# HPLC in Food Analysis

# FOOD SCIENCE AND TECHNOLOGY

## A SERIES OF MONOGRAPHS

### Series Editors

**George F. Stewart**
University of California, Davis

**Bernard S. Schweigert**
University of California, Davis

**John Hawthorn**
University of Strathclyde, Glasgow

### Advisory Board

**C. O. Chichester**
Nutrition Foundation, New York City

**J. H. B. Christian**
CSIRO, Australia

**Larry Merson**
University of California, Davis

**Emil Mrak**
University of California, Davis

**Harry Nursten**
University of Reading, England

**Louis B. Rockland**
Chapman College, Orange, California

**Kent Stewart**
USDA, Beltsville, Maryland

A complete list of the books in this series appears at the end of the volume.

# HPLC in Food Analysis

Edited by

**R. MACRAE**
Department of Food Science,
University of Reading,
Reading, England

**1982**

**ACADEMIC PRESS**

*A Subsidiary of Harcourt Brace Jovanovich, Publishers*

London   New York
Paris   San Diego   San Francisco   São Paulo
Sydney   Tokyo   Toronto

ACADEMIC PRESS INC. (LONDON) LTD.
24–28 Oval Road
London, NW1 7DX


U.S. Edition Published by

ACADEMIC PRESS INC.
111 Fifth Avenue
New York, New York 10003

*British Library Cataloguing in Publication Data*

HPLC in food analysis.—(Food and science technology)
  1. Food analysis     2. Liquid chromatography
  3. Beverages—Analysis
  I. Macrae, R.
  664'.07      TP372.5

  ISBN 0-12-464780-4
  CCCN 82-83666




Typeset by Preface Ltd., Salisbury, Wilts.
and printed in Great Britain by Thomson Litho Ltd., East Kilbride, Scotland.

# Contributors

**D. J. Folkes** The Lord Rank Research Centre, Lincoln Road, High Wycombe, Buckinghamshire HP12 3QR, U.K.

**E. W. Hammond** Unilever Research, Colworth House, Sharnbrook, Bedford, Bedfordshire MK44 1LQ, U.K.

**D. C. Hunt** Department of Industry, Laboratory of the Government Chemist, Cornwall House, Stamford Street, London SE1 9NQ, U.K.

**C. R. Loscombe** Laboratory of the Government Chemist, Cornwall House, Stamford Street, London SE1 9NQ, U.K.

**R. Macrae** Department of Food Science, University of Reading, Reading, Berkshire RG1 5AQ, U.K.

**R. Newton** Applied Chromatography Systems Ltd, Concorde House, Concorde Street, Luton, Bedfordshire LU2 0JE, U.K.

**P. J. van Niekerk** National Food Research Centre, Council for Scientific and Industrial Research, PO Box 395, Pretoria 0001, South Africa

**H. E. Nursten** Department of Food Science, University of Reading, Reading, Berkshire RG1 5AQ, U.K.

**K. Saag** Cadbury Schweppes Limited, Group Research, The Lord Zuckerman Research Centre, University of Reading, Whiteknights, Reading, Berkshire RG6 2LA, U.K.

**C. F. Simpson** Department of Chemistry, Chelsea College, University of London, Manresa Road, London, U.K.

**P. W. Taylor** The Lord Rank Research Centre, Lincoln Road, High Wycombe, Buckinghamshire HP12 3QR, U.K.

**A. P. Williams** Department of Basic Ruminant Nutrition, National Institute for Research in Dairying, Shinfield, Reading, Berkshire RG2 9AT, U.K.

# Preface

Chromatography has come a long way from being merely a separation technique even if, in a sense, the methodology has gone in a complete circle during the century or so in which it has been consciously recognized. The theoretical treatment was developed in classical physical chemistry style less than 50 years ago. The practical applications which have followed have taken an indirect route, from column to layer; qualititative to quantitative, preparative to picogram and with solid, liquid and vapour phases; all with an ever increasing diversity of application. Back to columnar form, albeit one which would have been unrecognized 100 years ago, liquid chromatography – high performance liquid chromatography or HPLC – today offers considerable advantages over some of the earlier chromatographic techniques. It is not limited by volatility or thermal instability. Food analysis covers a very wide range of sample substrates and the clean-up stage of the analysis can be very demanding. Chromatography can often be turned to advantage to accommodate both clean-up and the separation of the species of interest; and can then in addition give direct access to the normal analytical goal, quantification. The application of HPLC to vitamin analysis alone illustrates this well. The vitamins are a heterogeneous group of hydrophilic and hydrophobic compounds with virtually nothing in common chemically or physically. HPLC can cope with a large range, in a variety of complex foods; and is in some cases arguably the 'best' analytical technique available. Backed with theoretical, instrumental and practical introduction, the present volume deals with fats, amino acids and carbohydrates, their characteristic constituent molecular groupings and a wide range of additives both synthetic and natural; and will be of interest to the manufacturer, consumer and enforcement authority. All together, it presents a useful and important reference to the whole subject of HPLC in food analysis.

H. Egan
London SE1

August 1982

# Contents

# 1 Theory of Liquid Column Chromatography

## R. MACRAE
Department of Food Science, University of Reading, U.K.

## I INTRODUCTION

The tremendous growth of interest in liquid chromatographic methods over the last decade has paralleled that of gas chromatography in the preceding decade. Perhaps in many ways it is surprising that the rapid growth of gas chromatography did not stimulate a similar expansion in liquid chromatography simultaneously. The requirements for a liquid chromatograph with high resolving power, that is to say small particle size stationary phases and hence high inlet pressures, were elucidated in the classic paper of Martin and Synge (1941). However, despite this firm theoretical basis high resolution liquid chromatography was to lie dormant for some 20 years. Even the enormous success of high resolution amino acid analysers in the late 1950s (Spackman, Stein and Moore, 1958) did

1

not promote a general resurgence of interest in instrumental liquid chromatographs. In fact it was not until the early 1960s that modern liquid chromatography as it would be recognized today was born (Kirkland, 1969; Huber, 1969). It is interesting to note that the earliest applications of high pressure techniques were in the area of molecular size fractionation of polymers and not in the chromatography of simpler molecules, as might have been expected (Moore, 1964).

The debate as to the reasons for the laboured development of high resolution liquid chromatography, as compared with gas chromatography, must be left to students of the history of science, although it provides an interesting study in scientific development. The basic theory and instrumental requirements, together with at least some of the necessary technology, have been available for many years. It may have been simply that the rapid development of gas chromatography utilized fully the available resources of instrument manufacturers. This factor, coupled with belief that gas chromatography, with suitable sample derivitization where necessary, was a panacea for all chromatographic analyses, may be seen with hindsight as the reason for the slow development of high resolution liquid chromatography. In spite of its laboured early adolescence, or perhaps even as a direct consequence of it, liquid chromatography has seen an enormous increase in popularity over the last decade. Evidence of the successful maturation of liquid chromatography can be found both in the exponential growth of published scientific papers using the technique and also in the seemingly endless list of firms supplying instrumentation and supporting services. One result of this extraordinarily rapid expansion is that many analysts have become overwhelmed by the volume of available literature and have become perplexed by the apparent complexity of the technique. It is anticipated that an increasing understanding of chromatographic techniques will result from the inclusion of appropriate material in courses both at universities and other institutions of higher education, though to date there is little evidence of this happening.

The multitude of phraseology used to describe modern high resolution column chromatography has also contributed to the aura of mystic surrounding the technique. The terms high pressure, high performance and high speed liquid chromatography are all used interchangeably to describe modern high resolution column liquid chromatography. It is unfortunate that this last phrase, which in fact describes the technique most accurately, has not been universally adopted. However, in the present text the abbreviation HPLC will be used to cover all these expressions. There is a strong body of opinion that all liquid chromatography, irrespective of whether large or small particle size stationary phases are used, should be simply abbreviated to LC (Ettre, 1981). Theoretically this argument has much to commend it, but there are major differences in the instrumentation involved in the various types of liquid column chromatography and thus the term HPLC will be used to describe those column techniques employing small particle size stationary phases. In addition to the use of small particle size stationary phases, and the consequent requirement for high inlet pressures, HPLC systems usually employ sample injectors and continuous-flow detection, although the latter is often found in comparable low pressure systems. A basic HPLC chromatograph is a very simple instrument, the understanding and operation of which is well within the scope of all laboratory workers.

It is not the purpose of this chapter to provide a detailed account of the theory of the liquid chromatographic process but to describe in outline those factors which affect chromatographic performance, particularly those parameters which can be adjusted in the laboratory to optimize a given separation. A number of excellent texts have been published in which the interested reader may find a more complete treatise on the chromatographic process, for example those by Giddings (1965), Done, Knox and Loheac (1974) and Snyder and Kirkland (1979). Much progress has been made over the last two decades in the development of a complete theory for the chromatographic process but in many ways our present understanding is far from adequate. Liquid chromatography still remains to a large extent an empirical science. This may be contested by the theoreticians but for the practising chromatographer the empirical approach, guided by experience, is most commonly used to decide on chromatographic conditions. This is not to imply that a basic theoretical understanding is unnecessary for the efficient application of the technique but that theory, at its present state of development, will not provide all the answers to practical problems.


## II  THEORY OF LIQUID CHROMATOGRAPHY

### A  The Chromatographic Process

The basic chromatographic process consists of the partition of sample molecules between a mobile fluid and a stationary phase. In the case of gas chromatography the fluid is a gas and in liquid chromatography a solvent. In the case of column liquid chromatography the stationary phase is held in a column and the mobile phase is allowed to flow, or is pumped over it. In order that two components in a sample be separated they must pass down the column at different rates, that is to say they must partition to different extents between the mobile and stationary phases. An ideal separation is shown in Fig. 1.1(a). The factors which govern the partition ratio are

4                                    R. MACRAE

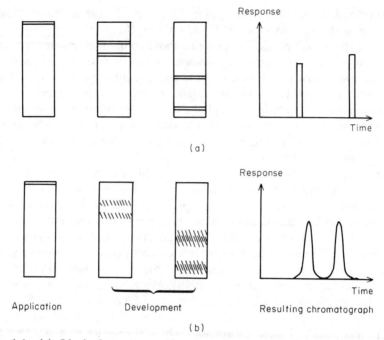

**Fig. 1.1** (a) Ideal chromatographic separation on a column with no band-spreading, (b) actual chromatographic separation on a column with band-spreading.

predominantly thermodynamic, so that sample retention on the column is thermodynamically controlled. In practice the idealized separation shown in Fig. 1.1(a) cannot be realized due to band-broadening. That is to say that as the components pass down the column diffusion takes place and the resulting chromatographic peaks are considerably broader than would have otherwise been the case. This process of band-broadening and its effect on the final chromatogram is shown in Fig. 1.1(b). The factors which affect diffusion are essentially kinetic in nature and thus can be considered separately from those thermodynamic factors which influence sample retention.

The entire *raison d'être* of column chromatography is the achievement of separations between components in a sample and an understanding of the factors that may influence this separation is important. Referring to Fig. 1.2 a term resolution ($R_s$) can be defined to quantify the resolution between chromatographic peaks:

$$R_s = \frac{2(t_{R2} - t_{R1})}{w_{b1} + w_{b2}},$$

Response

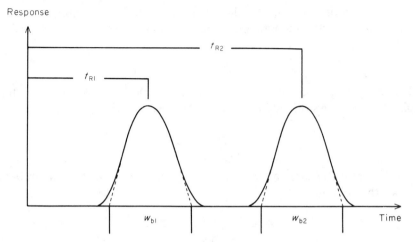

**Fig. 1.2**  Chromatographic resolution.

where $t_R$ is the retention time and $w_b$ is the peak width at the base of each peak expressed in the same units. From this expression it is evident that the resolution between peaks can be increased by either an increase in the difference between their retention times, which is thermodynamically controlled, or by reduction in the peak widths, which is dependent on kinetic factors. Both of these aspects are important and they will be considered separately but it is really the latter, the ability to produce narrow peaks, that has made HPLC such a powerful technique.

## B   Chromatographic Retention

The partition of sample molecules between the mobile and stationary phases may be considered as a dynamic equilibrium, both when the mobile phase is static and also during chromatographic elution. The concept of an equilibrium process taking place in a dynamic situation may be hard to accept but it is not unreasonable if it is appreciated that even a static equilibrium condition is in fact dynamic in the sense that sample molecules are continuously transferring between the phases. This equilibrium can be characterized by the capacity factor $(k)$, which is defined as the ratio of the amount of sample in the stationary phase $(a_s)$ to that in the mobile phase $(a_m)$, assuming here the sample to be one component:

$$k = a_s/a_m.$$

This factor can then also be expressed in terms of the mean residence times $(t)$ of the sample molecules in the stationary and mobile phases. The greater the proportion of time the molecules spend in the stationary phase the greater the chromatographic retention and hence the greater the value of $k$:

$$k = \bar{t}_s/\bar{t}_m.$$

The mean proportion of time spent by the sample molecules in the mobile phase $(R)$ follows directly from the above expression,

$$R = \frac{\bar{t}_m}{\bar{t}_m + \bar{t}_s} = \frac{1}{1 + k},$$

which is also the same as the relative migration rate of the sample band down the column:

$$R = \frac{u_{\text{sample band}}}{u_{\text{mobile phase}}} = \frac{1}{1 + k}.$$

The velocities $(u)$ of the sample band and mobile phase are inversely related to the elution times of the sample band $(t)$ and an imaginary band with zero retention $(t_0)$:

$$\frac{t}{t_0} = \frac{u_{\text{mobile phase}}}{u_{\text{sample band}}}.$$

A combination of the above expressions allows a definition of the capacity factor in terms of elution times:

$$k = (t - t_0)/t_0$$

This parameter is easily measured from an experimental chromatogram for the sample component of interest; $t_0$ is usually measured from the first disturbance on the detector base-line, often as a result of the sample solvent. The capacity factor is an attractive parameter for characterizing the retention of a sample component as it can be directly related to the equilibrium model in any theoretical treatment.

In an ideal equilibrium model the heat of transfer of molecules between phases can be related to the equilibrium constant by the Van't Hoff equation (Moore, 1963). Thus in the chromatographic model the capacity factor $(k)$ can be related to the heat change involved when sample molecules pass from the mobile to the stationary phase:

$$\frac{d(\ln k)}{dT} = \frac{-\Delta H}{RT^2},$$

where $\Delta H$ is the heat change per mole of solute and $R$ is the molar gas constant. In general the heat changes involved in liquid chromatography are small, only approximately a quarter of those encountered in gas chromatography and thus the effects of temperature on retention parameters would be expected to be far less significant. This is indeed found to be the case in practice, as shown by the widespread use of temperature programming in gas chromatography but not in liquid chromatography. However, it is unwise to dismiss temperature effects as unimportant, even though this is an experimental parameter that is rarely altered to change the degree of sample retention. The actual magnitude of the temperature effect will depend on the mode of chromatography employed (Chapter 3) and thus it is not feasible to quote a single figure for such temperature effects. The variation of the capacity factor for a component with changes in temperature will be different for adsorption or bonded-phase chromatography, reflecting the different phase changes involved.

An example of temperature effects for a bonded-phase system is shown in Fig. 1.3. Here the separation of glucose and galactose using an aminopropyl bonded phase was studied at various temperatures (Macrae, 1978 and unpublished data). As would be expected from the Van't Hoff equation, a plot of log $k$ against $T^{-1}$, where the temperature is expressed in kelvins, should produce a straight line, and this was found to be the result.

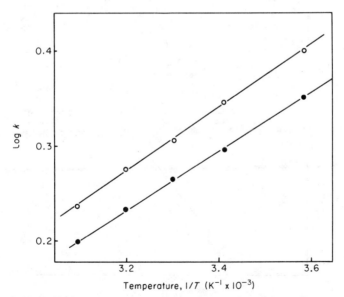

**Fig. 1.3**  Effect of temperature on capacity factors for glucose (●———●) and galactose (○———○). Column, Spherisorb 5-NH$_2$ (25 cm × 5 mm); mobile phase, 25% water in acetonitrile (v/v) with refractive index detection.

In this example the slopes are similar, although not identical, and thus changes in temperature result in only minor changes in column selectivity, with consequently a small effect on resolution. The fact that the heat changes are similar for these sugars is a direct consequence of their closely related structures. A more important effect, as far as retention is concerned, is that temperature changes are causing considerable alterations to retention times and thus for reproducible chromatography the column temperature should be maintained at a constant value. In many analyses a small shift in retention time will not in fact cause errors in peak assignment but will contribute significantly to errors in quantification. The effect of temperature changes on the kinetic aspects of the chromatographic process, which governs band-broadening, are more important and this will be referred to again subsequently.

A fundamental assumption that has been made in the above thermodynamic treatment is that the capacity factor is independent of concentration. If this is not in fact the case as the band passes down the column the higher concentration zone, that is to say the centre of the band, may travel faster or slower than the low concentration zones on the leading and trail-

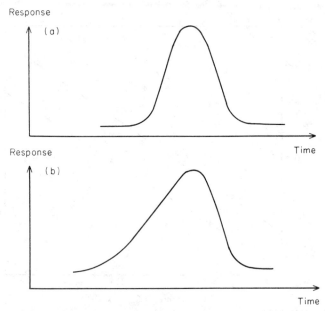

**Fig. 1.4** Effect of reduction in capacity factor at higher concentrations on peak shape: (a) capacity factor unaffected by concentrations, (b) capacity factor reduced at higher concentrations.

ing edges. The net effect of this phenomenon is to produce peaks with either poor leading or poor trailing edges. This is often due to the presence of sites of differing activity, which is a problem to a greater or lesser extent with all chromatographic phases. The result of a reduction in capacity factor at higher concentrations is shown in Fig. 1.4. This has been described by Knox, Laird and Raven (1976) as the breaking wave effect and is important at high sample loadings. Consider the situation in Fig. 1.5 showing the separating of two components. At low concentrations increases in sample loading will simply be observed as increases in peak heights, with no changes in retention or peak widths, and hence no change in observed resolution [Fig. 1.5(a)]. This is known as linear isotherm retention (Snyder and Kirkland, 1979). As the sample loading is further increased there will come a point where retention times are affected and peak tailing becomes apparent, with a deleterious effect on peak resolution [Fig. 1.5(b)] and when taken to an extreme situation resolution is almost

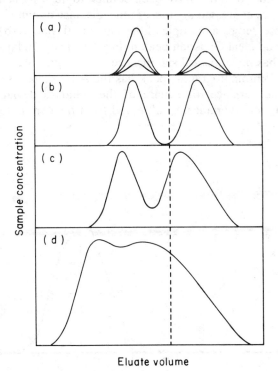

Fig. 1.5   Effect of sample loading on retention and resolution of two components. [Redrawn with permission from Snyder and Kirkland (1979).]

completely lost [Fig. 1.5(d)]. Sample loading thus becomes a critical factor
in controlling chromatographic resolution and is particularly important
when analytical separations are scaled up for preparative work.

The thermodynamic equilibrium theory can thus explain certain characteristics of the chromatographic process, for example the effect of temperature on retention. However, the theory only explains to a very limited
extent the process of band-broadening and attention must now be turned
to this.

## C  Band-broadening

The shape of chromatographic peaks is to a very great extent governed by
kinetic parameters and although it is not necessary to understand the
detailed theory involved it is at least desirable to be aware of those factors
that influence band-shape and its effect on chromatographic resolution.

As the sample passes down the column its components diffuse to form
Gaussian peaks under ideal conditions. The extent of this diffusion is critical and a number of terms have been defined to quantify band-spreading.
The peak width, expressed in units of time, distance or elution volume, can
also be represented by a multiple of the standard deviation ($\sigma$) of the curve
as shown for an ideal Gaussian peak in Fig. 1.6. The standard deviation in
this context has no statistical significance and results solely from the fact
that the ideal chromatographic peak has the same shape as a normal distribution. The variance, or square of the standard deviation, is often
quoted as the total variance, as observed from the chromatogram, and is

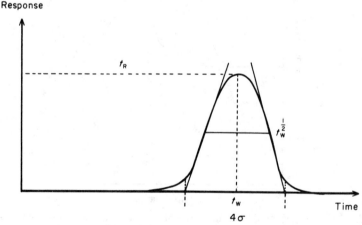

**Fig. 1.6**  Ideal Gaussian chromatographic peak.

the sum of the variance due to band-broadening on the column and that due to extra-column broadening. This will be referred to again later (p. 19).

The plate number ($N$) for a particular column, which is a measure of column performance or column efficiency, is defined in terms of the retention time ($t_R$) and the standard deviation, also expressed in units of time,

$$N = (t_R/\sigma)^2,$$

which may then be written directly in terms of base-line peak widths ($w_b$):

$$N = 16(t_R/4\sigma)^2 = 16(t_R/w_b)^2.$$

In practice it is often easier, and more precise, to measure the peak width at some other position, for example at half height. This may then be related to the above expressions by assuming an ideal Gaussian distribution:

$$N = 5.54(t_R/w_h)^2.$$

In the laboratory it is often found that the latter expression results in significantly higher plate numbers, presumably due to a combination of non-idealized band-shape and errors in measurement (base-line peak widths are difficult to measure precisely). Thus when comparing columns it is essential to adopt the same method. The same test sample should also be used as different compounds can give large differences in plate numbers. The plate height ($H$) is then simply defined as the column height per plate – also known as the height equivalent to a theoretical plate (HETP):

$$H = L/N,$$

where $L$ is the column length.

The concept of plate heights arose from the pioneering work of Martin and Synge, who recognized that the chromatographic process could be considered as analogous to distillation. This plate theory of chromatography allowed prediction of the Gaussian peak shape. The analogy is now largely irrelevant but the concept of plate height as a criterion of column efficiency is universally accepted. Plate height data recorded from peak widths at half height may not, however, reveal poor columns where the fault is due to peak trailing. In the extreme example shown in Fig. 1.7 column A and column B would give very similar plate heights, but it is quite clear that the latter is inferior chromatographically. To overcome this problem it is possible to define an asymmetry factor based on the ratio of the leading and trailing half-widths at various fractions of the peak height. In Fig. 1.7 the asymmetry factor calculated at one-tenth of the peak height clearly reveals the poor peak shape, whilst that recorded at one-half the peak height would have yielded a value very near to unity, thus not revealing the column defect.

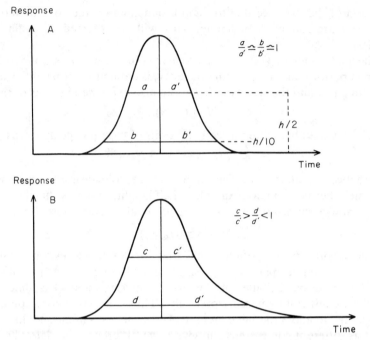

**Fig. 1.7**  Columns A and B have similar plate numbers but column B has poor tailing as revealed by the asymmetry factor at 10% peak height.

Chromatographic retention, as seen in Section II.B, can be readily explained in terms of a simple equilibrium model. However, the process of band-broadening is more complex and there are several mechanisms which contribute to the final observed peak shape. These mechanisms all result in a perturbation from the simple equilibrium model causing a "phase lag" between the concentration profile of the sample in the stationary phase to that in the mobile phase. This results in differences in the capacity factor throughout the band which means that the leading edge of the band will be moving faster than the trailing edge and band-broadening will result.

The mechanisms which contribute to this displacement between the concentration profiles, and hence deviation from the equilibrium model, are complex and despite much erudite theory several aspects of the chromatographic process remain poorly understood. These contributing mechanisms have been defined by Snyder and Kirkland (1979) and only a brief description will be attempted here. Theoretically it is possible to separate them but in reality there are probably interactions between them, making the overall chromatographic process more complex and less amenable to theoretical treatment.

The main contributing mechanisms to band-broadening are shown in Fig. 1.8. The first of these, eddy diffusion, is a direct consequence of the diverse pathways that sample molecules may take through the column. Where the solvent pathway is wide the flow will be fast and hence migration will be rapid, whereas in narrower solvent pathways the sample will be retarded in the slower solvent stream. The extent of eddy diffusion depends on both the nature of the chromatographic material and how well the column is packed. In general stationary phase material with spherical particles tend to give more uniformly packed columns than those with irregular particle shapes. Poorly packed columns that have not been well consolidated possess a very wide range of solvent pathways and hence the effects of eddy diffusion will be considerable. The contribution of this

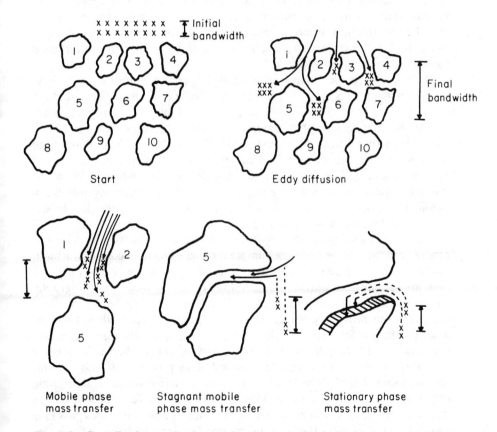

**Fig. 1.8**  Contributing mechanisms to band-broadening in liquid chromatography. [Redrawn with permission from Snyder and Kirkland (1979).]

process to the column plate height is found to be proportional to particle size of the packing material ($d_p$).

The second contributing process to band-broadening illustrated in Fig. 1.8 is known as mobile phase mass transfer. This is simply a consequence of the velocity gradient that exists across the diameter of any passage through which a fluid is flowing. The solvent and hence the dissolved sample near the sides of the passage, in this case next to the chromatographic material, will flow at a considerably lower velocity than that at the centre, or in midstream. This distribution of solvent velocities will result in further band-spreading and is dependent on the mobile phase velocity ($u$). Other factors which influence mobile phase mass transfer include the stationary phase particle size ($d_p$) and the rate of diffusion of the sample molecules within the mobile phase. A property of the solvent is thus directly affecting band-broadening and solvents which allow a high rate of sample diffusion are to be preferred, that is to say solvents with low viscosity. The contribution of mobile phase mass transfer to band-broadening is found to be proportional to $d_p^2 u / D_m$, where $D_m$ is the sample diffusion coefficient in the mobile phase.

Chromatographic columns contain a significant amount of mobile phase that is stagnant. This may be either in interstices between particles, that are not being swept by the solvent flow, or more importantly in deep pores within the chromatographic material. Sample molecules that enter such pores may either diffuse out quickly and hence not be significantly retarded or they may diffuse further into the pore, resulting in retardation and band-broadening. The factors influencing this process, which is known as stagnant mobile phase mass transfer, are similar to those encountered in mobile phase mass transfer and the contribution to the column plate height is found to be proportional to $d_p^2 u / D_m$.

In a manner analogous to stagnant mobile phase mass transfer a similar process can take place once the sample molecules have become associated with the stationary phase. (This association may be in the form of adsorption, ion-exchange or partition depending on the mode of chromatography employed, as discussed in Chapter 3.) Here it is not the pore depth that is important but rather the thickness of the chromatographically active material on the particle surface. The deeper the sample molecules penetrate into this layer the longer they will take to diffuse out and hence the greater the extent to which they will be retarded. This process can also be conceived as sample molecules transferring between adjacent chromatographic sites prior to diffusion back into the solvent stream. The concept of the stationary phase as having an active layer which sample molecules pass into and diffuse out of is probably well removed from reality and it is the interaction between sample molecules and chromatographic sites that should be considered. In this respect the contribution to band-broadening due to stationary phase mass transfer may also be explained by differences in activity between the various chromatographic sites. The expression for this contribution to band-broadening is similar to that for other mass transfer processes, that is to say proportional to $d_f^2 u / D_s$. In this case the particle size $(d_p)$ is replaced by the thickness of the conceptual stationary phase layer $(d_f)$ and $D_s$ is the diffusion coefficient of the sample molecules within it.

The final contribution to band-broadening that needs to be considered is that of simple diffusion. Whenever a concentration gradient exists in a fluid the process of diffusion will take place in an attempt to produce a uniform concentration throughout the volume available. This process of diffusion is non-specific and solute molecules will diffuse in all directions, irrespective of whether the solvent is flowing or not. Two consequences of diffusion can be observed in column chromatography. Firstly, longitudinal diffusion along the direction of solvent flow which results in symmetrical band-broadening. In practice diffusion of this kind will only lead to a significant reduction in chromatogram quality under those conditions where sample molecules have a long residence time on the column, that is to say when very low flow rates are used. Secondly, as the diffusion process is taking place in all directions simultaneously, the sample as applied to the top of the column will also tend to migrate towards the walls, this process is known as radial diffusion. Radial diffusion *per se* does not contribute to band-broadening but rather leads to secondary effects which do. There is much evidence to suggest that the sample molecules once they reach the column walls are subjected to significant retardation compared with those in the centre of the column (Knox *et al.*, 1976). Some of this broadening may be attributed to the slower solvent flow near the column walls, analogous to mobile phase mass transfer. It is also clear that there is an interaction between many solute molecules and the column wall material, giving rise to retardation and hence further broadening. The ideal situation is thus achieved when the sample passes down the column without approaching its walls; this is known as operating in the infinite diameter mode, that is to say as far as the sample is concerned the column has an effective infinite diameter. Whether this is possible or not will depend on the rate of radial diffusion which in turn will depend on the mobile phase, particularly its viscosity, and the column in use. Additionally the manner in which the sample is applied to the top of the column is important, as some injection systems cause considerable radial dispersion when the sample meets the column bed.

The contributions of these various processes to the column plate height can then be summarized as in Table 1.1. Assuming that they are non-

**Table 1.1**    Contribution of different band-broadening processes to column plate height $H$

| Process | $H_i$ |
|---|---|
| (i)   Eddy diffusion | $\left.\begin{array}{l} C_e d_p \\ C_m d_p^2 u / D_m \end{array}\right\} = Au^{0.33}$ |
| (ii)  Mobile phase mass transfer | |
| (iii) Longitudinal diffusion | $C_d D_m / u = B/u$ |
| (iv)  Stagnant mobile phase transfer | $C_{sm} d_p^2 u / D_m = Cu$ |
| (v)   Stationary phase mass transfer | $C_s d_f^2 u / D_s = Du$ |

*Note:* $C_d$, $C_e$, $C_m$, $C_s$, $C_{sm}$ = plate height coefficients; $d_p$ = diameter of packing particle; $d_f$ = thickness of stationary phase layer; $u$ = mobile phase velocity; $A, B, C, D$ = constants for a given column.
Reproduced with permission from Snyder and Kirkland (1979).

interactive they can be combined to give the total plate height (Giddings, 1965):

$$H = \frac{1}{1/C_e d_p + (1/C_m d_p^2 u / D_m)} + \frac{C_d D_m}{u} + \frac{C_{sm} d_p^2 u}{D_m} + \frac{C_s d_f^2 u}{D_s}.$$

This complex expression shows clearly the large number of parameters that are involved in band-broadening, and in particular the importance of particle size and diffusion coefficients. Fortunately for practical purposes it can be simplified by the introduction of reduced parameters (Giddings, 1963). The reduced plate height ($h_r$) is the number of particles per plate and regardless of particle size should be about 2 for a well packed column:

$$h_r = H/d_p.$$

In an analogous manner the solvent velocity can be treated as a reduced solvent velocity ($V$) which compares the rate of solvent flow down the column with the rate of molecular diffusion across the particle diameter:

$$V = u d_p / D_m.$$

All reduced parameters are dimensionless. They can then be substituted in the equation for the plate height which, after simplification, can be written in the form

$$h_r = B/V + AV^{0.33} + CV.$$

This simplification has also ignored the contribution due to stationary phase mass transfer, which is reasonable as this is small with most modern chromatographic materials. This was not the case, however, with earlier bonded-phase materials which possessed polymeric coatings with very poor transfer characteristics.

The reduced plate height, or more correctly the logarithm of the reduced plate height, can now be plotted against the logarithm of the reduced velocity and a plot such as that shown in Fig. 1.9 obtained. This is known as a Knox plot after its chief proponent, Professor J. H. Knox. The main role of the Knox plot, apart from its theoretical interest, is that it allows the question "is this a good or bad column?" to be answered. The various contributing terms to the reduced plate height have different relative importances as the reduced velocity is increased and these contributions are shown in Fig. 1.9. At low reduced velocities the term $B/V$ is most important; that is to say that at very low mobile phase velocities the process of longitudinal diffusion becomes significant, as would be expected. At high reduced velocities the term $CV$ predominates; that is to say that under these conditions there is a greater deviation from the equilibrium model, with slow equilibration of sample molecules between the mobile and stationary phases. From the graph in Fig. 1.9 it can be seen that there is a broad minimum at a reduced velocity of about 3, corresponding to a reduced plate height of about 1. In practice reduced velocities this low are not used, as they would lead to prohibitively long analysis times. More usual values would be in the range $10^2$–$10^3$, which corresponds to a reduced plate height of about 2. Typical values for the constants $A$, $B$ and $C$ for a well packed column of efficient material would be 1, 2 and $5 \times 10^{-2}$ respectively.

The main significance of the Knox plot is simply that all columns, irres-

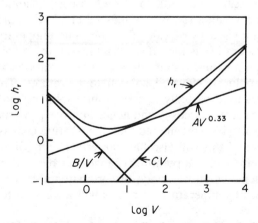

**Fig. 1.9** Knox plot showing various contributions to reduced plate height. [Redrawn with permission from Done *et al.* (1974).]

pective of their type or particle size, should yield curves of the same general shape with a minimum value for the reduced plate height in the same range. Gross deviations from the expected curve shape allow the chromatographer to decide whether the defect is in the stationary phase material, with poor sample molecule transfer characteristics, or whether the column is badly packed with poor flow characteristics, high eddy diffusion and poor mobile phase mass transfer. Poor chromatographic material is reflected in an increased $C$ value while a badly packed column is observed in an increased $A$ value. In practice, however, it is extremely difficult to produce a plot over a sufficiently wide range of reduced velocity to allow accurate quantification of these constants. Consequently most practical column testing procedures are confined to the determination of plate heights under standard conditions and a comparison of the experimental data with those obtained from previous columns, or from literature values.

## D    Diffusion and Temperature Effects

In the foregoing theoretical discussion the importance of the rate of diffusion of sample molecules in the mobile phase and the resulting effect on band-broadening can be seen. The diffusion of molecules in a liquid depends both on the nature of the molecules and that of the liquid. The two most important factors are the size of the molecules and the viscosity of the liquid. The first of these is largely out of the hands of the chromatographer, although conditions should be avoided that might lead to molecular aggregation. Solvent viscosities, however, differ widely and where a choice of solvents is possible that with the lower viscosity should be used. For example methanol with a viscosity of 0.54 cP at 25 °C would be preferred to ethanol with a viscosity of 1.08 cP at the same temperature. A further important consideration is the effect of temperature on solvent viscosity. The earliest high pressure systems used for size exclusion chromatography of polymers employed elevated column temperatures. This increase in temperature was used partially to assist solute solubility but more importantly to reduce solvent viscosity and hence improve peak shape and resolution. It should also be remembered that temperature increases will reduce retention times and this will also have an effect on resolution. The magnitude of the effect of temperature changes on band-widths and hence plate heights depends on the mode of chromatography employed. However, the effect of temperature increases will always be to reduce plate heights.

In ion-exchange chromatography, for example in amino acid analysis, the use of elevated temperatures to increase solute molecular diffusion is well established. In other modes of chromatography the effects of temperature changes have been less well studied although there is some evidence that such temperature effects may well be important. In the example shown in Fig. 1.10 the plate height for glucose on a partition column is shown as a function of temperature. There is a decrease in plate height of nearly 50% over the modest temperature range 10–40 °C (Macrae, unpublished data). Temperature effects can thus affect kinetic aspects of the chromatographic process (band-broadening) as well as thermodynamic aspects (retention) and hence should not be totally disregarded as an operating variable in liquid chromatography.

## E  Extra-column Band-broadening

The discussion of band-broadening so far has been concerned with broadening caused by the chromatographic column alone and the mechanisms of the processes involved. Unfortunately the column is not the only place in the chromatographic system where broadening can take place. In any chromatograph, however well designed, there will be a finite volume between the point of injection and that of detection, in addition to the volume of the column itself. This is known as extra-column volume ($V_{ext}$) and is made up of a volume between the effective injection point and the column inlet ($V_i$) and a volume between the column outlet and the point of

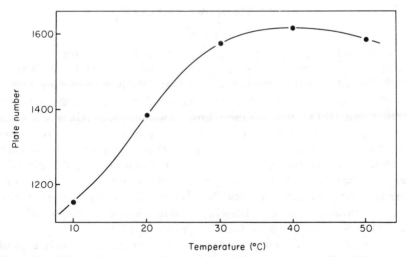

**Fig. 1.10**  Effect of temperature on plate height for glucose. Conditions as in Fig. 1.3

detection $(V_d)$, such that

$$V_{ext} = V_i + V_d.$$

It should be noted that this is not the same as dead volume, which would be an unswept volume, as the mobile phase will be flowing continuously through it. As the sample, dissolved in the mobile phase, passes through this extra-column volume broadening will take place simply as a result of the velocity gradient across the diameter of connecting tubing. The solvent stream near the walls, and hence that part of the sample in it, will move more slowly than that in midstream with resulting dilution of the sample and broadening.

The band-broadening observed from a chromatogram is a combination of extra-column and column band-broadening. These contributions can be summed in terms of their variances (p. 10) such that

$$\sigma_{total}^2 = \sigma_{column}^2 + \sigma_{extra\ column}^2$$
$$= \sigma_{column}^2 + \sigma_{injector}^2 + \sigma_{detector}^2.$$

In well designed systems the contribution from extra-column band-broadening should be very much smaller than that from the column. There should be no significant dead volume in the system, which in addition to band-broadening can give rise to memory effects, especially in injectors.

## III  QUALITATIVE ANALYSIS

The degree of retention of a compound, or its capacity factor, is characteristic of that compound under given chromatographic conditions. Thus a comparison of capacity factors of standards and unknown components in a mixture can theoretically lead to identification. However, it must be borne in mind that HPLC, just as other chromatographic techniques, does not determine structures but rather suggests compounds that behave in the same manner chromatographically under specified conditions. The concept of chromatographic similarity should be far more prevalent in the scientific literature than in fact it is. In many cases structures are assigned to compounds in samples from retention data alone without any supportive evidence, chromatographic or otherwise. It should also be remembered that capacity factors are altered at high sample loadings (p. 9).

It is essential that the chromatogram for the sample to be analysed and those for standards are recorded under identical conditions if comparisons of retention data are to be valid. The importance of certain operating parameters, such as column type and solvent, are self evident. However, the significance of other factors, for example column temperature, has been largely ignored, even though this can alter chromatographic data considerably. Small changes in column temperatures are unlikely to lead directly to incorrect peak assignments in simple systems but could seriously reduce the reliability of retention data stored over long periods. A further factor which must be taken into consideration is that of column ageing. Over extended periods of use retention characteristics for a given column may well change, due to stripping of bonded phases or irreversible binding of sample impurities to chromatographic sites. This is not a serious problem in the short term but again may reduce the reliability of stored retention data. One method of partially overcoming this problem is to use the capacity factor of a known compound as an internal reference and then compare the retention of other compounds to it. This assumes that the retention of the reference compound and the other compounds of interest behave in the same manner with changes in chromatographic conditions, which may or may not be the case in practice. A compound used in this way as an internal reference can also be used as an internal standard for quantification (p. 25). If there is any ambiguity that the unknown component and a particular standard do have the same retention time then the sample should be spiked with a small amount of the standard and the coincidence of the sample peak and the added material confirmed.

In many simple systems the identification of major components by comparison with standards leads to unambiguous assignments. Examples of this in the food area would be the identification of lactose in milk or caffeine in a coffee extract. However, even in these simple cases it must be realized that HPLC has not *proved* peak identification. In more complex systems where there is no evidence as to the nature of the components the process of structure identification may be very laborious. The first stage is a comparison with standards to give some idea of the nature of the unknown, although this may well come from experience. It is then possible to carry out a number of alterations to the chromatographic conditions which may provide further evidence about the compounds structure. This may involve studying the effect of pH on retention or the addition of ion-pair reagents or other solvent modifiers. A more powerful approach to peak identification, however, lies in the alteration of the detection conditions. If the compound absorbs in the ultraviolet or visible regions, the absorption at various wavelengths can be studied and thus related to structural features. A more complete picture can be obtained if the instrumentation available allows a spectrum to be recorded, while the component is trapped in the flow cell. In a similar manner the fluorescence characteristics of an unknown compound can be measured, again providing additional data as to its structural features. The alteration of detection, or chromatographic, conditions can also provide supportive evidence for tentative structural assignments. For example if the absorbances of an unknown peak in a sample and a standard alter in the same way with changes in detection wavelength, in addition to having the same retention time, there is an increased probability that the sample peak and the standard are in fact the same. The usual way to carry out this technique is to record absorbances of the unknown peak and standard at two or three wavelengths and then study the ratios of these absorbances for the two peaks (Shumaker and Yost, 1979). If the unknown and the standard are the same, these ratios should be the same. This technique can also be used to determine peak purity as any co-eluting material will disturb the absorbance ratios, provided that the impurity has a significantly different spectrum. The amount of identification work that can be carried out *in situ*, that is to say without isolation of the actual compound, is limited. The logical extension of such methodology would be the development of an efficient HPLC/MS system, in which eluted compounds could be introduced directly into the mass spectrometer source after removal of the chromatographic solvent. At present HPLC/MS systems are not widely used and their development is discussed in Chapter 13. Further identification of structure thus usually necessitates isolation of the compound from the column eluate, followed by conventional analytical procedures such as infrared spectrophotometry or nuclear magnetic resonance spectrometry. This may be extremely time-consuming and may involve the use of larger diameter columns to allow isolation of milligram quantities of material.

In the majority of published papers chromatographic similarity is used as the sole criterion for identity, but the limitations of this approach must be realized.

## IV    QUANTITATIVE ANALYSIS

The production of quantitative data from chromatograms involves the determination of either peak heights or peak areas. Peak heights can be determined relatively simply and this was the method of choice until the use of electronic integrators became widespread (Chapter 4). However, it is worth considering under what conditions each method will produce the more accurate results. Minor changes in mobile phase composition or column temperature will affect retention times with a consequent effect on peak widths and hence peak heights. The measurement of peak areas under these conditions would be preferable since there is some degree of compensation, as the peak width increases the height decreases and *vice versa*. On the other hand, changes in mobile phase flow rate will have a greater effect on peak areas than on peak heights, as most detectors are concentration dependent (Bakalyar and Henry, 1976).

In general, for accurate quantification, irrespective of the measurements to be used, complete chromatographic resolution is desirable, that is to say an $R_s$ value above 1.25. In practice this may not always be possible and it may prove necessary to quantify bands that are partially overlapping. In this situation, as depicted in Fig. 1.11, it may be that peak height measurements are less sensitive to the presence of overlapping peaks. In fact it can be shown (Snyder and Kirkland, 1979) that for $R_s = 1.0$ and an accuracy of ±3% it is possible to vary peak height ratios from 32 : 1 to 1 : 32, while for comparable accuracy peak areas can only vary from 3 : 1 to 1 : 3. Thus, depending on the resolution and the relative peak sizes, peak height measurements may be more accurate than peak areas.

A further requirement which is often overlooked in quantitative analyses is the linearity of the detection system. It is essential that the detector is shown to be linear over the entire concentration range to be used, or at least that the non-linearity is appreciated and a calibration plot used. In many cases this stage appears to be omitted in analyses with workers relying on original instrumental specifications. Even a small deviation from linearity as shown in Fig. 1.12 may have a significant effect on the final accuracy of results. Here with a one-point calibration the observed area would be correlated to a concentration value $B$, whereas the true value is $A$ which would have been determined from a complete calibration plot.

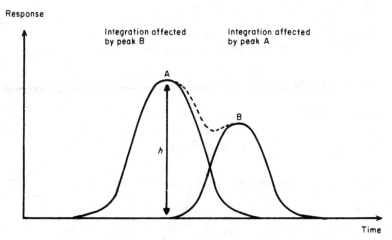

**Fig. 1.11**   Effect of overlapping bands or peak parameters.

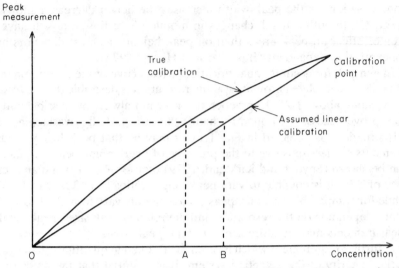

**Fig. 1.12**  Consequences of non-linearity on quantification when linearity is assumed.

There are many methods for manipulating peak measurement data to express analytical results, the most common of which are shown in Table 1.2. Percentage area calculations are the simplest and, provided that all the components respond equally to the detector, give good approximations to the relative amounts of each component. An example of the use of this method would be in the analyses of sugar mixtures, where each component has a similar refractive index response.

In the normalization procedure, also known as corrected area normalization, measured areas are corrected for variation in detector response and

**Table 1.2**  Common methods for quantification

1. Percentage area  Percentage of component $i = \dfrac{A(i)}{A} \times 100$

2. Normalization  Percentage of component $i = \dfrac{\text{abs } RF(i) \times A(i)}{(\text{abs } RF \times A)} \times 100$

3. Scale factor  Percentage of component $i = \dfrac{A(i)}{A} \times$ Scale factor $\times 100$

4. External standard  Amount of $i = A(i) \times \text{abs } RF(i)$

5. Internal standard  Amount of $i = \dfrac{\text{abs } RF(i) \times A(i)}{\text{abs } RF(IS) \times A(IS)} \times M(IS)$

thus better quantitative results are possible than with percentage area calculations. In this case it is necessary to determine the absolute response
factor (abs RF) from a standard chromatogram:

$$\text{abs RF}(i) = M(i)/A(i),$$

where $M(i)$ is the mass of component $i$ and $A(i)$ is the area under its peak.

A further method, 3 in Table 1.2, which is not frequently used, allows for
the fact that only part of the sample is detected. Here a scale factor is
employed to allow for this undetected portion of the sample. The method is
useful when selective detection, such as fluorescence, is used in liquid
chromatography.

The most important methods of quantification are those using external
or internal standards. In the former the peak of interest in a sample is
simply compared with the peak of a known amount of that compound in a
standard chromatogram, i.e. with its absolute response factor. In the internal standard method a known amount of a compound not naturally present
in the sample is added and quantification is then based on the ratio of that
standard to the peak of interest. It is necessary to know the absolute
response factors for the internal standard and the component of interest, or
at least their relative response factor. The internal standard should be as
similar as possible to the compound to be analysed but completely resolved
from it. The internal standard may be added at the beginning of the
analysis and so compensate for losses during the extraction procedure.
External calibration is simpler to carry out in practice and, in those cases

**Table 1.3**  Characteristics of methods of quantification

|  | Percentage area | Normalization | External standard | Internal standard |
|---|---|---|---|---|
| Amounts expressed as a percentage? | Yes | Yes | No | Opt |
| Amounts reported in any desired units? | No | No | Yes | Yes |
| Calibration table required? | No | Yes | Yes | Yes |
| Internal standard required? | No | No | No | Yes |
| Response factors used for detector response to different components? | No | Yes | Yes | Yes |
| Peaks are identified? | No | Yes | Yes | Yes |
| All peaks must be included in the calculation? | Yes | Yes | No | No |
| Control of injection volume is required? | No | No | Yes | No |

where the sample volume can be precisely controlled, for example with injection valves, is to be preferred. The internal standard method automatically compensates for variations in injection volume. The main features of the various methods of quantification are summarized in Table 1.3. All of the methods described above may be used with peak height as opposed to peak area measurements, which parameter is the most pertinent to use in any particular circumstances will depend on the factors already discussed and the instrumentation available.

## REFERENCES

Bakalyar, S. R. and Henry, R. A. (1976). *J. Chromat.* **126**, 327.
Done, J. N., Knox, J. H. and Loheac, J. (1974). *Applications of High Speed Liquid Chromatography*, John Wiley, London.
Ettre, L. S. (1981). *J. Chromat.* **220**, 29.
Giddings, J. C. (1963). *Analyt. Chem.* **35**, 1338.
Giddings, J. C. (1965). *Dynamics of Chromatography*, Part 1, Marcel Dekker, New York.
Huber, J. F. K. (1969). *J. chromatogr. Sci.* **7**, 85.
Kirkland, J. J. (1969). *J. chromatogr. Sci.* **7**, 7.
Knox, J. H., Laird, G. R. and Raven, P. A. (1976). *J. Chromat.* **122**, 129.
Macrae, R. (1978). *Lab. Pract.* **27**, 719.
Martin, A. J. P. and Synge, R. L. M. (1941). *Biochem. J.* **35**, 1358.
Moore, J. C. (1964). *J. Polymer Sci. A* **2**, 835.
Moore, W. J. (1963). *Physical Chemistry*, 4th edn, Longmans, London, p. 180
Shumaker, W. E. and Yost, R. W. (1979). In *Liquid Chromatographic Analysis of Food and Beverages* (G. Charalambous, ed.), Academic Press, London, pp. 1.
Snyder, L. R. and Kirkland, J. J. (1979). *Introduction to Modern Liquid Chromatography*, 2nd edn, John Wiley, New York.
Spackman, D. H., Stein, W. N. and Moore, S. (1958). *Analyt. Chem.* **30**, 1190.

# 2 Instrumentation for HPLC

## R. NEWTON

Applied Chromatography Systems Ltd, Luton, U.K.

## I INTRODUCTION

The instrumentation used in high performance liquid chromatography (HPLC) has changed considerably since the introduction of the first commercial instruments in the late 1960s. The purpose of this chapter is to review and discuss the apparatus which is currently in use. This discussion of HPLC equipment will concentrate on that used for analytical work, although some brief reference will be made to preparative and semi-preparative HPLC.

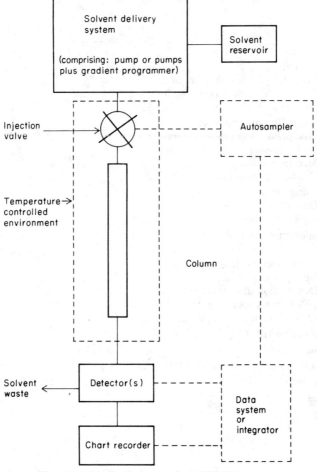

**Fig. 2.1**  Block diagram of an HPLC system.

The individual components of a liquid chromatograph can be described with reference to the block diagram shown in Fig. 2.1. The solvent delivery system, consisting of a pumping system, possibly with a gradient device, is used to supply mobile phase from the solvent reservoirs through the injection valve to the column. The injection valve is used to apply the sample mixture to the head of the column. After passing through the column, the eluate passes through the detectors, which detect the separated components of the sample mixture and supply a signal to the chart recorder where the chromatogram is recorded. As an option to this basic system it is possible to add an autosampling device for the automatic presentation of samples to the column.

A further module, or group of modules, regarded by many analysts as essential, is a computer-based data handling system, where raw data from the detector can be taken, stored and processed to provide automatic quantitative calculations on the sample being analysed.

Computer-based systems may be used with automatic sampling devices so that a large number of samples can be conveniently analysed while the liquid chromatograph is left unattended. Automatic HPLC is discussed fully in Chapter 4 and no further reference to it will be made here.

One final option, which can be useful for certain separations, is to enclose the column and injection valve in a temperature-controlled environment.

Before proceeding further to discuss the individual components of this system in detail, one very important point must be emphasized. The HPLC column is at the heart of any system as it is here that the chromatographic separation takes place. Unless the column is of the correct material, is efficiently packed and is used with a compatible mobile phase, no matter how sophisticated the instrumentation high quality chromatograms will *never* be obtained.

## II HPLC PUMPS

The main criteria for the design of an ideal HPLC pump can be summarized as follows:

(1) It should be manufactured from materials resistant to attack by the mobile phases used in HPLC. This is also a feature which must be satisfied by all the other parts of the system which come into contact with the mobile phase, i.e. pipework, fittings, detector flow cells, etc. From a practical point of view this means the use of the highest quality stainless steels, inert polymers such as polytetrafluoroethylene (PTFE) and the use of materials

**Table 2.1**   Resistance of 316 stainless steel to attack
by HPLC mobile phases and acids

| Mobile phase | Corrosion resistance |
|---|---|
| Acetic acid | Resistant |
| 880 ammonia | Resistant |
| Ammonium hydroxide | Resistant |
| Dichloromethane | Resistant |
| Hydrochloric acid | Some corrosion |
| Phosphoric acid | Resistant |
| Nitric acid | Resistant |
| Sulphuric acid | Some corrosion |
| Hexane | Resistant |
| $HNO_3/H_2SO_4$ mixtures | Resistant |

such as ruby and sapphire for many components. Table 2.1 shows the resistance to attack of 316 stainless steel by a wide variety of chemicals. Since silica is the basis of most of the current column packings this means that the pH range of HPLC mobile phases is restricted to less than pH 8.0. Table 2.1 indicates that the only common material to avoid using in HPLC eluents is any compound capable of producing chloride ions in solution or sulphuric acid.

(2) It should be capable of continuously supplying large volumes of mobile phase without attention.

(3) It should have low internal volume so that the changing of mobile phase can be achieved rapidly and conveniently.

(4) It should not contribute towards any detector noise.

(5) It should be capable of being used in a gradient elution system.

(6) It should be capable of providing flow rates at the pressures developed with HPLC columns. In practice, with packed microporous analytical columns of internal diameter approx. 5 mm and length up to 30 cm, this means pressures up to 6000 p.s.i.[†] and analytical flow rates of between 0.5 and 3.0 ml min$^{-1}$.

(7) Flow from the pump should be precise so that variation in peak retention time does not occur.

These features are difficult to achieve in practice, although modern pumps do approach this ideal extremely closely.

---

[†]Throughout this chapter pressures have been given in pounds force per square inch (p.s.i.) since this unit is still in common use in HPLC instruments. 1 p.s.i. = 6.8948 kPa.

## A Constant Pressure Syringe Pumps

A schematic diagram of this type of pump is shown in Fig. 2.2. The pump consists of a piston which is caused to move inside a pump chamber by the action of a pneumatic actuator. The pump chamber is filled by the force of compressed air acting downwards on the diaphragm of the pneumatic actuator. The piston is driven downwards, the outlet check valve closes and the inlet check valve opens drawing mobile phase into the pump chamber.

At the end of the fill stroke the path of the compressed air in the actuator is diverted via the gas valve, so that it now acts upwards on the diaphragm. This forces the piston to move upwards, closing the inlet check valve, opening the outlet check valve and forcing the mobile phase through the column. Depending on the pump design, the gas valve can be operated either manually or automatically. The pump chamber volume is normally between 2 and 60 ml. A main feature of constant pressure syringe pumps is that because the cross-sectional area of the actuator diaphragm is very much larger than that of the piston, a pressure amplification is obtained within the system. Thus, it is possible to generate very high mobile phase pressures, up to 10 000 p.s.i., with relatively low gas input pressures of the order of 80–100 p.s.i.

There are two main advantages of this type of pump. During the delivery stroke the delivery of solvent through the column is perfectly smooth and

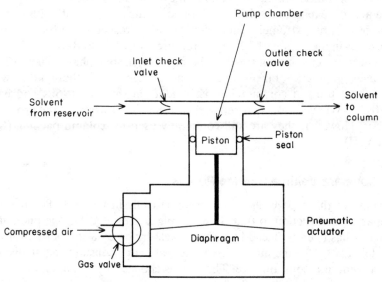

**Fig. 2.2** Schematic diagram of a gas-pressurized syringe pump.

uninterrupted. Hence the pump does not contribute to detector noise. Secondly, pumps of this type are simple in construction and are comparatively inexpensive and generally extremely reliable and easy to maintain.

They do, however, suffer from certain disadvantages. The fact that they are constant pressure devices means that at a given input pressure anything which causes a change in column back pressure, e.g. changes in ambient temperature or column deterioration, will cause a change in flow from the pump. This in turn will change the retention times of the peaks in the chromatogram. This can occur slowly over a period of time or very rapidly, particularly if great care is not taken to remove undissolved particles from samples prior to injection.

Changes in retention times of this nature can make interpretation of the resulting chromatograms difficult and this must be regarded as a serious disadvantage of constant pressure pumps. On the fill stroke the pump stops, this in turn causes a disturbance on the detector base-line. The degree of disturbance is related to the speed of the fill stroke which in turn is related to the pump chamber volume.

Even with small chambers of the order of 2 ml and very rapid fill times, base-line spikes can be observed particularly with flow-sensitive detectors such as refractive index or electrochemical detectors. It should also be noted that too rapid a refill stroke can create a partial vacuum within the pump chamber resulting in cavitation and disturbance to the flow.

During method development in HPLC it is important that changing of the mobile phase should be simple and economic, requiring only small volumes of expensive solvents. Syringe pumps, even with 2 ml pump chambers, are extremely inefficient in this respect and pumps with capacities in the region of 50–60 ml require many hundreds of millilitres of solvent to purge the final trace of the previous mobile phase. These disadvantages, in practice, far outweigh the advantages and, although they were commonly used in commercial instruments in the late 1960s and early 1970s, they have now been replaced for analytical work by reciprocating pumps. However, they are still commonly used for column packing (Section VIII).

## B Constant Volume Syringe Pumps

A pump of this type is shown diagrammatically in Fig. 2.3. The pump is similar in construction to that shown in Fig. 2.2 except that the pneumatic actuator has been replaced by a screw-jack driven by a stepper motor. On the delivery stroke the motor drives the piston at a constant speed, displacing solvent at a constant rate. The flow is changed by changing the motor speed and is normally dialled directly by controls on the front of the

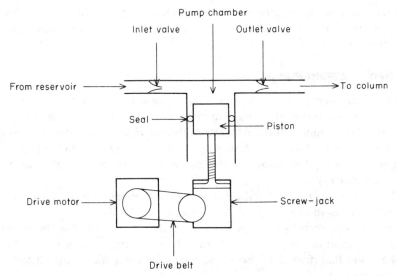

**Fig. 2.3** Schematic diagram of a constant volume syringe pump.

instrument. These pumps have many similar characteristics to constant pressure syringe pumps but they do differ in a number of important aspects.

First of all they deliver solvent at a constant rate regardless of column back pressure. Thus peak retention times can be expected to remain constant with these pumps, provided that the solvent composition is not changed. In this respect, they represent a considerable improvement over constant pressure devices.

The fill cycle of the pump, however, is now more complicated than its gas-pressurized equivalent. Simple reversal of the drive motor mechanism is totally unsatisfactory because of the slow reverse speed involved. If a constant volume syringe pump of capacity 60 ml were operated at a forward flow of 1 ml min$^{-1}$ simple reversal of the direction of the drive motor would mean that the pump chamber would take an hour to refill.

Several mechanisms have been used to solve this problem: (1) the use of very large pump chambers up to 500 ml – this obviously makes solvent changing more difficult; (2) a gas pressure-assisted fill cycle; (3) the use of rapid refill motors in conjunction with slow speed delivery drive motors operated via a clutch mechanism; (4) a combination of (1), (2) and (3).

These design features, which are essential for the satisfactory operation of these pumps, are unfortunately expensive. Their resultant high cost combined with the problems associated with changing solvents has made constant volume syringe pumps as unpopular as their constant pressure counterparts. However, these pumps may be of considerable importance for capillary columns.

## C Reciprocating Pumps

Rather than having a pump head volume measured in millilitres and compression stroke times of many minutes, reciprocating pumps commonly have internal volumes of the order of 100–200 $\mu$l which are swept several times per minute. This immediately overcomes the major disadvantage of syringe pumps so that mobile phases can now be changed simply, quickly and economically.

The piston is driven by either a d.c. or a stepper motor, via a fixed eccentric or, as shown in Fig. 2.4, via a cam.

The flow in a reciprocating pump may be changed by altering the stroke or displacement of the piston, while maintaining a fixed motor speed. The stroke – and therefore the flow – is normally set via a micrometer fitted to the pump drive.

It must be noted that with pumps of this type the resultant flow will also depend on the back pressure applied to the outlet valve. In chromatography this is normally the back pressure on the column, which will depend on the viscosity of the mobile phase being pumped. Thus at a fixed stroke setting different flows will be obtained with different mobile phases. The

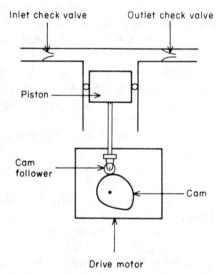

**Fig. 2.4** Schematic diagram of a reciprocating pump.

relationship between flow and back pressure with these pumps will depend on the design of the pump head. The reason for this is the compression or compliance of the mobile phase and the materials used in the pump construction. Liquids are very slightly compressible, albeit only to the extent of a few per cent under normal HPLC conditions. Fluorinated hydrocarbon polymers, the most commonly used pump seal material, are also quite compliant. The degree to which the mobile phase and seal are compressed depends upon the pressure exerted upon them.

The forward stroke of the piston can therefore be represented by the simple equation

$$S = C + D, \tag{2.1}$$

where $S$ = total piston stroke, $C$ = that part of the stroke during which the system is compressed, $D$ = that part of the stroke during which mobile phase is delivered from the pump. Hence, at a fixed value of $S$ (stroke or displacement) the $C$ (compression) part of the stroke will be higher at higher pressures and hence the delivery or flow will be lower. The relationship between pressure and flow will depend on a variety of parameters such as pump volume, total pump displacement and seal design. Thus reciprocating pumps, in which the stroke is varied to vary the flow, can never be strictly regarded as constant flow devices.

However, for routine analytical work where no solvent changing is desired and only small variations in column back pressure are encountered, this type of pump represents an inexpensive reliable method of pumping for HPLC.

As an alternative to varying the pump stroke it is possible to fix the pump stroke and to vary the drive motor speed. Thus by doubling the motor speed the flow from the pump is doubled and a reasonably linear relationship between motor speed and flow exists which is independent of column back pressure. These pumps can be regarded as constant flow pumps, although constant speed devices are normally used with the drive motors to maintain reasonable flow precision.

## 1 Pressure and Flow Pulsations

Consider a reciprocating pump with a displaced volume of 100 $\mu$l. At a flow rate of 1 ml min$^{-1}$ the piston will have 10 output strokes and 10 input or fill strokes per minute. As in the case of syringe pumps the fill cycle produces a fluctuation in flow. The frequency of these fluctuations will, however, be greater with reciprocating pumps and therefore a greater problem. The actual flow–time curve depends on the shape and design of the cam or eccentric driving the piston. These flow variations are known as pulsations and occur in the case of single piston reciprocating pumps, controlled by varying either the stroke or the motor speed.

In fact to describe some reciprocating pumps as "constant flow" is not strictly correct, they are "constant mean flow" or "constant volume" pumps. Pump pulsations cause serious problems in the form of extraneous detector noise and therefore must be removed.

## 2 Pulse Dampeners

Various methods have been used to remove these pressure pulsations (Johnson and Stevenson, 1978; Ventura and Nikelly, 1978). They all consist of hydro-pneumatic mechanical devices containing large volumes, which are placed between the pump and the column. They can either be placed in-line or in a T-configuration. Their function is to absorb the flow variations of the pump. Various designs of pulse dampener have been described elsewhere, but in their simplest form they consist of a length, normally coiled, of narrow bore stainless steel or Teflon tube. A "teed-in" pressure gauge, such as a Bourdon tube is also very effective at removing pulsations. A combination of the above methods can also be used.

Pulsations tend to be worse at lower pressures and at pressures of 1000 p.s.i. or above the chromatographic column itself acts as a reasonably effective pulse dampener.

The disadvantage of using pulse dampeners is that because of their large dead volumes, normally of the order of 10 ml, changing mobile phases again becomes tedious and uneconomic. In particular, in a T-configuration a drain point is extremely important since the dampener contains an unswept dead volume. This could have undesirable effects on the chromatography, particularly if immiscible mobile phases are involved in the solvent change, e.g. the change from water to dichloromethane. This would normally be carried out by flushing the water-filled pump with methanol, followed by dichloromethane. If the "teed" pulse dampener were not drained it would still contain water, which could leak out into the dichloromethane stream, with which it is immiscible, creating spurious noise in the detector.

## 3 Twin Piston Reciprocating Pumps

As an alternative to the use of pulse dampeners, a system involving the use of two pistons operating 180° out of phase has been developed. Such a pump is shown schematically in Fig. 2.5. The flow–time curve for a twin piston pump is shown in Fig. 2.6. Fractionally before piston A begins its fill stroke, piston B begins its compression stroke, so that the loss in flow from one piston is always taken up by a corresponding increase in flow from the

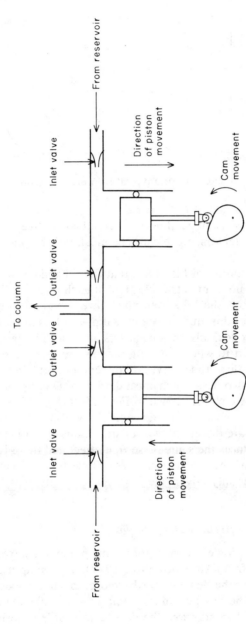

**Fig. 2.5** Schematic diagram showing the operation of a twin piston reciprocating pump.

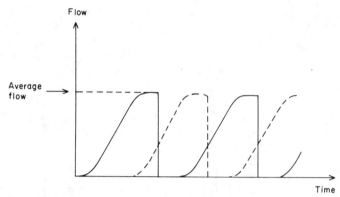

**Fig. 2.6**  Flow–time curve for a typical twin piston reciprocating pump.

other. The resultant average flow is constant and, except for the most flow-sensitive of detectors, e.g. electrochemical, may be regarded as pulse free.

One important feature of Fig. 2.6 is that the fill and compression cycles of each piston are not symmetrical: although each piston cycle is identical to that of its partner, fill stroke must always be more rapid than the compression stroke; this must be the case if piston B is to be able to begin its compression stroke as piston A begins its fill cycle. This fact is important and will be referred to when gradient systems are being considered. The exact shape of the flow–time curve will depend entirely on the pump cam profile. There are two basic mechanical designs of twin piston pumps used by manufacturers which are shown in Fig. 2.7. In Fig. 2.7(a) the two pistons are driven by one cam mounted centrally between the pistons. In Fig. 2.7(b) the pistons are driven by two separate cams operating side by side. Either design produces the same pulse-free effect and there is little difference in the cost of the two alternatives. Three-headed pumps are also available but their only advantage is in the generation of low pressure gradients (Section II.C.1).

## 4   Other Pulse-free Reciprocating Pumps

In the mid 1970s a single piston reciprocating pump was introduced which is relatively pulse free (Altex Scientific, 1977). The pump uses a variable speed motor to alter the flow but, furthermore, the motor speed is controlled so that during the first part of the compression stroke the motor speed is rapidly increased. In addition, the drive cam profile is designed so that the speed of the input stroke is increased. In this way the total fill cycle is reduced to less than a second, which in turn reduces the size of the flow and

(a)

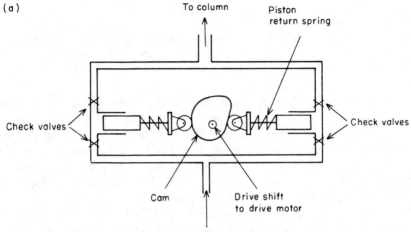

To column

Piston
return spring

Check valves

Check valves

Cam

Drive shift
to drive motor

From reservoir

(b)

To column

Check valves          Check valves

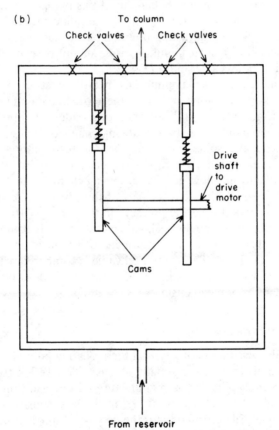

Drive
shaft
to
drive
motor

Cams

From reservoir

**Fig. 2.7**  Schematic diagrams of commercial twin reciprocating pumps: (a) side
view of single cam system; (b) top view of twin cam system.

pressure fluctuations within the pump. This pump design is remarkably effective and has been used by a number of other manufacturers. At high detector sensitivities, however, an in-line pulse dampener is necessary to remove the residual pulsations.

More recently a totally different design of pulse-free, single piston reciprocating pump has been introduced (Applied Chromatography Systems, 1981). It operates on the principle of fixing the drive motor speed, but instead of driving the piston at rates of 10–30 strokes per minute, the piston speed is increased to 23 strokes per second. The pump chamber volume is now reduced to less than 10 $\mu$l. Thus the pump delivers an extremely rapid series of low volume pulses. At this pumping speed the detectors, pressure gauges and recorders used in HPLC are incapable of detecting the pump pulsations and so the pump can be regarded as being totally pulse free. A schematic diagram of this pump is shown in Fig. 2.8. The drive motor is operated at constant speed and the flow is changed by altering the stroke, by adjusting the angular displacement of the arm A, which brings about a corresponding displacement of arms B and C which are pivoted at points 1, 2 and 3 (Fig. 2.8).

When arm A is at its maximum downward deviation to the horizontal this gives maximum piston stroke, and therefore maximum flow. The pump in this form, however, is still not operating under conditions of constant volume delivery, since the stroke is being changed to alter the flow. It is necessary to make some compensation for the compressibility effects occurring within the pump head. This has been accomplished by incorporation of an electronic flow compensation mechanism. In this system the exact point at which the delivery part of the piston stroke commences is determined by monitoring the pressure in the chamber (the pressure falls when the outlet valve opens). This delivery stroke is compared with the total piston stroke and from a knowledge of the head geometry the true flow can be calculated. This computed value is compared with the selected flow rate and if they are not identical the flow rate is adjusted automatically via a feedback mechanism.

## 5   Pump Construction

The internal construction of a typical reciprocating pump head is shown in Fig. 2.9. The piston seal is normally of glass-filled Teflon fixed within the pump body which in turn is manufactured from a chemically resistant stainless steel. The use of glass filling in the seal construction is to help prevent mechanical deformation of the polymer. Similar compression seals are used within the inlet and outlet unions to prevent leakage of solvents around the unions. The piston itself is normally sapphire rather than stain-

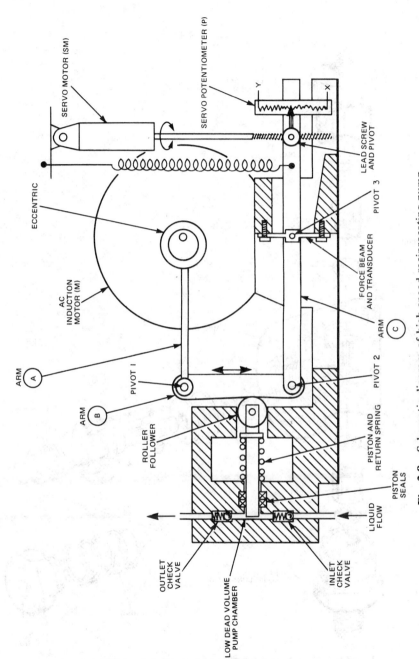

**Fig. 2.8** Schematic diagram of high speed reciprocating pump.

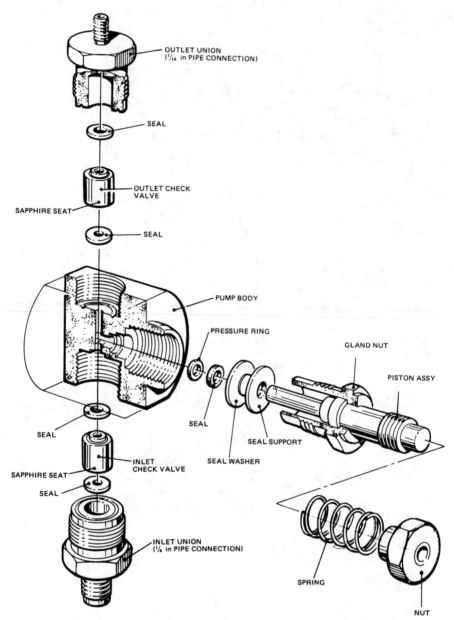

**Fig. 2.9** Internal construction of a reciprocating pump head.

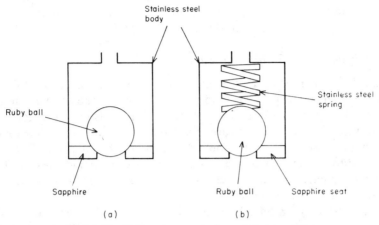

**Fig. 2.10**   Internal construction of check valves.

less steel, since even the slightest corrosion of a metal piston could cause permanent damage to the piston seal. Sapphire, however, is totally resistant to attack by all the common HPLC solvents. In a typical check valve a ruby ball sits on a sapphire seat forming the seal; pressure within the pump chamber causes the ball to open or close depending on whether it is functioning as an inlet or an outlet valve.

Two types of valve are commonly used (Fig. 2.10). In the first [Fig. 2.10(a)] the ball operates under gravity and is only closed by the operation of the pump. In the second [Fig. 2.10(b)] the valve is spring-loaded, in this case the spring keeps the valve closed until the pressure developed by the pump forces it to open. The spring is used to prevent check valve leakage at low pressure, which can occur with gravity-operated valves. For this reason gravity valves are often used in pairs, particularly when operating as outlet valves.

### 6   Advantages of Reciprocating Pumps

The reasons for the success of reciprocating pumps in analytical HPLC can be summarized as follows:

(1) They have theoretically infinite solvent reservoirs.
(2) They can readily be made pulse free.
(3) Their low internal dead volumes make solvent changing convenient. This cannot be overemphasized in method development.
(4) They are generally reliable and simple to use and maintain.
(5) They are compact and relatively inexpensive.

These points correspond extremely closely to the ideal requirements for an HPLC pump cited in the introduction to this section.

### 7 Maintenance of Reciprocating Pumps

It must be remembered that reciprocating pumps are mechanical devices and that their life expectancy and reliability will depend on the care they receive during routine use. The following comments can be regarded as general hints on the routine care and maintenance of pumps. Most reciprocating pumps are fitted or supplied with in-line solvent filters. These are designed to prevent small particles entering the pump and causing blockages and damage to seals and check valves. It is extremely important that in-line filters are always used.

When in-line filters have become blocked they act as a resistance to flow and as such can cause cavitation within the pump head. This can be prevented by ensuring that solvents are dust and particle free by pre-filtering. It is also useful to ensure that in-line solvent filters are frequently cleaned, preferably ultrasonically. If solvents containing buffers are used it is important that they are not allowed to stand within the pump for long periods of time, e.g. overnight. If any buffer should crystallize out of solution, while the pump drive is off, the crystals will almost certainly cause damage, either to the check valves by scratching the moving surfaces or to the seals by cutting the seal surface when the pump is restarted. This can be prevented by flushing the pump thoroughly with water after use; alternatively, the flow can be reduced to a minimum, e.g. 0.1 mm min$^{-1}$, to ensure that the solvent is not stationary within the pump.

## III   GRADIENT ELUTION SYSTEMS

Gradient elution in liquid chromatography involves changing the composition of the mobile phase continuously during the chromatographic run. In order that retention times are reproducible, it is extremely important that the apparatus used for gradient work should be capable of reproducing given gradients accurately. Furthermore, gradient elution is frequently used for the separation of complex mixtures of compounds which may be structurally very similar. It is therefore an advantage for gradient devices to be capable of producing a large number of gradient profiles so that the analyst has a high degree of flexibility in his choice of mobile phase conditions.

There are two ways of forming gradients. The solvents can be mixed prior to the pump, i.e. low pressure gradient systems; alternatively, the solvents can be mixed after passing through the pumps, i.e. high pressure gradient systems. These will be considered separately and are also outlined in Chapter 4 (Fig. 4.2).

## A  Low Pressure Gradient Systems

In low pressure gradient systems the mobile phase is being constantly supplied to the pump and thus the only type of pump which can be used is a reciprocating pump. The mixing chamber can theoretically be very simple, e.g. two or more reservoirs connected by suitable plumbing and feeding each other by gravity. Systems of this kind have been reported in the literature (Perrett, 1976); however, they do not always work with all the combinations of mobile phase likely to be used by the chromatographer, they do not always give good reproducibility and they can be inconvenient to use. Consequently they are not offered commercially. Commercial low pressure gradient elution systems are all based on the apparatus shown in Fig. 2.11.

The inlet line of the pump is connected to two or sometimes three switching valves. In turn the switching valves are connected to the solvent reservoirs. The opening and closing of the switching valves is controlled by the gradient programmer, which is purely an electronic device used to control the gradient profile.

During the gradient run the length of time during which valve B is allowed to remain open relative to valve A will be increased at a rate which will depend on the gradient profile. The length of the time cycle of the switching valves is critical. If it is too long serious mixing problems will occur leading to excessive "mixing" noise on the detector base-line. If, on the other hand, it is too short, it will be very difficult to operate at low proportions of one solvent. This is because it takes a finite time for the switching valves to open, which in turn means that there is a minimum quantity of solvent which can be mixed. In practice cycle times of 1–2 s are commonly used. Under normal conditions this allows the accurate addition of approximately 5% of one solvent to another.

One further complication must be considered with this type of gradient system. In the case of twin piston pumps the fill stroke is more rapid than the compression stroke. Thus the pump is not taking solvent at all times during its cycle of operation. Therefore at low concentrations of solvents A or B, when one switching valve is open for only a very short period, it is possible that this may coincide with a time interval during which no solvent is being taken by the pump. Consequently an error occurs in the gradient profile which can seriously affect the mobile phase composition. This is obviously much more serious with single piston pumps but does not occur

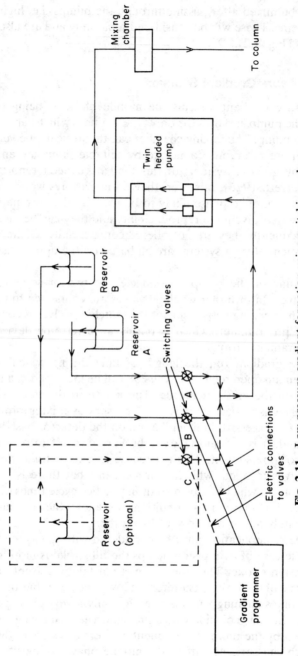

**Fig. 2.11** Low pressure gradient formation using switching valves.

at all with three-headed pumps, since in this case there is never a time when solvent is not being taken by the pump.

There are two procedures which have been devised to overcome this problem. The phenomenon is statistical and its effects can largely be eliminated by using large mixing chambers (5–10 ml) in-line between the pump and the injection valve. This does have the disadvantage that it introduces a large volume which increases the regeneration time of the system and has to be taken into account when the gradient profile is being designed. In an alternative method the pump cam position can be electronically monitored and the switching valves only opened when the pump heads are in their fill cycle. This does overcome the problem completely but can add significantly to the cost of the system.

A more complete asessment of low pressure gradient systems has been carried out by Billiet, Keehnen and De Galan (1979) who compared a number of commercially available chromatographs.

The main advantage of low pressure programming is the cost, since it involves the use of only a single pump. This is offset to a degree by the problems associated with the interaction between the pump fill cycle and the switching valves discussed above, which can only be solved either by increasing volumes within the chromatographic system or by increasing the cost. Low pressure gradient systems do have an added advantage over their high pressure counterparts in that by using a third switching valve it is possible to carry out ternary gradients using three solvents. This does, however, further increase the cost and is only of marginal benefit in practice.

## B  High Pressure Gradient Systems

Liquid chromatographs designed to operate using high pressure mixing techniques use two pumps and the gradient programmer now controls the pumps directly rather than operating on switching valves. The total flow desired by the analyst is selected via controls on the programmer, e.g. 1 ml min$^{-1}$, and if 100% A is desired pump A operates at 1 ml min$^{-1}$ and B does not operate at all. If the composition is changed to 75% A and 25% B , pump A is driven at a flow of 0.75 ml min$^{-1}$ and pump B at 0.25 ml min$^{-1}$, the total flow remaining at 1 ml min$^{-1}$. Thus gradients can be constructed in a simple, straightforward manner. Despite the cost of the extra pump, high pressure mixing of this kind is favoured by many analysts. This is largely because it is simple, the solvent mixing problems are less than with low pressure systems (although mixing chambers are normally still required see below), and accurate reproducible gradients can be obtained even at low concentrations of one solvent.

There is also an advantage in using two pumps in that when not in use as a gradient chromatograph the two pumps can be used separately as two isocratic systems. This is particularly true of modular HPLC systems and many users consider high pressure gradient systems to be more flexible than low pressure systems because of this.

High pressure systems have been designed using either syringe or reciprocating pumps.

True ternary gradients are, however, less straightforward, although they can be achieved by starting with mixed solvents, e.g. solvent A could be water and B could be a 50 : 50 mixture of acetonitrile and methanol.

## C  Mixing Chambers

The physical and chemical properties of any two solvents likely to be used in gradient elution are unlikely to be identical. For instance, a common combination of solvents is water and methanol. Even the purest grade of methanol will absorb ultraviolet light quite strongly at wavelengths below 240 nm, whereas water at the same wavelengths is still relatively transparent. Thus if these two solvents are inadequately mixed during a gradient run portions of solvent pass down the column and into the detector, some of which are methanol rich and some of which are water rich. This produces a sinusoidal base-line of the type which can be seen in Fig. 2.12. Base-line "mixing" noise of this type is an obvious disadvantage and may be removed by the use of an in-line mixing chamber. For solvents whose physical properties are very similar it is often sufficient to use a short (10 cm) tube packed with glass beads, in this respect the column itself also acts as an aid to mixing.

With less miscible mobile phases mechanical mixing is required, in commercial liquid chromatographs this is normally a small magnetic stirrer.

In the case of the high speed pump described in Section II.C.4, no mixing chamber is required. This is because the extremely low volumes delivered at each pump stroke mix readily within the volumes of the pipework and column of the chromatograph.

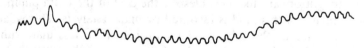

**Fig. 2.12**  Detector noise as a result of poor mixing. Conditions of experiment were as follows: solvent, 60% methanol in water; flow rate, 2 ml mm$^{-1}$; detector, ultraviolet at 240 nm; sensitivity, 0.2 AUFSD.

## D  Gradient Programmers

Gradient programmers are electronic control devices which change the mobile phase composition with respect to time either by acting directly on the pumps (high pressure system) or on input switching valves (low pressure system). In its simplest form the electronic system is pre-programmed with a series of composition versus time curves, of which one is selected. However, greater flexibility may be achieved with the so-called multi-linear system, in which the gradient profile is constructed of a number of linear gradients. This allows constant composition sections in the curve and negative gradients permit efficient re-establishment of initial solvent conditions.

The use of microprocessor-based programmers also allows other parts of the system, e.g. autosamplers and integrators, to be controlled and these aspects, together with examples of the various programmers, are discussed further in Chapter 4.

## IV  INJECTION DEVICES

The injection of the sample to the head of the column is one of the most critical parts of the chromatographic process. Ideally the sample should arrive at the column as a discreet spot. Any phenomenon which causes extra diffusion at the injector will result in band-broadening and an associate loss in chromatographic resolution (Kelsey and Loscombe, 1979).

## A  On-column Injection

The simplest way of carrying out on-column injection is to inject through a septum using a micro-syringe. A suitable design is shown in Fig. 2.13. Alternatively the glass beads may be replaced by some other inert material such as quartz wool or a porous PTFE plug. It is important, however, that the septum should be lined with PTFE to prevent attack by organic solvents. This method of injection is inexpensive and because it involves the use of micro-syringe offers a great deal of flexibility in terms of the sample volumes which can be injected. In practice, there are serious problems associated with the use of septa. In the author's experience they are impossible to use at pressures greater than 1000 p.s.i. and, even at pressures lower than this, they tend to leak unless changed very frequently, preferably at least once a day. In addition their low capital cost can be offset by the cost of damage to syringes, which can be considerable.

Another way of producing direct on-column injection is to use a stop-

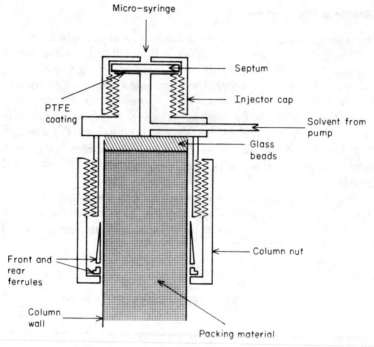

Fig. 2.13  Septum injection for HPLC.

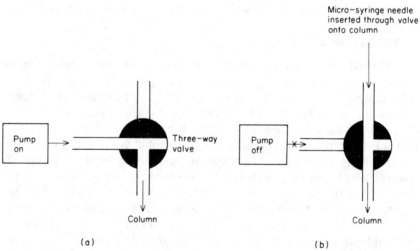

Fig. 2.14  On-column injection using a three-way manual valve: (a) system
running; (b) injection.

flow technique. The top of the column is packed with glass beads or some other suitable material as before but no septum is used; instead the method involves the use of manual valves as shown in Fig. 2.14. To carry out stop-flow injection the pump is first of all switched off, and the injection valve opened. The needle is inserted through the open valve and the sample injected. The valve is then closed and the pump switched on.

Stop-flow injection retains the flexibility allowed by the use of a micro-syringe and has the advantage that it can be used at pressures up to 6000 p.s.i. Furthermore, the lack of a septum within the injector greatly prolongs syringe life. However, it can be inconvenient to have to stop and start the pump for each injection.

## B  Injection Valves

Despite the risk of the band-spreading which can occur with off-column injection, the convenience associated with the use of high pressure injection valves makes them a very attractive proposition. In fact, modern valve designs do not contribute significantly to band-broadening with packed columns. There are two types currently in use:

(1) *Internal loop valves*    These valves consist of a rotor mounted within the valve body, as shown schematically in Fig. 2.15. The sample loop consists of a calibrated groove cut into the rotor surface. The sample is loaded into the loop via a 1 or 2 ml syringe and as the rotor is turned during the injection the sample is moved into the solvent stream and hence injected onto the column. The volume of sample which can be injected in this way varies from 10 $\mu$l upwards. Although simple to use and reliable in operation, internal loop valves suffer from the disadvantage that it is impossible to change the volume of sample being injected without changing the valve.

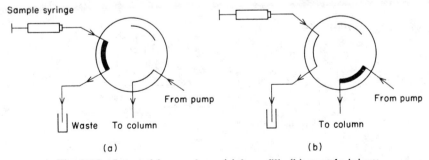

**Fig. 2.15**  Internal loop valves: (a) loop fill; (b) sample inject.

(2) *External loop valves*  These valves (Fig. 2.16) are very similar in construction to internal loop valves except that the sample loop is now a length of calibrated capillary tubing mounted externally. It is therefore now possible to change the sample volume by changing the loop.

Neither external nor internal loop valves of this type have the flexibility involved in using micro-syringes since they tend to have large internal volumes associated with them. Consequently, the total sample required also tends to be quite large, generally 0.5 ml or greater.

In recent years another type of valve has become available in which these dead volumes have been reduced to a minimum. The design is typified by the Rheodyne 7125 injection valve shown in Fig. 2.17. The valve is very similar to that shown in Fig. 2.16 except that the sample is introduced via a micro-syringe, directly into the sample loop. The syringe needle passes through a PTFE needle seal and butts against a hole in the rotar seal which forms the entrance to the sample loop. In order to prevent damage to the rotar seal it is essential that a flat-tipped syringe needle is

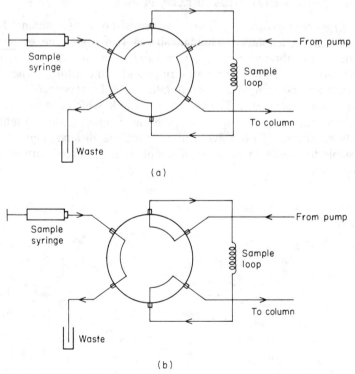

**Fig. 2.16**   External loop injection valves: (a) loop fill; (b) inject.

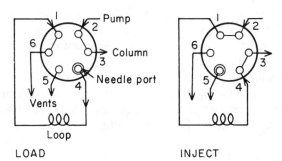

**Fig. 2.17** The Rheodyne 7125 injection valve. (Reproduced by courtsey of the Rheodyne Corp.)

used. There is now no dead volume on the sample loop inlet and hence the sample loop can be either totally or partially filled with a high degree of accuracy. In the author's experience with this valve, in the partial fill mode it is possible to carry out repeat injections with a precision of better than 1% (coefficient of variation) as measured by peak height. Additionally, it is possible to change the total loop volume from 10 $\mu$l up to several millitres. Automatic valves of this kind are also available.

For packed column HPLC this type of valve probably represents an optimum in that it combines the flexibility of using a micro-syringe with the ease of operation of a high pressure valve. It is, however, still important to use low volume tubing, preferably 0.010 inches (254 $\mu$m) i.d. or less, to connect the valve to the column in order to minimize band-broadening.

## V TEMPERATURE CONTROL

Maintaining a constant temperature within the column is less important in HPLC than it is in gas–liquid chromatography (GLC). This is because HPLC separations are less affected by changes in ambient temperature. There are some exceptions, however; for example, in gel permeation chromatography where the polymeric materials under analysis may be only partially soluble at room temperature. Some ion-exchange and reversed phase separations are also affected by temperature changes.

In these cases it is important that the temperature is controlled accurately and reproducibly. This can be done either by placing the column within a water jacket (Macrae, 1978) or by using an air-circulating oven.

If a circulating water bath/water jacket combination is used, the temperature control achieved is dependent on the temperature stability of the water bath. Ovens have an advantage in that injection valves can be mounted within the oven cabinet. This can be important, particularly if sample solubility is a problem, since it prevents the precipitation of sample within the injector.

The use of air-circulating ovens does present a safety hazard if inflammable solvents are being used. It is therefore essential to have the oven fitted with an efficient nitrogen purge to prevent fire or explosion if solvent leaks should occur.


## VI  DETECTORS


Detectors can, in many respects, be described as the weak link in the HPLC chain, and it is a fact that no totally universal, sensitive detector exists. This is despite the efforts of many workers over the last decade to develop the HPLC equivalent of the flame ionization detector (FID) commonly in use in GLC. The main reason for this is that in HPLC the mobile phase is not an inert material as in the case of the carrier gas in GLC. The mobile phase in HPLC is normally an active participant in the separation process and almost all of the solvents used for HPLC have some organic content. Since the technique is normally used for the separation of small amounts of organic compounds, the presence of relatively large amounts of organic solvent in the mobile phase complicates the problem of finding suitable detection systems. In addition all flow-through detectors must conform to a number of other design criteria. Poppe (1980) has looked at the design and characterization of liquid phase flow-through detectors for HPLC and other applications. He suggests that the following detector characteristics are the most important and they should be known and understood by the analyst for any detector under study:

(1) *Calibration*  The relationship between response and solute concentration should be known; for preference this should be linear. If it is not linear, throughout its total working response range, the range of its linearity should be established. Most detectors currently in use for HPLC do have a reasonable linear working range.

(2) *Dynamic behaviour of the detector*  In HPLC this can be summarized as peak-broadening effects. Band-spreading in HPLC detectors normally occurs as a result of having large volumes within the flow cell or the connecting pipes. Detector response time is also important since a very slow response time can artificially distort the detector signal. Response

times for the detectors in common use in HPLC vary from 0.2 to 10 s but normally fall within the range 0.2–2 s.

(3) *Selectivity*   The dependence of the detector response on the compound type must be known.

(4) *Noise and detection limit*   The signal to noise ratio is important, particularly when trace analysis is involved.

Broadly speaking the detectors which have been developed for HPLC can be classified into three types:

(1) Solute-specific detectors, in which some specific property of the solutes, which is not shared by the mobile phase, is continuously monitored. Post-column reaction detectors can be considered to be a form of specific detector.

(2) Bulk property detectors, in which some bulk property of the mobile phase is monitored and the changes associated with the presence of the solute are used as the detection principle.

(3) Desolvation detectors, in which the mobile phase is removed by evaporation, leaving the solute behind. The solute is then detected separately by another technique.

## A  Solute-specific Detectors

### 1  Ultraviolet Light Detectors

Detectors based on the absorption of ultraviolet (UV) light by the solute are by far the most common. Figure 2.18 shows the general layout of a UV detector for HPLC. It consists of a light source, some method of wavelength selection, a flow-through cell and a photodetection system, which may be a photomultiplier, a photodiode or some other light-

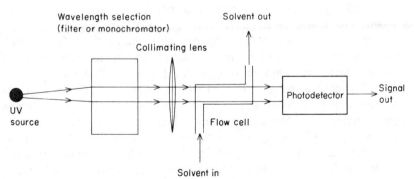

**Fig. 2.18**  Schematic diagram of an ultraviolet detector for HPLC.

measuring device. The two main sources of UV light in common use in HPLC detectors are mercury and deuterium lamps.

Mercury lamps have a strong emission line at 254 nm and detectors employing mercury lamps are inexpensive, very stable and sensitive, provided that the compounds under analysis absorb UV light appreciably at 254 nm. A filter is also used in conjunction with mercury lamps to remove other unwanted emissions. Since mercury lamps also display weak emissions at wavelengths higher than 254 nm, e.g. 313, 334 and 365 nm, it is possible to monitor UV absorption at these wavelengths. By using a phosphor in conjunction with the lamp, absorption at 280 nm can also be measured.

For many compounds these wavelengths are perfectly adequate. However, a large number of organic compounds do not absorb light in these regions of the spectrum. A significant number of substances only show UV maxima at wavelengths lower than 254 nm, e.g. many pesticides, and it is impossible to use detectors based on mercury lamps at these low wavelengths. Deuterium lamps, on the other hand, emit a continuum from approx. 190 to 500 nm, and instruments based on these lamps offer a great deal more flexibility to the analyst.

In deuterium lamp detectors wavelength selection is achieved by one of two methods. In continuously variable wavelength detectors a monochromator is used. These detectors resemble UV spectrophotometers. There are, however, a number of significant differences between HPLC UV detectors and UV spectrophotometers. Most UV spectrophotometers are double beam instruments with a chopper mechanism used to produce a reference beam of light which passes through a cuvette containing the solvent, while the analytical beam passes through a cuvette containing the sample dissolved in the solvent. The spectrophotometer ratios the two signals and displays the true absorbance of the sample.

In the case of a UV detector for HPLC the reference cell, when fitted, is normally filled with air and the reference beam impinges on a reference photocell. The reference signal is now used to balance out variations in lamp output and hence reduce detector noise. Filling the reference cell with solvent serves no useful purpose and in most cases increases noise levels by reducing the amount of light reaching the photocells.

Many UV detectors are not fitted with a reference cell, but a quartz beam splitter is used to produce a reference beam which is then used as before to reduce the effects of changes in lamp intensity.

The band-pass of UV spectrophotometers is normally of the order of 1 nm or less, and they are seldom used at sensitivities lower than 0.5 absorbance units full scale deflection (AUFSD). UV detectors on the other hand are required to operate at sensitivities down to 0.005 AUFSD. At a band-pass of 1 nm the amount of light reaching the photodetector is relatively low and the resulting noise at this sensitivity would render the detector unusable. Consequently, the band-pass of commercial variable wavelength detectors is normally set between 5 and 10 nm. This has two effects, first of all the detector is now less specific than a monochromator, since a set wavelength of 240 nm at a band-pass of 10 nm will also allow the passage of a significant amount of light having a wavelength between 235 and 245 nm. Using a band-pass of 10 nm also means that deviations from Beer's law can be expected, since Beer's law is only valid for monochromatic light. To a certain extent this can be corrected electronically but most UV detectors can be expected to be non-linear at low sensitivity settings.

The range of non-linearity will vary from model to model and it is important that the linear range of the detector is known for each compound under investigation, since some substances disobey Beer's law even in monochromatic light.

As an alternative to using a monochromator with a deuterium lamp it is possible to select the wavelength by means of interference filters. This can be of particular advantage at wavelengths below 240 nm, where stable high sensitivity operation is required. Although the emission from a deuterium lamp is considerable in the region 190–220 nm, even the highest grade of solvents will absorb light strongly at these wavelengths. Furthermore, very high light losses can occur within the relatively long light path of monochromator optics. These two factors combine to give poor noise and drift characteristics with some monochromator-based detectors, particularly at high sensitivities. Alternatively, if a high transmission (approx. 18%) narrow band-pass (10 nm or less) interference filter is placed between the deuterium lamp and flow cell, the overall optical path length from lamp to photodetector can be reduced. This results in high light levels passing through the flow cell despite absorption by the solvent. Detectors using this simple design are now commercially available.

Although mercury and deuterium lamp sources are by far the most commonly used, it is also possible to buy detectors using zinc lamps for operation at 214 nm and electrodeless gas discharge lamps for operation at 206 nm.

The design of the flow cell in UV detectors is also critical. The cell volume must be small and is normally 10 $\mu$l. In detectors used for analytical work the path length is commonly 8–10 mm. It is important that if air enters the cell it is removed automatically by the force of the mobile phase and does not stay trapped within the cell, causing spurious noise. Most UV detectors available at the present time do satisfy this criterion very well. Refractive index effects can also occur within the cell, particularly during gradient elution. Some detectors have flow cells designed in such a way as to minimize these refractive index effects.

Before leaving discussion of UV detectors, it is worthwhile mentioning some other possibilities which are available. The first of these is monitoring two different wavelengths simultaneously. The simplest method of doing this is to operate two detectors in series. This is perfectly satisfactory provided that low volume connections are used, and can be a useful technique where peak overlap is suspected.

As a further aid to peak indentification it is also possible to carry out stop-flow scanning (Readman, Brown and Rhead, 1981) on a peak if a good quality variable wavelength detector is used. For this technique to be useful it is important that the band-pass of the detector is as low as is practically possible to obtain good spectral resolution. Secondly the background spectrum of the solvent must be removed. Some instruments designed specifically for stop-flow work automatically subtract the solvent background absorption.

Instruments controlled by microprocessors can also change wavelength automatically at pre-specified times during the chromatographic run. This can be useful since it means that the optimum wavelength for each compound present can be used rather than either using detectors in series or selecting a compromise wavelength for a particular mixture.

For the future there is no doubt that the ability to monitor several wavelengths or even the whole spectrum simultaneously has many attractive features. This can already be achieved by placing the monochromator on the exit side of the flow cell and monitoring all wavelengths by using either a diode array (James, 1981) or a vidicon tube. Such a detector is discussed in Chapter 4.

In food analysis many components can be analysed with UV detection and many examples will be found in later chapters in this book. Detection limits for samples with good UV characteristics are 1–10 ng.

## 2   Fluorescence Detectors

The number of compounds which fluoresce is smaller than that which show only absorption of light, thus fluorescence is a more specific technique. Furthermore, it is also generally more sensitive. HPLC detectors based on the measurement of fluorescent radiation are available and have been used extensively in trace analysis. In particular, in the determination of polynuclear aromatic hydrocarbons, aflatoxins and, after derivatization, amino acids, see later chapters. With good detector design, it is possible to achieve detection limits in the picogram range. The light sources which have been used in fluorescence detectors include xenon, mercury and deuterium lamps. Flow cells are similar to those used in UV detectors except that the emitted light is measured at right angles to the incident beam. It is impossible to obtain meaningful fluorescence measurements in the same plane as the incident light beam. Monitoring the light at 90° to the incident beam can, however, result in lower sensitivities than might otherwise be expected. Fluorescent radiation is being emitted at all angles around the flow cell; the photodetector, however, will only see a relatively small amount of that emitted light. This situation can be improved by using a larger flow cell up to 70 μl in volume. The extra volume and bandspreading which this introduces tends to be less important in fluorescence detection because it is generally highly specific. One manufacturer (Schoeffel Instruments, 1975) has overcome this particular problem in another way by using a parabolic reflector (see Fig. 2.19) which collects the fluorescent radiation over 180° and focuses it on the end window of a photomultiplier tube.

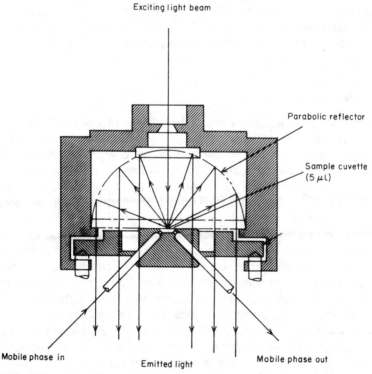

**Fig. 2.19** Schematic diagram of the Schoeffel SF 970 fluorescence detector. (Reproduced by courtesy of Schoeffel Instruments Corp.)

Wavelength selection is carried out by using either optical filters or monochromators. This has resulted in the evolution of three types of fluorescence detector:

(1) All filter detectors, in which both excitation and emission wavelengths are chosen by selecting the appropriate filter. These detectors are reliable, relatively inexpensive and for many applications perfectly adequate. They can lack flexibility for some research applications.

(2) Instruments using monochromators to select the excitation wavelength and cut-off filters to select the emission.

(3) Instruments using monochromators to select excitation and emission wavelengths.

The two last-mentioned types offer much greater flexibility.

More recently, a fluorescence detector which uses a radioactive source rather than a lamp has been reported (Malcolm-Lawes and Warwick, 1980). In this β-induced fluorescence detector, β-particles from the source interact with the mobile phase and the solute, producing fluorescence. Although not yet commercially available this offers some advantages since a radioactive source has predictable life and emission characteristics, unlike optical lamps, and, furthermore, no stabilized power supply is required.


## 3  Electrochemical Detectors

Polarography or voltammetry is a well established technique in the field of metal ion estimation. However, this technique is not specific to metals and can be applied to a large number of organic compounds in solution. It therefore follows that by placing the appropriate electrodes in a flow cell it can be used for HPLC detection. The component parts of such a flow cell are shown in Fig. 2.20. In practice three electrodes are used, the working electrode, to which the working potential is applied and through which the measured current flows, the reference electrode, which completes the electrical circuit, and an auxillary electrode. The use of an auxillary electrode is important so that the potential at the reference electrode does not change, since by definition it is being used as an electrical reference point to measure the working potential. This condition will only apply if no current passes through the reference electrode. In a system using only two electrodes current will pass through the reference electrode when redox reactions begin to occur at the surface of the working electrode. The auxillary electrode ensures that only a small amount of current flows through the reference electrode so that the correct working potential is applied to the working electrode.

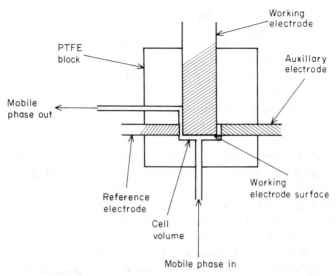

**Fig. 2.20** Schematic diagram of a typical electrochemical detector.

In practice, the chromatographer simply applies a fixed potential to the working electrode equal to or greater than the half-wave potential of the compound under analysis. When that compound elutes from the HPLC column the appropriate electrochemical reaction takes place and the resulting increase in current is measured and recorded as a chromatographic peak. The flow cells used in electrochemical detectors are simple in construction (Kissinger, Refshange, Dreiling and Adams, 1973; Conac, 1979). The electrodes are mounted in a PTFE block and the total cell volume is normally of the order of 2 $\mu$l or less. Several different materials can be used for the working electrode, the most common being either glassy carbon or carbon paste. The reference electrode is either a standard calomel electrode or of silver/silver chloride, with the auxillary being either of carbon or platinum.

Rather than using a d.c. method it is also possible to use a differential pulse detector (Mayer and Greenberg, 1979) which makes use of the differential pulse polarographic wave-form. This method offers enhanced selectivity when compared with d.c. instruments. Electrochemical detectors are simple, inexpensive and very sensitive (picograms or less) provided that the compound under study is electrochemically active. Most published data on the electroactivity of organic compounds refer to half-wave potentials measured when mercury has been used as the working electrode. Carbon is preferred to mercury as the working electrode in HPLC detectors because of the difficulties associated with the construction of a flow cell utilizing dropping mercury. Unfortunately the half-wave potentials observed with mercury are lower than those obtained with carbon by as much as several hundred millivolts. This means that in order to determine the correct operating potential for a particular compound with carbon the analyst must have access to a polarograph fitted with carbon electrodes from which the half-wave potential can be obtained. Alternatively a series of chromatograms can be run using different working electrode potentials to determine the optimum value. It is also important to remember that electrochemical detectors can only be used with mobile phases which are themselves electrolytes. This restricts their use to reversed phase and ion-exchange chromatography. Although electrochemical detectors are sensitive devices, they are also very sensitive to flow changes and therefore can only be used with pulse-free pumps. One final point is that the mobile phases used with electrochemical detectors must be thoroughly degassed as dissolved oxygen interferes with the redox reactions occurring within the cell. Electrochemical detectors have been widely used in the field of clinical chemistry, in particular for the analysis of catecholamines (Loullis, Hingtgen, Shea and Aprison, 1980); they can, however, theoretically be used for the detection of a large number of compounds, including amines, phenols, acids, esters, ketones and aldehydes, although few references appear in the literature.

## 4   Infrared Detectors

In the same way that it is possible to use the principle of the absorption of UV and visible light by organic compounds, it is also possible to utilize the absorption of infrared radiation (IR). Some compounds absorb very strongly in the infrared, for example those compounds containing carbonyl groups.

The construction of IR detectors is similar to that used for UV detection. They consist of an IR source, a flow cell and photodetector. Wavelength selection is carried out by a monochromator or a combination of monochromator and optical filter. The major limitation on their use is that it can be difficult to find a suitable mobile phase which does not absorb in the same region of the IR spectrum as the solute. For example, tetrahydrofuran has very few optical windows within the IR unless the cell path length is reduced to less than 0.1 mm, which can limit sensitivity. Similarly acetonitrile, a common HPLC solvent, has few optical windows in the IR unless a flow cell of path length lower than 1 mm is used. Nevertheless, IR detectors can be very useful for some substances which are difficult to detect by other means, e.g. some lipids (Dupont Instruments, 1977).

## 5  *Derivatization and Post-column Reaction*

The solute-specific detectors described in the previous sections can be very sensitive provided that the solute absorbs UV, visible or IR radiation, or fluoresces, or is electrochemically active. In some cases, if the particular compound cannot be detected by these methods, it may be possible to convert it into a derivative which can be detected by a solute-specific detector.

Derivatization prior to injection of the sample can be carried out, for

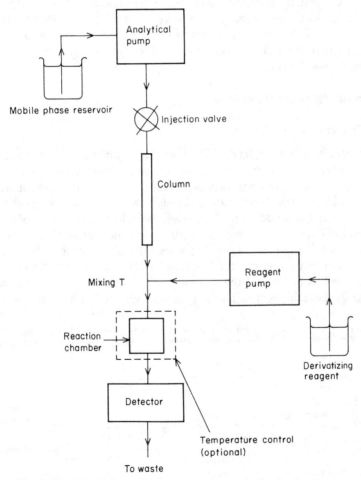

**Fig. 2.21**  Block diagram of a post-column derivatization system.

example the preparation of the *o*-phthalaldehyde (Lindroth and Mopper, 1979) derivatives of amino acids. However, this can complicate the chromatography since the derivatives will almost certainly possess different chromatographic properties to the original compounds. Post-column derivatization, therefore, offers considerable advantage in this respect (Fig. 2.21). The reagent pump supplies the derivatization reagent to the reactor where it is mixed with the column eluate and the chemical derivative is produced and passes subsequently to the detector, which is commonly either a UV or fluorescence detector. The reactor can be heated to increase the rate of the derivatization reaction. Three types of reactor have been used, coiled tube reactors (Frei, 1979) packed bed rectors (Schwedt and Reh, 1981) and segmented flow reactors (Scholten, Brinkmann and Frei, 1981). For post-column reaction techniques to be useful it is not essential that the derivatization is carried to completion, but the extent of reaction must be reproducible.

## B  Bulk Property Detectors

### 1  Refractive Index Detectors

The refractive index (RI) of liquids is a relatively easy parameter to measure both in static systems and in flow-through cells. Two basic types of RI detector are commercially available. In the deflection instrument (Fig. 2.22) the light beam passes through two matched triangular shaped flow cells and is reflected back through the cells from the face of a mirror placed behind the cells. One cell is fitted with the mobile phase (the reference cell) and the column eluate passes through the other (the analytical cell). When both cells are filled with mobile phase the refraction occurring in each cell is equal and the system is at optical zero. The presence of a solute in the analytical cell will cause the reflected light beam to deviate from optical zero, producing a signal which can be passed to a chart recorder. In detectors utilizing a Fresnel prism (Fig. 2.23) two light beams

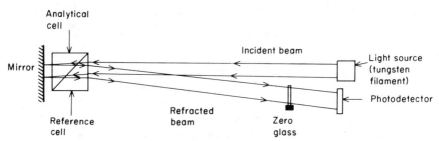

**Fig. 2.22**  Schematic diagram of a deflection RI indicator.

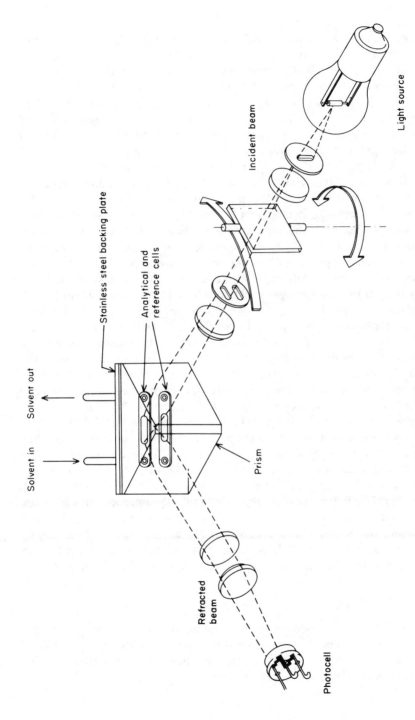

**Fig. 2.23** A Fresnel prism RI detector.

are focused through the prism on the reference and analytical cells. The flow cells are formed by placing a pre-shaped PTFE gasket between the back face of the prism and the front face of a stainless steel backing plate, through which the inlet and outlet pipe connections of each cell also pass. With mobile phase in both analytical and reference cells the two light beams are refracted equally through the prism, causing light beams of equal intensity to fall on the photodetector. With solute in the analytical cell the light beams are refracted by different amounts and different amounts of light from the two beams fall on the photodetector. This difference is measured and recorded as the solute peak. The cell volume in RI detectors is small, of the order of 5 $\mu$l, so that band-spreading is not a serious problem. RI detectors can be regarded as universal in that most compounds affect the refractive index of the liquid. However, both positive and negative changes in RI can be observed depending on the nature of the solutes, which can produce chromatograms showing positive and negative peaks. This can cause difficulties with integrators and data systems. There are a number of disadvantages of RI detectors. In common with other bulk property detectors they are sensitive to changes occurring in the environment, in particular, temperature. In fact the effect on RI of a 1 °C temperature change is much greater than the effect of the presence of a small amount of a dissolved solute. Thermostating the cell helps to reduce the thermal drift which occurs with all RI detectors; however, the detection limit tends to be rather poor (c. 1–10 $\mu$g), largely as a result of poor thermal stability. RI detectors also tend to be pressure and flow sensitive, although not as sensitive to pulsations as electrochemical detectors. The use of pulse-free pumps with RI detectors is therefore an advantage. They cannot be used with gradient elution. Theoretically gradients are possible provided that the flow through the analytical and reference cells is perfectly matched. In practice this is very difficult to achieve even in isocratic operation. It is in fact common practice to fill the reference cell with static mobile phase and to seal the inlet and outlet tubes of the reference with dead-end connectors. This prevents drift due to evaporation of the mobile phase in the reference cell. In food analysis the most common application of RI detectors is in the separation of carbohydrates (see Chapter 6).

## 2  Viscometric Detectors

Detectors based on viscosity measurements are not very common; they have, however, been used successfully in the analysis of polymeric materials (Arlie, 1974). Like RI detectors they are temperature sensitive and have relatively poor detection limits.

### 3  Conductivity Detectors

The conductance of a solution depends upon the concentration of ions present. Thus flow-through conductivity meters can be used as HPLC detectors to monitor the elution of ions from a chromatographic column. The construction of the flow cells used in these detectors is simple, usually consisting of two stainless steel tubes held firmly against a PTFE gasket. A hole through the gasket forms the cell with the tubes forming two electrodes to which a small d.c. voltage is applied. The eluate from the column passes through the cell and out to waste. The current passing between the two electrodes is dependent upon the conductivity of the eluate.

Using conductivity detectors concentrations of inorganic anions such as $Cl^-$, $Br^-$ and $SO_4^{2-}$ can be determined at concentrations in the region of $1-10~\mu g~kg^{-1}$. Similar detection limits are obtained for organic acids. In common with other bulk property detectors conductivity detectors are very sensitive to temperature changes which is a major cause of base-line drift. This can be minimized by siting the detector in a draught-free environment. As in the case of RI detectors, both positive and negative peaks can be observed. This occurs as the result of a number of complex on-column matrix effects, which depend largely on the background conductivity of the mobile phase (Pohl and Johnson, 1980).

## C  Desolvation Detectors

Since the mobile phase is the cause of many of the problems in HPLC detection, there is considerable attraction in the possibility of removing the mobile phase and analysing the sample separately. Many attempts to build detectors operating on this principle have been made. These have largely been directed towards producing devices in which the column eluate is transferred to some transport medium which in turn passes through an evaporation chamber in which the solvent is removed, leaving the solute on the transport medium. The solute, on the transport material, then passes into a pyrolysis chamber where the solute is converted to gaseous organic residues which can be detected by means of a flame ionization detector (FID).

This type of system is typified by the moving wire detector (Scott and Lawrence, 1970) (Fig. 2.24) which is the only transport detector to have been built commercially, with the exception of those systems used in conjunction with on-line HPLC/mass spectrometry systems (see Chapter 13).

The carrier wire passes through a heated quartz tube where any trace of organic material adhering to the wire is burnt off. It then passes around a pulley in a coating block where eluate from the column is applied. The wire

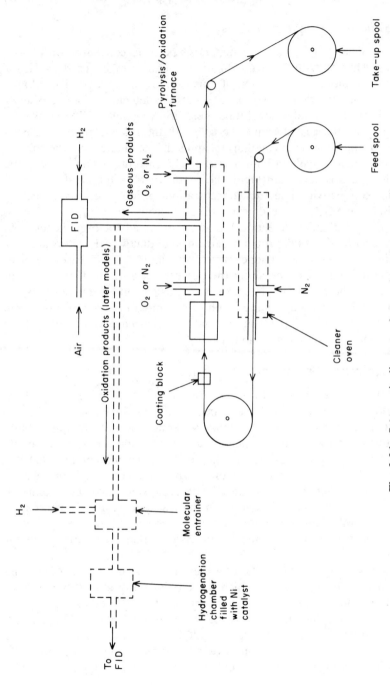

**Fig. 2.24** Schematic diagram of a moving wire detector.

passes into a heated evaporation chamber where the mobile phase is removed. The wire plus solute moves through a high temperature furnace operating at approximately 650 °C where thermal decomposition of the solute occurs. Finally the wire is fed via a system of pulleys to a wind-up spool. In early versions of the dectector the pyrolysis of the solute was carried out in an atmosphere of nitrogen and the gaseous decomposition products were fed directly to an FID. In later models (Fig. 2.24) the pyrolysis is carried out in an oxygen atmosphere and the reaction products, mainly $CO_2$, are fed into a molecular entrainer, where they are mixed with hydrogen, then into a reactor containing a nickel catalyst where the carbon dioxide is converted to methane before passing into the FID. The conversion to methane gives a much better linear dynamic range, a more predictable response and a better sensitivity for those compounds which give poor yields under simple nitrogen pyrolysis.

The moving wire detector, originally manufactured by Pye Unicam, of Cambridge, England, is no longer commercially available. Its failure as a commercial entity was almost certainly due in part to the fact that it is mechanically complicated, and has poor sensitivity as the wire is capable of adsorbing less than 1% of the total column eluate. Despite its drawbacks the moving wire system does possess some advantages, mainly that it is possible to use it with gradient elution techniques; this can be of great advantage with some complex carbohydrate and lipid mixtures.

The use of transport material other than wire has been reported, including belts of steel spring wound around a steel core (Stolyhwo, Privett and Erdahl, 1973), alumina discs (Szakasits, 1974) and metal discs and grids (Dubsky, 1973). No commercial devices have appeared as the result of this work.

Nephelometry following removal of the mobile phase can also be used to detect HPLC solutes. This detector, described either as an evaporative analyser (Charlesworth, 1978) or as a mass detector, is shown schematically in Fig. 2.25. The column eluate enters the dectector via a nebulizer and the resultant atomized solvent spray is forced down a heated tube, approximately 40 cm long, by the flow of nebulizer gas. The temperature of the heated tube can be adjusted so that the solvent is evaporated by the time it reaches the tube base.

The optical system consists of a primary light beam and a photomultiplier tube set at an angle of 120° to the primary beam, mounted at the bottom of the heated tube. If mobile phase only is eluting from the column the optical chamber is filled with solvent vapour, and light passed into a light trap positioned at 180° to the incident beam.

When a mixture of solute and solvent enters the detector the solvent is removed by evaporation as before, leaving behind a cloud of fine solute

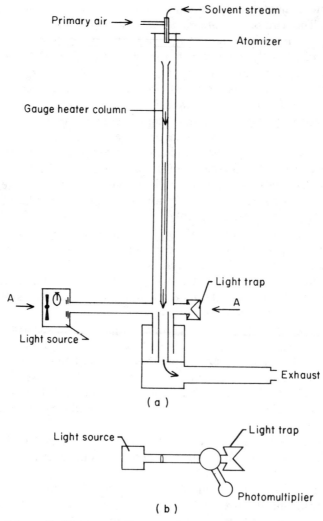

**Fig. 2.25**  Schematic diagram of the mass detector or evaporative analyser. (b) shows the cross-section A–A in part (a) from above.

particles. This particulate cloud enters the optical chamber and light from the primary beam is scattered by the particles to be detected by the photomultiplier. For polymers, involatile and thermally stable molecules it is mass responsive. Thus calibration for these materials is straightforward. For smaller and less stable solutes it is not mass responsive and must be calibrated for each compound of interest.

As in the case of the moving wire detector, it can be operated with gradients, e.g. in the analysis of carbohydrates (Macrae and Dick, 1981). It also appears to be more sensitive than RI detectors by approximately an order of magnitude, although this depends on the nature of the solute. It cannot, however, be operated with mobile phases containing dissolved salts and non-volatile buffers.

## D  HPLC/Mass Spectrometry

Mass spectrometry (MS) is a very powerful analytical tool which can be applied to the identification of unknown organic compounds. The combination of gas chromatography and mass spectrometry has been of particular value in the analysis of complex mixtures. When used in combination with chromatography, mass spectrometry can be considered to be a universal detector with the further advantage that it can be used to identify and quantify unknown peaks in a mixture. It is therefore a logical step to combine mass spectrometry with HPLC and this is discussed in detail in Chapter 13.

## E  Other Detection Systems

The detectors described in the previous section can be regarded as the most common currently used in liquid chromatography. The search for more sensitive, more universal or more specific detectors still continues, and reports of new detection devices continue to appear in the literature. It is beyond the scope of this chapter to review fully all the detection methods which have been described. A number of general reviews devoted to the topic of detectors have appeared elsewhere (Hein, 1980; McKinely, Popovich and Layne, 1980).

Some of the more recent and interesting systems which have been reported include the combination of HPLC with nuclear magnetic resonance (Buddrus, Herzog and Cooper, 1981), chemiluminescence techniques (Veazey and Nieman, 1980), flame emission (McGuffin and Novotny, 1981) and inductively coupled argon plasmas (Gast, Kraak, Poppe and Maessen, 1979). Other detectors have included a flame aerosol detector (Wise, Mowery and Juvet, 1979), a laser-induced photoacoustic detector (Shohei and Tsugo, 1981), a liquid ionization detector (Tsuchiya et al., 1981), a photoconductivity detector (Popovich, Dizon and Ehrlich, 1979) and a phosphorus-sensitive detector involving the use of molecular emission techniques (Cope, 1980). Piezoelectric crystals (Konash and Bastiaans, 1980) have also been used and properties such as optical activity and circular dichroism (Westwood, Games and Sheen, 1981) have been utilized.

One other technique which should be mentioned is the use of on-line radioactivity detectors (Peng, Horrocks, Donald and Alpen, 1980) which can be extremely valuable in trace analysis, for example in the study of drug or pesticide metabolites.

## VII  PREPARATIVE HPLC

Liquid chromatography has an advantage over other separation techniques in that preparative work is very simple since most of the detectors used are non-destructive. This means that any particular peak in the chromatogram can simply be collected after passing through the flow cell (Verzele and Geeraert, 1980). The type of apparatus used in preparative HPLC is determined by the amount of compound required to be collected. This in turn depends on the purpose for which the compound is required, possibly for further analysis by a secondary technique, e.g. mass spectrometry, or as a pure standard for clinical trials. Table 2.2 shows the approximate amount of sample required for a number of end-user applications. Provided that the compound is reasonably well separated from other peaks in the mixture for quantities less than 5 mg analytical columns (4.5 mm i.d. and 25 cm long) and therefore analytical pumps and detectors can be used. For quantities between 1 and 100 mg columns of 9 or 10 mm i.d. (semi-preparative) are required, and for quantities over 100 mg a column of diameter 20 mm i.d. (preparative) or above is necessary.

The optimum flow for an analytical column is approximately 2–3 ml min$^{-1}$, and for a preparative column it will be several hundred millimetres per minute.

Table 2.2   Approximate amount of sample required for some end-user applications

| Purpose | Amount required |
| --- | --- |
| Analytical standards | ~500 mg |
| Organic synthesis | 100 mg–10 g |
| NMR spectroscopy | 10 mg |
| Fourier transform NMR | 10 $\mu$g |
| Infrared spectroscopy | 1 mg |
| Fluorescence spectroscopy | <1 mg |
| UV/visible spectroscopy | <1 mg |
| Elemental analysis | 10 mg |
| Mass spectrometry | <1 mg |

Most reciprocating pumps for analytical HPLC have a maximum flow of 10 ml min$^{-1}$; they can, however, be readily converted for semi-preparative work by replacing the pump head and piston with one of a large capacity. Many manufacturers supply kits so that this conversion can be carried out by the analyst. Similarly injection valves used in analytical work are also suitable for semi-preparative work provided that the injection loop is increased in size. This may not always be necessary, since often it is more convenient to increase the sample concentration and maintain a low injection volume.

The detection of peaks in semi-preparative liquid chromatography is far less critical than it is in trace analysis since sensitivity is no longer a problem. UV and RI detectors are commonly used although most analytical RI detectors will not cope with the pressures generated at flows greater than 2–3 ml min$^{-1}$ and tend to leak unless a stream splitter is used. Analytical UV detectors on the other hand tend to be too sensitive for preparative work. Their sensitivity can be reduced by replacing the flow cell with a cell of shorter path length. Again many manufacturers will supply preparative cells for their detectors. As an alternative to replacing the flow cell UV detectors can be "detuned" for particular compounds by operating them at wavelengths where the compounds have reduced absorbance. With many compounds this means using wavelengths greater than 300 nm. For full preparative HPLC it is difficult to modify the equipment used for analytical work to satisfy the higher flows and sample volumes involved. It is therefore better to use a purpose-built preparative chromatograph with fully integrated pumps, injectors, column systems and detectors. Currently two commercial instrument are available for full preparative HPLC (manufactured by Waters Associates and Jobin-Yvon).

Fraction collectors are also extremely useful accessories for preparative HPLC, most of the currently available models being perfectly satisfactory.

### VIII  COLUMN PACKING

A wide range of different types of column packing materials are currently available and, whereas the price of commercially packed columns is still high, the capital cost of the apparatus required for column packing is relatively low. Therefore, for those laboratories with a high column throughput, there is an economic advantage to be gained by packing columns "in-house".

A typical column packing rig is shown in Fig. 2.26. It consists of a constant pressure pump connected by suitable plumbing to a slurry reservoir

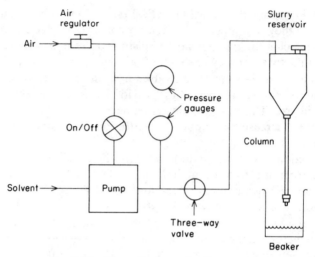

**Fig. 2.26**   Flow diagram of a slurry packer.

which in turn is connected to the empty column. Constant pressure pumps are preferred for column packing since during the initial stages of packing a column they are capable of supplying a high flow rate. Practice shows that this produces more efficient columns. The slurry, consisting of the packing material and a suitable solvent, such as methanol or acetone, made up in the approximate ratio 4–6 g of packing to 50 ml of solvent, is placed in the reservoir. The manufacturers' instructions should be consulted for each packing material as to the best slurry solvent, optimum slurry composition and optimum packing pressure. The pump previously primed is then switched on, and the three-way valve opened. This forces the slurry from the reservoir into the column. The column is then kept attached to the packing rig while several hundred millilitres of solvent are pumped through.

The optimum amount of solvent required to be pumped through a freshly packed column should be determined by experiment, although 300 ml should be adequate. The column is then removed and tested.


## IX   CAPILLARY AND MICROBORE HPLC

There is currently a growing interest in the use of microbore, micro-packed capillary and open tubular columns in HPLC. The advantages of these columns are numerous. They allow faster more efficient separations and because they necessitate the use of lower flow rates capillary techniques are more economical in their use of expensive, high purity solvents.

The high efficiency and low flow rates associated with capillary columns are unfortunately largely incompatible with current instrument design. This has been emphasized by a number of workers active in the field of capillary column research (Reese and Scott, 1980; Novotny, 1980; Ishii and Takeuchi, 1980; Knox, 1980). Narrow bore HPLC techniques are discussed further in Chapter 13.

## X CONCLUSION

HPLC can now be regarded as an off-the-shelf technique. The purpose of this chapter has been to provide a general introduction to current instrumentation and to suggest likely future developments. For further general reading and in particular more specific discussion of some detectors, the interested reader is referred to the books in the bibliography in addition to the literature cited.

## REFERENCES

Altex Scientific (1977). Product literature, Altex Scientific Inc., Berkeley, Calif., USA.
Applied Chromatography Systems (1981). Product literature, Applied Chromatography Systems Ltd, Luton, England.
Arlie, J. P. (1974). Paper presented to the Fourth National Colloquium on Gel Permeation Chromatography, Lyons.
Billiet, H. A. H., Keehnen, P. D. M. and De Galan, L. (1979). *J. Chromat.* **185**, 515.
Buddrus, J., Herzog, H. and Cooper, J. W. (1981). *J. Magn. Reson.* **42**, 453.
Charlesworth, J. M. (1978). *Analyt. Chem.* **50**, 1414.
Conac, M. (1979). *Lab. Pharmac. Probl. Tech.* **27**, 873.
Cope, M. J. (1980). *Analyt. Proc., Lond.* **17**, 273.
Dubsky, H. (1973). U.K. Patent no. 1 323 840.
Dupont Instruments (1977). Data sheet no. E19719, Dupont Instruments, Wilmington, Del., USA.
Frei, R. W. (1979). *Proc. analyt. Div. chem. Soc.* **16**, 289.
Gast, C. H., Kraak, J. C., Poppe, H. and Maessen, F. J. M. J. (1979). *J. Chromat.* **185**, 549.
Hein, H. (1980). *Chem. Lab. Betr.* **31**, 559.
Ishii, D. and Takeuchi, T. (1980). *J. chromatogr. Sci.* **18**, 462.
James, G. E. (1981). *Can. Res.* **13**, 39.

Johnson, E. L. and Stevenson, R. (1978). *Basic Liquid Chromatography*, Varian Associates Inc., Walnut Creek, California, p. 256.
Kelsey, R. J. and Loscombe, C. R. (1979). *Chromatographia* **12**, 713.
Kissinger, P. R., Refshange, C., Dreiling, R. and Adams, R. N. (1973). *Analyt. Letters* **6**, 465.
Knox, J. H. (1980). *J. chromatogr. Sci.* **18**, 453.
Konash, P. L. and Bastiaans, G. J. (1980). *Analyt. Chem.* **52**, 1929.
Lindroth, P. and Mopper, K. (1979). *Analyt. Chem.* **51**, 1667.
Loullis, C. C., Hingtgen, J. N., Shea, P. A. and Aprison, M. H. (1980). *Pharmac. Biochem. Behav.* **12**, 959.
McGuffin, V. L. and Novotny, M. (1981). *Analyt. Chem.* **53**, 946.
McKinley, W. A., Popovich, D. J. and Layne, T. (1980). *Am Lab.* **12**, 37.
Macrae, R. (1978). *Lab. Pract.* **9**, 719.
Macrae, R. and Dick, J. (1981). *J. Chromat.* **210**, 138.
Malcolm-Lawes, D. J. and Warwick, P. (1980). *J. Chromat.* **200**, 47.
Mayer, W. J. and Greenberg, M. S. (1979). *J. chromatogr. Sci.* **17**, 614.
Novotny, M. (1980). *J. chromatogr. Sci.* **18**, 473.
Peng, C. T., Horrocks, D. L. and Alpen, E. L. (1980). *Liquid Scintillation Counting: Recent Applications*, Academic Press, New York, p. 141.
Perrett, D. (1976). *J. Chromat.* **124**, 187.
Pohl, C. A. and Johnson, E. L. (1980). *J. chromatogr. Sci.* **18**, 442.
Popovich, D. J, Dizon, J. B. and Ehrlich, B. J. (1979). *J. chromatogr. Sci.* **17**, 643.
Poppe, H. (1980). *Analyt. chim. Acta* **114**, 59.
Readman, J. W., Brown, L. and Rhead, M. M. (1981). *Analyst, Lond.* **106**, 122.
Reese, C. E. and Scott, R. P. W. (1980). *J. chromatogr. Sci.* **18**, 479.
Savage, M. (1979). *Am. Lab.* **11**, 19.
Schoeffel Instruments (1975). Product literature, Schoeffel Instruments Corp., Westwood, N.J., U.S.A.
Scholten, A. H. M. T., Brinkmann, U. A. T. and Frei, R. W. (1981). *J. Chromat.* **205**, 229.
Schwedt, G. and Reh, E. (1981). *Chromatographia*, **14**, 123.
Scott, R. P. W. and Lawrence, J. G. (1970). *J. chromatogr. Sci.* **8**, 65.
Shohei, O. and Tsugo, S. (1981). *Analyt. Chem.* **53**, 471.
Stolyhwo, A., Privett, O. S. and Erdahl, W. L. (1973). *J. chromatogr. Sci.* **11**, 263.
Szakasits, J. J. (1974). US Patent no. 3 788 479.
Tsuchiya, M., Kawabe, K., Toyoura, Y., Taira, T., Tanaka, S., Saite, Y. and Otake, W. (1981). *Nippon Kagaku Kaishi* **1**, 145.
Veazey, R. L. and Nieman, T. A. (1980). *J. Chromat.* **200**, 153.
Ventura, D. A. and Nikelley, J. G. (1978). *Analyt. Chem.* **50**, 1017.
Verzele, M. and Geeraert, E. (1980). *J. chromatogr. Sci.* **18**, 559.
Westwood, S. A., Games, D. E. and Sheen, L. (1981). *J. Chromat.* **204**, 103.
Wise, S. A., Mowery, R. A. and Juvet, R. S. (1979). *J. chromatogr. Sci.* **17**, 601.

## BIBLIOGRAPHY

Bristow, P. (1976). *Liquid Chromatography in Practice*, HETP Components for Liquid Chromatography, Wilmslow, Cheshire.

Brown, P. R. (1973). *High Pressure Liquid Chromatography: Biochemical and Biomedical Applications*, Academic Press, New York.

Deyl, Z., Macek, K. and Janak, J. (eds) (1975). *Liquid Column Chromatography: A Survey of Modern Techniques and Applications*, Elsevier, Amsterdam.

Done, J. N., Knox, J. H. and Loheac, J. (1974). *Applications of High Speed Liquid Chromatography*, John Wiley, London.

Huber, J. F. K. (ed.) (1978). *Instrumentation for High-performance Liquid Chromatography*, Elsevier, Amsterdam.

Johnson, E. L. and Stevenson, R. (1978). *Basic Liquid Chromatography*, Varian Associates Inc., Walnut Creek, California.

Kremmer, T. and Boross, L. (1980). *Gel Chromatography: Theory, Methodology and Applications*, John Wiley, Chichester.

Lawrence, J. F. and Frei, R. W. (1976). *Chemical Derivatisation in Liquid Chromatography*, Elsevier, Amsterdam.

Paris, N. A. (1976). *Instrumental to Liquid Chromatography: A Practical Manual on HPLC Methods*, Elsevier, Amsterdam.

Perry, S. G., Amos, R. and Brewer, P. I. (1972). *Practical Liquid Chromatography*, Plenum, New York.

Sawicki, E., Mulik, J. D. and Wittgenstein, E. (eds) (1978). *Ion Chromatographic Analysis of Environmental Pollutants*, John Wiley, Chichester.

Scott, R. P. (1977). *Liquid Chromatography Detectors* (Journal of Chromatography, Vol. 11.) Elsevier, Amsterdam.

Snyder, L. R. and Kirkland, J. J. (1979). *Introduction to Modern Liquid Chromatography*, 2nd edn, John Wiley, New York.

Unger, K. . (1979). *Porous Silica: Its Properties and Use as Support in Column Liquid Chromatography*, Elsevier, Amsterdam.

Weinstein, M. J. and Wagman, G. H. (eds) (1978). *Antibiotics: Isolation, Separation and Isolation*, Elsevier, Amsterdam.

# 3   Separation Modes in HPLC

## C. F. SIMPSON

Department of Chemistry, Chelsea College, University of London, U.K.

# I  INTRODUCTION

The separation of the components of a mixture by any chromatographic procedure is a function of two separate and distinct processes which can take place within the system. [In the ensuing discussion only methods of chromatographic separation in which the stationary phase is contained within a column (of varying dimensions) and the mobile phase is restricted to a liquid medium will be considered.] Separation may be considered to be a function of (1) how the solute bands are moved apart and (2) how the individual bands disperse. Figure 3.1 exemplifies the situation: suppose that there are two bands which are insufficiently separated [Fig. 3.1(a)], then it is possible to achieve a complete resolution ($R_s = 1.5$) of the bands either by changing the thermodynamic parameters of the column so that the bands are moved apart [Fig. 3.1(c)] or by improving the kinetic processes within the column (i.e. how well and with what the column has been packed) [Fig. 3.1(b)]. The second parameter is a function of many interrelated processes connected with the construction of the column, the particle size of the packing, its distribution, how well the column has been packed and extra-column parameters, all of which contribute to the band dispersion observed. A detailed account of these factors will be found in Chapter 1. Here consideration will be given to the various thermodynamic and mechanistic processes which control separation. It will be assumed that columns of adequate efficiency are available so that this factor can, in effect, be ignored.

In the above paragraph the term "thermodynamics of the separation process" was used. In fact, thermodynamics is not able to yield very much information *per se*. It can on occasion be of use in liquid/liquid chromatography when the distribution coefficient may be used to predict a separation system. Unfortunately, such information is rarely available. A more rewarding view of the way a separation will progress may be gained from consideration of the solute–solvent and/or surface interactions on a molecular level; hence it would be useful to consider the forms these interactions may take.

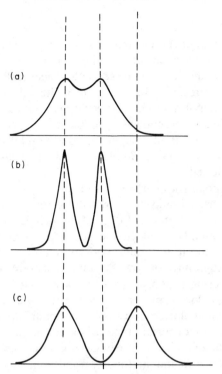

**Fig. 3.1** Dual nature of resolution: (a) unresolved peaks; (b) peaks resolved by improving the kinetics of the system; (c) peaks resolved by changing the thermodynamics of the system.

## II  MOLECULAR INTERACTIONS

There are three basic molecular interactions which need to be considered. These arise from ionic, polar and dispersive interactions between the solute molecules and the stationary and mobile phases.

### A  Ionic Forces

These result from interactions between charged species, e.g. ions, and occur when molecules have a nett positive or negative charge. For example, if it is desired to separate anions chromatographically then the stationary phase should contain cations which are capable of interacting selectively with the sample anions (e.g. acids). The converse situation also holds (e.g. with bases).

## B  Polar Forces

These arise from an uneven electron distribution over the molecules and are brought about by electron withdrawing or donating groups or elements present in the molecule. They are essentially "electrical" in nature, but no formal charges are present. The presence of the substituent groups produces permanent dipoles in the molecules containing them, and these dipoles are capable of interaction with dipolar groups present in the stationary phase or in the solvent and, dependent on the relative strengths of the stationary phase/mobile phase dipolar groups, a competition will occur which can be exploited.

In practice, if a polar (dipolar) series of solutes is to be separated, then a polar stationary phase would be used with a non-polar or weakly polar mobile phase. Polar compounds of this nature include alcohols, aldehydes, etc., and, in compounds in which the active polar group is capable of hydrogen bonding, would be preferably separated in a polar substrate in partition chromatography or on a polar substrate in bonded-phase chromatography, using a non-polar or weakly polar, mobile phase, e.g. hexane containing chloroform, ethyl acetate, etc. (see later).

However, not all molecules possess permanent dipoles; even so, the molecular construction of certain types of molecule (e.g. aromatics) results in them being polarizable, i.e. they can have a temporary dipole induced in them by the presence of the stationary/mobile phase, and these induced dipoles can provide a basis for separation. Here similar stationary phases as indicated above (including silica) would be used, but the mobile phase would be less polar, e.g. pure hexane or hexane containing a weakly polar constituent.

## C  Dispersive Forces

These are not quite so simple to comprehend; they are essentially "electrical" in nature, but do not result from a nett charge or a permanent dipole. They are more tenuous in character and occur because the electrons surrounding the molecule are not necessarily homogeneously distributed about the molecule; this inhomogeneity results in a short term "dipole" which is capable of inducing another short term "dipole" in an adjacent molecule, thus causing a weak nett attractive or repulsive force between the molecules. Separations involving dispersive interactions are usually found in "reversed phase" chromatography, either in the true partition or bonded-phase modes. Here, the stationary phase is non-polar in character while the mobile phase is polar (e.g. water or water plus an organic medium such as methanol, acetonitrile or tetrahydrofuran). Separation occurs through the dispersive interactions between the non-polar stationary phase and the non-polar or weakly polar solute molecules. However, quite polar solutes can also be separated by reversed phase chromatography, but there the mechanism depends upon the effect of the adsorption of an organic modifier on to the surface of the reversed phase, and hence, purely dispersive forces only exert a minor influence on the separation mechanism.

On the basis of the above synopsis of the molecular interactions which can occur in liquid chromatography, it will be clear that the situation is complex. However, it is possible to allocate the various mechanisms outlined above into the sub-divisions which are recognized as being "individual" liquid chromatographic modes, although it must be emphasized that often there is appreciable overlap between the methods indicated. These modes include the following:

(1) Adsorption chromatography (AC), in which the stationary phase is a polar substrate such as silica gel or alumina and the mobile phase is essentially non-polar (e.g. hexane or heptane), but often with a slightly or moderately polar medium added.

(2) Liquid/liquid partition chromatography (LLPC), in which the stationary phase is polar (and this is adsorbed onto a matrix such as silica gel or celite – which should play no part in the separation process) with a non-polar or weakly polar mobile phase. This is so-called "normal" chromatography. Alternatively, the stationary phase may be non-polar (adsorbed on silanized silica gel) and the mobile phase polar – so-called "reversed phase" chromatography.

(3) Bonded-phase chromatography (BPC), in which the stationary phase (which may be polar or non-polar) is actually "bonded" to an insoluble matrix and the mobile phase is either non-polar or polar respectively.

(4) Ion-exchange chromatography (IEC), in which the stationary phase contains ionic moieties, cationic or anionic in nature, and the mobile phase is a suitably buffered ionizing medium such as water or water–alcohol, etc.

(5) Gel permeation chromatography (GPC), a technique which does not, or ideally should not, include any of the mechanisms outlined above. Rather this technique involves the separation of the solute molecules in terms of their molecular dimensions by using a series of stationary phases with a variety of defined pore sizes which, in series, are capable of including or excluding molecules of appropriate size. This is often termed "molecular sieving", and is used to separate the sample components broadly on the basis of size.

The techniques outlined above are not mutually exclusive (except possibly GPC) and often there is overlap between the methods. However, they will be considered separately for the purposes of this chapter.

## III  ADSORPTION CHROMATOGRAPHY

### A  The Nature of Silica Gel

Silica gel is by far the most important adsorbent in use today in liquid chromatography. Indeed, silica gel forms the substrate in LLPC and in the preparation of many bonded phases. Hence, it would be useful to consider briefly the surface structure of silica gel, not only because of its use as an adsorbent, where clearly this knowledge is important, but also because of its other applications. For an in-depth study of this subject, the interested reader is referred to reviews by Ilher (1979), Unger (1979) and Scott (1982a).

Silica gel is a porous solid with a surface area which is inversely related to the pore size of the particles. Techniques are now available to control the pore size distribution of silica closely during manufacture, and typical sizes available range between 4 and 400 nm.

In adsorption chromatography a compromise is reached between pore size and surface area by using silica of 6 nm pore size with a surface area of about $200 \ m^2 \ g^{-1}$ with particle sizes in the range 5–10 $\mu$m. (Note, however, that 3 $\mu$m silicas are now becoming available.)

The surface of silica gel is complex and is covered with hydroxyl groups attached to the surface silicon atoms in a random manner, and it is in the nature of these surface hydroxyl groups which confer upon silica its adsorptive capability.

Depending on the separation between these surface silanols, they can exist either as free or as hydrogen-bonded hydroxyls (Davydov, Zhurvelev and Kiselev, 1964). Alternatively, the fully hydroxylated silica may contain paired hydroxyl groups (geminal or vicinal) as well as free hydroxyls (Peri and Hensley, 1968). In addition, because of its nature, associated with each surface silanol is water of hydration, and it may be shown that there are three distinct forms of water on the surface of silica gel (Scott, 1982a): (1) loosely bound water which may be removed by heating the gel to 110 °C or by solvent extraction; (2) strongly held water which is evolved by heating to 250–650 °C; (3) water which is evolved by heating to 1100 °C and results from dehydration occurring between adjacent hydroxyls, giving a siloxane (Fig. 3.2). The process of water removal by (1) and (2) is reversible, whereas (3) is a non-reversible process.

It is the water which is removed by heating to 110 °C which is particularly important chromatographically in so far as it is a variable quantity, loosely associated with the more strongly bound water [type (2)] and confers upon the silica gel its variable adsorptive nature and necessitates careful control of the water content of the mobile phases used in order that repeatable separations can be achieved.

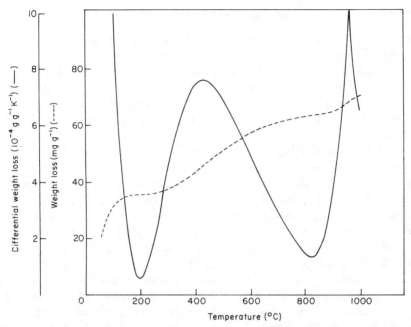

**Fig. 3.2** Curves relating weight loss (– – – –) and differential weight loss (——) of silica to heating temperature (taken from a thermogravimetric curve). [Redrawn with permission from Scott and Traiman (1980).]

Thus, in order to obtain reproducible separations on a day to day basis, it is recommended (Scott and Kucera, 1973a) that the loosely bound water be removed by solvent extraction using a solvent series, e.g. ethanol, acetone, ethyl acetate, ethylene dichloride and heptane. Solvent extraction may be carried out in the column by progressively passing four to five column volumes of each solvent through the column (about 20–25 ml for a 250 mm × 4.6 mm i.d. column). It is important to remember, however, that the activity obtained from a given batch or supplier of silica gel will not necessarily be identical with the activity obtained from another batch or supplier. Although on a batch basis, because of the careful control exercised during manufacture, differences should be small.

Thus, the silica surface which it is preferable in practice is one obtained as indicated above, and consists of a monolayer of water hydrogen-bonded to the surface silanols. But it will be clear from the above discussion that considerable care must be exercised when attempting to reproduce a separation given in the literature.

## B The Effect of the Mobile Phase

A wide variety of mobile phases may be used in adsorption chromato-graphy ranging from non-polar (dispersive) solvents to polar materials (often hydrogen bonding in nature), and, depending on the nature of the phase system used, then the nature of the surface exposed will be modified in its adsorption capability. When silica gel is in contact with dispersive solvents, it has been shown that a layer of the solvent (e.g. butylchloride or chloroform in heptane) is adsorbed onto the surface (Scott, 1976). Thus, any solute chromatographed would not necessarily be interacting with the silica surface but with the adsorbed solvent layer. Alternatively, the solute may displace the adsorbed solvent and hence interact with the exposed silica surface. If, however, a polar solvent is used, e.g. ethyl acetate, or an alcohol in heptane, a double layer is formed on the surface by successive addition (Scott, 1982a). The first layer is strongly bound to the surface silanols by hydrogen bonding, while the second layer, is more weakly bound. Thus, any solute being chromatographed under these conditions will be interacting with the primary layer of adsorbed polar solvent and not with the silanols on the surface. Effectively, the surface of the silica has been deactivated and, unless the solute is very polar, it will not displace this bound layer of polar solvent. In these circumstances, it is very difficult to distinguish between true adsorption and a partitioning mechanism.

## C Interaction of Solutes when the Silica Surface is in Contact with a Solvent

In view of the above conclusions, it is necessary to consider in more detail what is happening when a solute is present in the system, and how the solute can interact with the silica surface and/or the adsorbed solvent layer(s) on the surface. There are two cases to consider:

(1) *Solvent molecules are weakly held* The solute molecules may dis-place the solvent from the surface and interact directly with the hydrated silanols on the surface (Scott, 1982b). Figure 3.3 illustrates this phenom-emon and shows that increasing the sample size of anisole results in a reduc-tion in the concentration of the mobile phase moderator (butyl chloride) on the stationary phase, with a concomitant increase in the adsorbed solute (anisole), i.e. the anisole is displacing the butyl chloride initially adsorbed on the "silica" surface.

(2) *Solvent molecules are strongly adsorbed* In this case ethyl acetate is the moderator in heptane solution at 0.35% m/v. In an evaluation of four different solutes, (i) anisole, (ii) 2,4-dinitrobenzene, (iii) m-dimethoxybenzene and (iv) benzyl acetate, with each of the four solutes

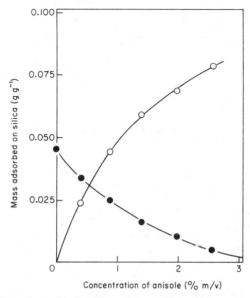

**Fig. 3.3** The adsorption of anisole (O—O) on silica gel from a solvent mixture of butyl chloride and $n$-heptane and the desorption of butyl chloride (●—●). [Redrawn with permission from Scott (1980).]

being added (independently) in increasing amounts, the curves shown in Figs 3.4 and 3.5 were obtained. Figure 3.4 demonstrates that in the case of anisole (i) $k = 2.4$ and 2,4-dinitrobenzene (ii) $k = 4.7$ no displacement of the ethyl acetate occurs (horizontal lines marked $E_s$ and $E_m$) but the two solutes associate with it (lines marked $S_s$ and $S_m$) (the subscripts s and m denoting that the solute or solvent was present in the stationary and mobile phases, respectively), i.e. the solute molecules are being adsorbed on the surface of the adsorbed ethyl acetate. Figure 3.5 demonstrates that a similar effect occurs with $m$-dimethoxybenzene ($k = 10.5$), but with benzyl acetate ($k = 27$), which has a similar polarity to ethyl acetate, the curves show that the addition of benzyl acetate brings about an increase in the ethyl acetate concentration in the mobile phase, i.e. the solute is displacing the adsorbed ethyl acetate and is interacting directly with the hydrated silanol surface.

Thus, to summarize the information presented above with regard to the mechanism of adsorption chromatography, it is clear that the silica surface contains various forms of hydroxyl function which have strong hydrogen-bonding capability and permit multilayer formation with solvents which readily undergo hydrogen bonding. The silica surface has three layers of

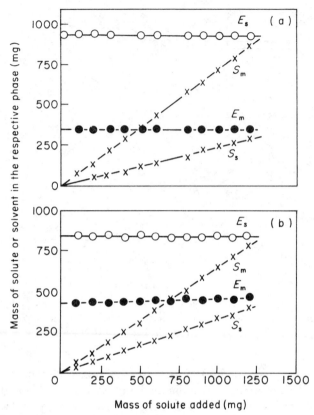

**Fig. 3.4** Curves relating mass of solute and solvent in the two phases to total mass of solute added. $E_m$ = mass of ethyl acetate in the mobile phase; $E_s$ = mass of ethyl acetate on the silica gel; $S_m$ = mass of solute in the mobile phase; $S_s$ = mass of solute on the silica gel. Concentration of ethyl acetate, 0.35% m/v; volume of mobile phase, 100 ml. (a) Solute, anisole; $k$ = 2.4; mass of silica gel, 10.04 g. (b) Solute, nitrobenzene; $k$ = 4.7; mass of silica gel, 10.28 g. [Redrawn with permission from Scott and Kucera (1978).]

water associated with it. The first layer appears to be water strongly hydrogen-bonded to the surface silanol groups and is not entirely removed by heating until the temperature is in excess of 600 °C. Associated with this primary layer by hydrogen bonding are the other two water layers (which also interact between themselves) and these layers are much more easily removed by heating to 120 °C or by washing with anhydrous solvents.

For preference the silica gel used in chromatography should only carry the adsorbed monolayer in order to ensure reproducible activity. This silica

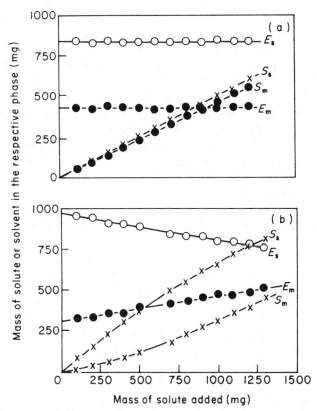

**Fig. 3.5** Curves relating mass of solute and solvent in the two phases to the total mass of solute added. (a) Solute, $m$-dimethoxybenzene; $k = 10.5$; mass of silica gel 10.23 g. (b) Solute, benzyl acetate; $k = 27.0$; mass of silica gel, 10.17 g. All other details as in Fig. 3.4. [Redrawn with permission from Scott and Kucera (1978).]

surface, when in contact with non-polar solvents, adsorbs a monolayer on the hydrated silanols and, in chromatographic terms, the mechanism is one of association and/or displacement, with the solute molecules interacting directly with the hydrated silanols (e.g. the separation of aromatics). In the presence of low concentrations of a polar moderator in heptane, e.g. ethyl acetate at 0.5% m/v, the surface adsorbs a monolayer of the moderator, and as the concentration of the moderator is increased a bilayer is formed. At low moderator concentrations, the solute molecules interact only with the adsorbed monolayer and it is not until the solute has a polarity equivalent to, or greater than that of the moderator that the displacement mechanism again occurs.

Thus it is possible to produce an interactive surface which is constant in its polarity over a reasonable range of moderator concentrations. For example, in the range 3–15% m/v of a polar moderator in heptane, the surface will be completely covered by a monolayer, and even at 15% m/v roughly only 10% of the second weakly adsorbed layer will be formed and hence interaction will be with the primary layer of polar solvent. Thus, any changes in retention observed within this range of concentrations will be as a result of changes in the interactions of the solute with the mobile phase rather than with the surface of the modified silica.

The understanding of the method of solute interaction with the stationary phase as outlined above requires further development if predictions about the changes in retention volume with respect to changes in mobile phase composition are to be made.

Scott and Kucera (1975) have shown that under conditions in which the silica surface is covered with an adsorbed monolayer of polar moderator

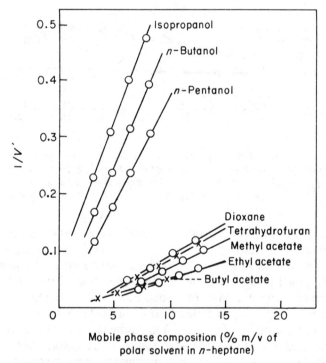

**Fig. 3.6** Graphs relating the reciprocal of the corrected retention volume of phenylmethylcarbinol to the composition of the mobile phase containing different polar solutes in *n*-heptane. Column, 25 cm × 4.6 mm i.d.; column packing, Partisil 10. [Redrawn with permission from Scott and Kucera (1975).]

then the adjusted retention volume $V'_R$ is related to the concentration of polar moderator $C_P$ by the equation

$$1/V'_R = A + BC_P, \qquad (3.1)$$

where $A$ and $B$ are constants and are related to the surface area of the adsorbent $A_s$ and the distribution coefficient $K$ because

$$V'_R = KA_s. \qquad (3.2)$$

The validity of equation (3.1) has been established for a series of polar moderators using phenylmethylcarbinol as the test solute. Figure 3.6 illustrates the linear relationships that are obtained.

The conclusions stated above have been confirmed by independent work on mixed solvent theory by McCann *et al.* (1982) which also casts doubt on the only other theory available, attributable to Snyder (1968, 1974). Snyder's theory is based on a highly approximated theoretical approach and will not be considered here. However, the interested reader might care to compare and contrast the two theories and draw his own conclusions.

## IV LIQUID/LIQUID PARTITION CHROMATOGRAPHY

### A Introduction

LLPC was introduced by Martin and Synge (1941). They were attempting to carry out a liquid/liquid extraction, with little success, when they realized that it was unnecessary to pass two liquids countercurrently, but that they could hold one liquid stationary on a suitable medium, pack it into a column, and then, on passing the second solvent through the bed so obtained, achieve efficient separations.

Thus LLPC may be viewed as being an extension of adsorption column chromatography, discussed above, except that a relatively heavy loading of an appropriate stationary phase is used. The resulting system is a pure partitioning one, in which conventional distribution equilibria occur, and hence extraction data can provide a useful guide to the choice of phase system applicable in a chromatographic situation.

LLPC has been largely ignored in recent years, mainly because of the problem of stationary phase loading stability. Nevertheless, it is a powerful technique and should be considered when appropriate problems arise.

### B Liquid/Liquid Distribution Equilibria

The rate of migration of a solute in an LLPC system depends upon the equilibrium distribution of the solute between the stationary liquid phase

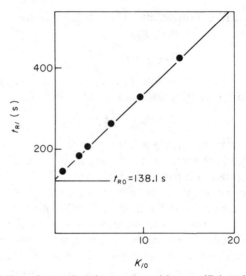

**Fig. 3.7** Correlation of retention time and partition coefficient for carbamate pesticides. Mobile phase, dibutyl ether; stationary phase, ODPN. [Redrawn with permission from Huber and Vodenik (1976).]

and the mobile phase. It is possible to use this equilibrium model to decide whether the mechanism operating is in fact pure partition or a mixture of partition and adsorption. This involves demonstrating the linearity between retention time and partition coefficients for a range of similar compounds, for example carbamate pesticides, as shown in Fig. 3.7 (Huber and Vodenik, 1976).

Clearly this uncertainty as to the separation mechanism is important and the reasons why a dual mechanism may occur need explanation. This may be found by considering the surface structure of silica gel (the most usual support in LLPC). It was stated above (Section III.A) that the silica surface is derived from the pore structure present, and it is the nature of these pores which to a large extent control the separation mechanism. The requirement for a pure partition mechanism is that the surface of the support is evenly coated with the partitioning solvent. Should this criterion not be achieved, and an uneven coating result, then a mixed mechanism may well occur. The sizes of the pores in the support contribute to uneven coating, and small pores have been shown to result in a mixed mechanism.

Huber (personal communication) recommends that the specific surface area of the support should not exceed $30 \ m^2 \ g^{-1}$, which implies a mean pore size of $>80$ nm.

## C Preparation of Columns for LLPC

### 1 *The Support*

For the normal elution mode, in which the stationary phase is polar, the silica surface provides an admirable support in that its surface area and pore size can be carefully controlled. It is most probable that the liquid stationary phase is a multi-layer built up through hydrogen bonds or polar interactions with the hydrated silanols on the silica surface (Section III.A).

In the "reversed phase" mode, where the stationary phase is non-polar, the nature of the silica surface requires modification so that the non-polar phase can associate with it through dispersive interactions. This is readily achieved by reacting the surface silanols with trimethylchlorosilane or hexamethyldisilazane in dry heptane or toluene. A longer chain alkyldimethylchlorosilane may also be used. Care has to be exercised in this procedure because of the presence of hydrated silanol groups on the surface, and it may be preferable to heat the silica gel to at least 200 °C prior to silanization.

### 2 *The Stationary Phase*

LLPC offers the analyst the widest range of solvent pairs for performing analyses. In practice, however, the choice is narrowed appreciably through the requirements of the system, i.e. the stationary phase and the mobile phase must be immiscible to a large extent (but note that a small, finite degree of miscibility is necessary, otherwise partitioning would not occur; and indeed it is rare, if not impossible, to achieve complete immiscibility between solvents). The viscosity of the phases ideally should be low, otherwise the diffusion and mass transfer rates may become unacceptably small, leading to band-broadening within the column, and concomitant loss in operating efficiency.

### 3 *Coating the Support and Packing*

Three methods may be distinguished:

(1) *The solvent evaporation technique* This procedure is equivalent to the method of preparing packings in gas chromatography. The appropriate amount of stationary phase is dissolved in a suitable solvent, preferably volatile, and added to a known weight of silica gel, previously heated to 200 °C. The volatile solvent is removed in a rotary evaporator to provide a dry, free-flowing powder. The packing is dry-packed into a suitable column. This technique is only applicable to particle sizes of $\geq 20$ $\mu$m, particles of smaller size are more suitably packed using a slurry technique. Care must

be exercised here because the slurry medium may dissolve the stationary phase. However, if through prior experiment the equilibrium level of stationary phase is included in the slurry medium, successful columns may be packed using particles of 5–10 $\mu$m.

(2) *In-situ coating technique* In this method, 5–10 $\mu$m silica gel, preferably activated by heating to 200 °C, is packed using normal slurry methods into an appropriate column; the slurry medium is then removed from the column using mobile phase saturated with the desired stationary phase until a constant capacity factor for a test solute is obtained. The level of stationary phase loading so obtained will remain constant provided that the same mobile phase composition is maintained and the column solvent reservoir, etc., are thermostated. Higher stationary phase loadings may be obtained by injecting aliquots of the pure stationary phase into the flowing mobile phase through the column prior to passing mobile phase saturated with stationary phase.

(3) *Precipitation technique* Here the stationary phase is precipitated within the pores of the support matrix. A high concentration of the stationary phase in an appropriate solvent is pumped through the packed bed and this solvent is then displaced by an eluent in which the stationary phase is immiscible, which results in a "precipitation" of the stationary phase on the surface of the support.

## D Operation of LLPC Columns

By their very nature, partitioning columns require special care in their use; however provided that care is taken, excellent performance can be obtained. The principle problem is to ensure that the phase ratio remains constant; should it vary, then capacity factors will also vary. High mobile phase flow rates can strip the stationary phase from the support by frictional forces, and hence low ($<2$ ml min$^{-1}$) flows are mandatory. Similarly, any change in operating temperature will alter the phase ratio, hence it is preferable to thermostat all parts of the system. Obviously, a change in the nature of the mobile phase will be deleterious to repeatability, but careful control of the composition of the mobile phase regulates this effect.

## E Ternary Phase Systems

Huber (1971) has demonstrated that ternary phase systems can be used very successfully to provide invariant stationary and mobile phase compositions. A typical ternary phase diagram for the system comprising 2,2,4-trimethylpentane (iC$_8$), ethanol and water is shown in Fig. 3.8. The characteristic of these systems is that any point in the two-phase region has characteristic compositions of the separate layers, e.g. at any point along

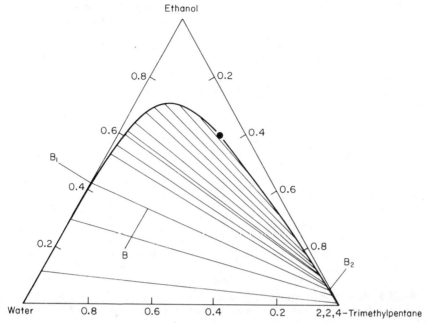

**Fig. 3.8** Liquid–liquid equilibria curve of the ternary system water, ethanol and 2,2,4-trimethylpentane. ●, Plait point (extrapolated); $T = 25$ °C. [Redrawn with permission from Huber (1971).]

the line B will produce two discrete phases with molar fraction compositions: $B_1 = 0.575$, $H_2O$; 0.42, EtOH; 0.004, $iC_8$ and $B_2 = 0.008$, $H_2O$; 0.100, EtOH; 0.892, $iC_8$. In other words $B_1$ is a polar phase and $B_2$ is non-polar. Clearly these may be used as stationary and mobile phases respectively, or vice versa.

## F Toasted Columns

Another procedure which has been used very successfully to produce stable stationary phases, particularly with high surface area supports, has been reported by Vermont, Delevil, De Vries and Guillemin (1975). Spherosil (460 m² g⁻¹) is first coated with an increasing amount of stationary phase by the solvent evaporation procedure and then "toasted" for 10 h. For ODPN, the temperature was maintained at 110 °C and for Carbowax at 250 °C.

Figure 3.9 shows the effect of "toasting" the $\beta\beta$-oxydiproprionitrile (ODPN) on Spherosil compared with a conventionally coated support while Fig. 3.10 illustrates the change in mechanism on increasing the

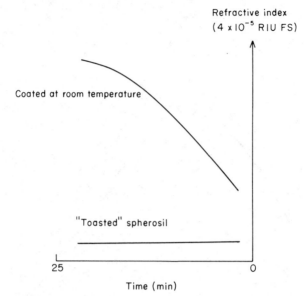

**Fig. 3.9**  Base-line recording of Spherosil XOA 400 coated with ODPN (88 g per 100 g). Hexane flow rate, 1 ml min⁻¹. [Redrawn with permission from Vermont *et al.* (1975).]

ODPN load. Clearly this technique merits further investigation, and is similar to a technique used by Ilher (1955). The technique probably involves the chemical bonding of a multi-layer to the surface of the silica by the reaction

$$\equiv Si-O-H + H-O-C\lessgtr \longrightarrow \equiv Si-O-C\lessgtr + H_2O$$

to form a silicate which would be stable to non-polar (or weakly polar) mobile phases. The present author has also investigated these materials and found them to be very satisfactory (Simpson, 1969, 1974). They may also be used in gas chromatography.

## V  BONDED-PHASE CHROMATOGRAPHY

### A  Introduction

In Section IV.F "toasted" phases were mentioned and the indication is that they are a form of bonded phase. The idea of chemically bonding a liquid

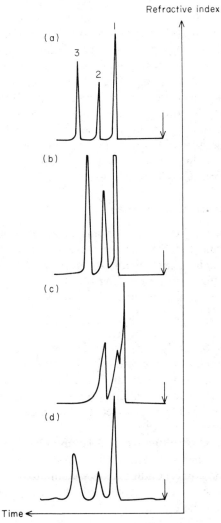

**Fig. 3.10** Chromatogram showing the effect of the amount of stationary phase. Column, 25 cm × 4 mm packed with Spherosil coated with ODPN and toasted before packing; eluant, hexane. Peaks: 1, benzene; 2, naphthalene; 3, anthracene. (a) Partition: 120 g ODPN per 100 g Spherosil; 460 m$^2$ g$^{-1}$. (b) Partition: 88 g ODPN per 100 g Spherosil. (c) Adsorption/partition: 20 g ODPN per 100 g Spherosil. (d) Adsorption: 0 g ODPN. [Redrawn with permission from Vermont *et al*. (1975).]

to a solid has been known for some time. Ilher and Pinksney (1947) bonded long chain hydrocarbon to silica to make it more soluble in rubber. Stewart and Perry (1968) suggested silanizing silica gel with a long chain chlorosilane. But the first demonstration of the use of chemically bonded phases in liquid chromatography was made simultaneously by Halász and Sebastion (1969) and Simpson (1969). Since that time, the number of bonded phases has proliferated, Indeed it has been estimated that about 80% of current analytical problems are handled by BPC. The reason for their popularity, which has caused the virtual eclipse of LLPC, lies in their rapid equilibration times, and, in the case of "reversed phase" BPC, the cheapness and wide availability of the mobile phases used.

However, BPC has limitations as well as advantages; here the practical aspects of bonded phases will be discussed.

## B  Preparation of Bonded Phases

In Section III.A the nature of the silica surface was described and shown to contain surface silanols, and these are present with a density of about 8 $\mu$mol m$^{-2}$; however, it is not possible to react all of the available groups because of steric effects of the bonding species.

The first type of bonded phase to be prepared was the silicate ester type:

$$\geq Si-O-H + H-O-R \longrightarrow \geq Si-O-R + H_2O,$$

where R is ODPN or Carbowax 400, etc. These materials are stable in organic solvents, and may be prepared simply in the laboratory by warming the silica with the appropriate modifying agent using a Dean and Stark apparatus. However, when R = hydrocarbon (from octadecyl alcohol) the Si – O – C bond is readily hydrolysed by the aqueous solvents used in reversed phase chromatography.

The second method, and one which is commonly used in manufactured bonded phases is to react the surface silanols with mono-, di- or trichloro-organosilanes or the corresponding alkoxy compound.

When the reaction is carried out with a monochloroalkyl silane, a monomeric layer results, e.g.

$$\geq Si-OH + Cl-SiR_2'R'' \longrightarrow \geq Si-O-SiR_2'R''.$$

With di- or trichlorosilanes in the presence of a limited amount of water a complete series of reactions can occur leading to multi-layers and/or cross-linked species:

$$(1) \quad \geq Si-OH + Cl_2SiR_2 \xrightarrow{H_2O} \geq Si-O-(\underset{R}{\overset{R}{Si}}-O)_n H$$

(2) $\equiv$Si—OH + Cl$_2$SiR$_2$ $\longrightarrow$ $\equiv$Si—O—$\overset{\displaystyle R}{\underset{\displaystyle R}{\text{Si}}}$—Cl

$\searrow$ H$_2$O

$\equiv$Si—O—$\overset{\displaystyle R}{\underset{\displaystyle R}{\text{Si}}}$—OH

$\swarrow$ R″R′$_2$SiCl

$\equiv$Si—O—$\overset{\displaystyle R}{\underset{\displaystyle R}{\text{Si}}}$—O—$\overset{\displaystyle R'}{\underset{\displaystyle R'}{\text{Si}}}$—R″

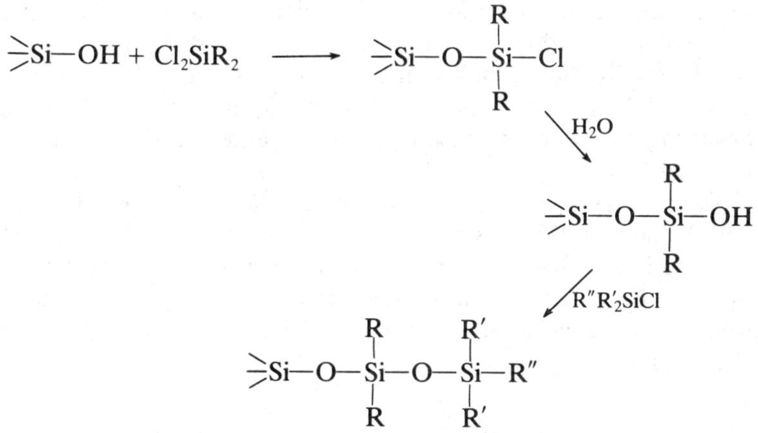

(3) $\equiv$Si—OH + Cl$_3$SiR $\longrightarrow$ $\equiv$Si—O—$\overset{\displaystyle Cl}{\underset{\displaystyle Cl}{\text{Si}}}$—R

$\swarrow$ H$_2$O

$$-\overset{\vdots}{\text{Si}}-\text{O}-\overset{\vdots}{\text{Si}}-\text{R}$$
$$\underset{\displaystyle -\text{Si}-\text{O}-\text{Si}-\text{R}}{\overset{\displaystyle |\qquad\quad|}{\underset{\displaystyle \vdots\qquad\quad\vdots}{\text{O}\qquad\text{O}}}}$$

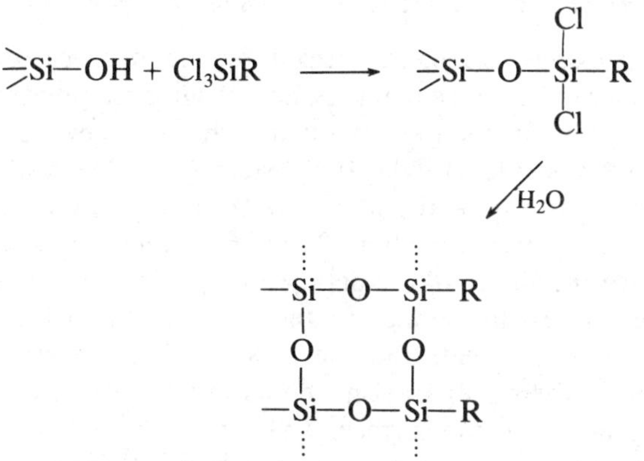

Linear polymeric layers [as formed by reactions of type (1)] can have poor mass transfer characteristics, and this type is to be avoided. However, the bulk polymeric layer [as formed by reactions of types (2) and (3)] or mixtures of the two mechanisms are the most usual forms offered. If reaction (3) does not go to completion, new silanols can be left on the surface.

## C  Polar Bonded Phases

Polar surfaces can be formed with a variety of terminal groups, the most common having cyano, nitro, amino or diol functionalities, and they are commonly marketed under these names. Basically, the process of bonding these various moieties to the silica surface deactivates the surface in the same way that ethyl acetate absorbed on silica gives surface deactivation. Bonded phases, however, show fewer problems than solvent-modified silicas, e.g. chemisorption, tailing and possibly catalytic activity. But the separations obtained are in general similar to those obtained with classical silica columns modified with polar moieties. An added advantage lies in the rapid equilibration times obtained on changing solvent compositions. Table 3.1 lists some of the wide variety of polar bonded phases currently available in order of increasing polarity together with applications of these materials.

Table 3.2 gives a comparison of capacity factors for three bonded phases and the silica gel from which they are derived. The data clearly show the increase in selectivity obtained on increasing the degree of substitution of hydroxyl by amine. The maximum retention is obtained with the "hydroxylamine" terminal group, and a beneficial improvement in selectivity is also obtained.

## D  "Reversed Phase" Bonded-phase Chromatography

Reversed phase packings are characterized by the hydrocarbon chain bonded to the surface of the silica matrix. Chain lengths ranging from $C_1$ to $C_{22}$ are available, but by far the most popular is the octadecyl ($C_{18}$) chain, which gives rise to the octadecylsilyl (ODS) packings, and secondly the $C_8$ chain. In addition, cyclohexyl and phenyl bonded phases are available.

At the time of writing there are about 24 reversed phase packings available, and of these over half are of the octadecyl type. While in general the majority of these phases are similar in their selectivities with aqueous solvents, they are not identical, and the differences occur through the chemistry involved in their preparation, the nature of the silica used and the hydrocarbon loading on the surface. These differences become more marked with the use of non-aqueous solvents. Clearly, therefore, care has to be exercised to ensure that the same packing is used when attempting to reproduce a separation quoted in the literature. Table 3.3 lists the majority of the packings currently available and illustrates the diversity of these materials and reinforces the necessity for care.

The chain length of the bonded hydrocarbon has a significant effect on the capacity factors of solute. This is illustrated in Fig. 3.11 and shows that the capacity factor decreases linearly with decrease in hydrocarbon chain length (Melander and Horváth, 1980). Thus, the speed of analysis is a function of chain length at a given mobile phase composition.

The nature of the bonded phase, i.e. whether it is a monomeric (brush) or polymeric type, also has an effect on the majority of mobile phase equilibrations. It has been shown (Scott and Simpson, 1980) that equilibration is rapid with the polymeric type of phase whereas the monomeric type have significant equilibration times.

Reversed phase BPC is the most widely used form of chromatography at

**Table 3.1** Characteristics of some polar bonded phases

| Polarity | | Name | Functionality | $d_p$ ($\mu$m) | Application or comments |
|---|---|---|---|---|---|
| Weak | { | Hi-eff Micropart–ester | Ester | 5 | Subject to hydrolysis in aqueous solutions |
| | | Nucleosil-NMe$_2$ | Dimethylamino | 5, 10 | Also weak anion exchanger |
| | | LiChrosorb diol | Diol | 10 | Tetracyclines analysis, and exclusion chromatography of proteins |
| | | Fe-Sil-X-1 | Fluoroether | 13 ± 5 | Carbonyl compounds |
| Medium | { | Nucleosil-NO$_2$ | Nitro | 5, 10 | Less polar than silica |
| | | Cyano-Sil-X-1 | Alkyl nitrile | 13 ± 5 | Can be used in reversed phase also |
| | | $\mu$-Bondapak-CN | Nitrile | 10 | Selective for double bonds, particularly in ring systems |
| | | Spherisorb-CN | Nitrile | 5 | |
| | | Chromex HEMA | Polyhydroxyethyl methacrylate | 11 ± 1 | Contains both hydrophilic and hydrophobic sites, therefore useful in both normal and reversed phases |
| Strong | { | Amino-Sil-X-1 | Alkylamino | 13 ± 5 | Selective for nitro compounds and aromatics |
| | | LiChrosorb-NH$_2$ | Amino | 10 | For sugars and peptides |
| | | Micropak-NH$_2$ | Aminopropyl | 10 | Carbohydrates and nucleotides |
| | | $\mu$-Bondapak-NH$_2$ | Amino | 10 | As above |

Data abstracted from Majors (1980), where complete information may be found.

**Table 3.2** Capacity and selectivities of polar solutes on different bulk modified silica column packings using a constant mobile phase composition

Terminal group: $S$ value

| Test solute | SiO$_2$ base material 400 m$^2$ g$^{-1}$ | | —CH(OH)CH$_2$OH 394 m$^2$ g$^{-1}$ | | —CH(OH)CH$_2$NH$_2$[†] 401 m$^2$ g$^{-1}$ | | —CH(OH)CH$_2$NH$_2$[‡] 430 m$^2$ g$^{-1}$ | |
|---|---|---|---|---|---|---|---|---|
| | $k$ | $\alpha$§ | $k$ | $\alpha$§ | $k$ | $\alpha$§ | $k$ | $\alpha$§ |
| 2-Chlorophenol | 0.21 | 3.81 | 0.25 | 3.4 | 0.46 | 2.30 | 0.72 | 1.50 |
| Phenol | 0.80 | 0.86 | 0.85 | 0.94 | 1.06 | 1.16 | 0.14 | 1.54 |
| 4-Chlorophenol | 0.69 | 1.23 | 0.80 | 1.48 | 1.23 | 1.84 | 1.75 | 5.19 |
| 4-Nitrophenol | 0.85 | 1.75 | 1.18 | 0.85 | 2.26 | 1.00 | 9.08 | 0.45 |
| Aniline | 1.49 | — | 1.00 | 1.30 | 2.26 | 2.04 | 4.08 | 2.65 |
| α-Picoline | Large | | 1.30 | | 4.60 | | 10.8 | |

[†]Molar ratio of OH/NH$_2$ = 6.
[‡]Molar ratio of OH/NH$_2$ = 1.
§Values of $\alpha$ were calculated by the author between successive values of $k$ using the relationship $\alpha = t'_{R_2}/t'_{R_1}$.
The mobile phase composition was dichloromethane/acetonitrile, 95 : 5 v/v.

**Table 3.3** Characteristics of commercially available reversed phase packings

| Chain length | Name | Functionality | Substrate | Particle size ($\mu$m) | Comments |
|---|---|---|---|---|---|
| Long | Micropak CH | Octadecylsilane | LiChrosorb Si | 10 | Polymeric layer, 22% hydrocarbon |
| | Micropak MCH | Octadecylsilane | LiChrosorb Si | 10 | Monomeric layer, 8% hydrocarbon |
| | Partisil ODS | Octadecylsilane | Partisil | 10 | 5% loading, monomeric |
| | Partisil ODS-2 | Octadecylsilane | Partisil | 10 | |
| | Partisil ODS-3 | Octadecylsilane | Partisil | 10 | 16% loading, polymeric |
| | Zorbax ODS | Octadecylsilane | Zorbax | 6–8 | 5% loading |
| | $\mu$-Bondapak $C_{18}$ | Octadecylsilane | $\mu$-Porasil | 10 | 10% loading |
| | Hypersil-ODS | Octadecylsilane | Hypersil | 7 | 8% carbon |
| | LiChrosorb $C_{18}$ | Octadecylsilane | LiChrosorb Si | 5, 10 | Monolayer, 22% loading, 1–9 pH |
| | Spherisorb ODS | Octadecylsilane | Spherisorb Si | 5, 10 | Spherical, pH 8 max. |
| | Nucleosil $C_{18}$ | Octadecylsilane | Nucleosil 100Å | 5, 10 | Capacity twice $C_8$ |
| Medium | Zorbax $C_8$ | Octylsilane | Zorbax | 6–8 | For polar compounds |
| | LiChrosorb $C_8$ | Octylsilane | LiChrosorb Si | 5, 10 | Monolayer or 14% load, pH 1–9 |
| | Nucleosil $C_8$ | Octylsilane | Nucleosil 100Å | 5, 10 | General purpose |
| | Partisil $C_8$ | Octylsilane | Partisil | 10 | General purpose |
| | Chromegabond $C_8$ | Octylsilane | LiChrosorb Si | 10 | $C_1$ and $C_{18}$ also available |
| | Chromegabond | Cyclohexene | LiChrosorb Si | 10 | Low polarity |
| | $\mu$-Bondapakphenyl | Phenyl | $\mu$-Porasil | 10 | Low–intermediate polarity |
| Short | Hypersil SAS | Short alkyl chain | Hypersil | 7 | For ion-pair chromatography |
| | LiChrosorb $C_2$ | Ethyl | LiChrosorb Si | 5, 10 | For polar compounds |
| | Zorbax TMS | | Zorbax | 10 | For polar compounds |

Data abstracted in part from Melander and Horváth (1980).

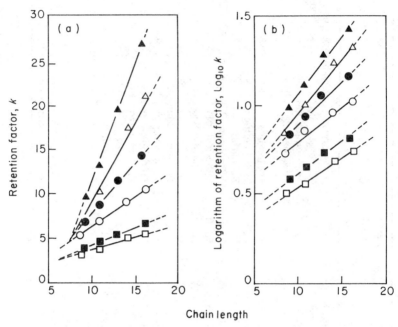

**Fig. 3.11** Dependence of retardation factor (a) and log (retardation factor) (b) on the carbon load of various alkylsilicas. Four different ligates ($C_8$, $C_{11}$, $C_{15}$, $C_{18}$) were bound to 10 $\mu$m LiChrosorb Si 100 having a BET surface area of 282 $m^2\,g^{-1}$. The open and closed symbols represent data obtained with bonded phases prepared with alkyltrichlorosilane and alkylmethyldichlorosilane respectively. The mobile phase was 27.6% v/v water in methanol at 23 °C. The eluates are 1,4-dibromobenzene ($\square$, $\blacksquare$), 1,2,3,4-tetrachlorobenzene ($\bigcirc$, $\bullet$) and 4,4-dibromobiphenyl ($\triangle$, $\blacktriangle$). [Redrawn with permission from Melander and Horváth (1980).]

the present time; about 80% of all published applications use the technique. The reason for this popularity is threefold: first, the ready availability in a pure state of the mobile phases used; secondly, the rapid equilibration times when changing the mobile phase composition; thirdly, the mobile phases used are transparent in the ultraviolet, down to as low as 190 nm if specially purified, and hence confer flexibility on the monitoring wavelength, which can be of significant assistance when analysing substances which only absorb weakly at 254 nm.

The major disadvantage of reversed phase BPC lies in the non-equivalence between various manufacturers' products, and so problems can be experienced when attempting to reproduce a published separation. However, rigid quality control from certain manufacturers has provided products which will reproduce a given separation to within an accuracy of

10% in the worst cases, and this deviation may often arise from inferior solvents or variation in equipment design, not from the packing.

## VI ION-EXCHANGE CHROMATOGRAPHY

Ion-exchange chromatography (IEC) and related techniques are used to separate ionized or ionizable compounds. The reaction, which occurs on the surface of the substrate, is a simple ionic equilibrium and two forms may be distinguished for monovalent ions:

(1) Anionic exchange,

$$\overline{B^+Y^-} + X^- \underset{}{\overset{K_{XY}}{\rightleftharpoons}} \overline{B^+X^-} + Y^-;$$

(2) Cationic exchange,

$$\overline{A^-M^+} + NH^+ \underset{}{\overset{K_{NM}}{\rightleftharpoons}} \overline{A^-NH^+} + M^+.$$

The bar indicates the stationary phase, and $K_X$, $K_{NM}$ are the equilibrium constants (selectivity coefficients) of the exchange reactions.

It will be clear from the above that in ion-exchange chromatography the ionized forms of solutes are adsorbed to fixed ionic sites on the stationary phase of opposite charge, while counterions ($Y^-$ or $M^+$) are released into the mobile phase to preserve electroneutrality of the overall system.

Effectively the ion-exchange process can be regarded as a competition between the solute ions and the counterions present in the mobile phase to pair with the oppositely charged fixed sites on the substrate, i.e. the solute ions in the mobile phase have to displace at least one counterion which is normally paired with a fixed functional group before they can be adsorbed themselves (c.f. Section III.B). Clearly, therefore, one means of adjusting retention in IEC is to vary the concentration of the counterion (which will vary the equilibrium constant of the reaction); namely for anionic exchange

$$K_{XY} = \frac{[B^+X^-][Y^-]}{[B^+Y^-][X^-]}, \tag{3.3}$$

and for cationic exchange

$$K_{NM} = \frac{[A^-NH^+][M^+]}{[A^-M^+][NH^+]}. \tag{3.4}$$

Clearly IEC can only occur when the solute is in a charged form, and the presence of a charge depends upon the $pK_a$ of the species concerned, which in turn depends upon the pH of the mobile phase. For acids,

$$HX \xrightleftharpoons{K_a} H^+ + Y^-; \qquad K_a = [H^+][Y^-]/[HX]. \qquad (3.5)$$

For bases,

$$NH^+ \xrightleftharpoons{K_a} N + H^+; \qquad K_a = [N][H^+]/[NH^+]. \qquad (3.6)$$

Note that

$$K_a = K_w/K_b. \qquad (3.7)$$

If it is assumed that only charged forms are present, i.e. the pH is adjusted so that all species present are fully charged, then it is possible to write down the distribution coefficient $D_i$ of species $i$ as the ratio of the concentration of solute in the stationary phase to the total concentration of charged and uncharged forms in the mobile phase. (Note: $D$ describes the distribution constant, rather than the more familiar "$K$" so as not to confuse it with the dissociation constant used above.) In anionic exchange,

$$D_X = \frac{\overline{[B^+X^-]}}{[X^-] + [HX]} = K_{XY} \frac{\overline{[B^+Y^-]}}{[Y^-]} \cdot \frac{1}{(1 + [H^+]/K_a)}, \qquad (3.8)$$

and in cationic exchange

$$D_N = \frac{\overline{[A^-NH^+]}}{[NH^+] + [N]} = K_{NM} \frac{\overline{[A^-M^+]}}{[M^+]} \cdot \frac{1}{(1 + K_a/[H^+])}. \qquad (3.9)$$

These equations follow from equations (3.3)–(3.7) above.

Equations (3.8) and 3.9) demonstrate that the distribution coefficient is controlled by the number of available functional groups present on the substrate, and is inversely proportional to the counterion concentration in the mobile phase together with the $pK_a$ and pH of the solute and mobile phase respectively.

These two equations enable the distribution coefficient (and hence the separation obtainable) to be varied by modifying the mobile phase composition, and, most importantly, in a predictable manner.

The simple analysis of the process presented above neglects several other important parameters which can also have an effect on this ion-exchange process. These are given below with their effects on retention.

## A    Ion-exchange Capacity of the Substrate

This is easy to comprehend; ion-exchange capacity is defined as the maximum amount of solute in milliequivalents that can be adsorbed per gram (or unit volume) of dry ion-exchanger. Clearly the larger the ion-exchange capacity, the larger the distribution coefficient and also the capacity ratio, $k$. The ion-exchange capacity depends upon the number of available charged sites and this is a function of the pH of the mobile phase, and also the type of functional group available (see later).

Further, while to a first approximation ion-exchange capacity only affects $k$, in practice secondary selectivity effects may also become apparent which can change the order of elution.

## B   pH of the Mobile Phase

pH is a particularly valuable parameter to vary, and is especially valuable with weak acids or bases. For strong acids and bases, the distribution coefficient only starts to drop at very low or very high pH values respectively; but with weak acids and bases, a dramatic change in $D_i$ occurs when pH $= pK_a$. This is a particularly valuable effect when separating amphoteric substances, e.g. proteins.

pH can also have a profound effect on the ion-exchange capacity of the substrate, depending on the nature of the bound functional groups. For ion-exchangers containing the $- SO_3^-$ or $- N^+R_3$ groups, i.e. strongly acidic or basic groups, $D_i$ starts to decrease slowly when the pH value equals that of the $pK_a$ of the functional group. But for weakly acidic ( $- COO^-$) or basic ( $- NR_2$ and $- NR$) groups, $D_i$ passes through a maximum which depends on the $pK_a$ of the solute and the bound functional group and drops to zero at high and low pH.

## C   Effect of the Nature and Concentration of the Counterion

It was shown earlier [see equations (3.8) and (3.9)] that for monovalent ions the distribution coefficient, $D_i$, is inversely proportional to the concentration (molality) of the counterion. This was an oversimplification, but correct in a simple system in which only one type of counterion was considered. If the nature of the counterion is changed, then other factors have to be considered also, e.g. the charge on the ion, the size of the hydrated ion (ionic radius) and its polarizability. Furthermore, the effect of the counterion is increased if its chemical nature (of the organic part) is similar to the substrate.

Counterions may be arranged in terms of their increasing affinity for ion-exchangers as follows:

Anions   $F^- < OH^- < CH_3COO^- < HCOO^- < Cl^- < SCN^- < Br^-$
$< CrO_4^{2-} < NO_3^- < I^- <$ oxalate $< SO_4^{2-} < ClO_4^-$
$<$ citrate

Cations    $Li^+ < H^+ < Na^+ < NH_4^+ < K^+ < Rb^+ < Cs^+ < Ag^+ < Mn^{2+}$
$< Mg^{2+} < Cu^{2+} < Ca^{2+} < Sr^{2+} < Ba^{2+} < 3^+$ cations $< Ce^{3+}$

This order holds, in general, for both strong and weak ion-exchangers except that, in the case of the latter group, $H^+$ and $OH^-$ have the strongest elution strength.

## D   Influence of Temperature

The column temperature (including the mobile phase, which should be pre-equilibrated to the same temperature) plays a significant role in IEC for two main reasons. First, the exchange equilibrium rate is increased with increasing temperature; this leads to a decrease in solute retention. Secondly, but possibly more important, the decrease in viscosity of the mobile phase has a beneficial effect on mass transfer and hence on column kinetics; this is shown by an increase in column efficiency. An added advantage in using elevated temperatures lies in the decreased pressure drop across the column for a given flow rate, which can be beneficial if ion-exchangers are used which are susceptible to compression at high pressure.

## E   Effect of the Addition of Organic Solvents

The addition of organic solvents to the mobile phase can have a profound effect on the selectivity and retention of ionized organic substances. It was mentioned earlier that secondary effects can play a part in the ion-exchange mechanism. The effects are adsorption (other than by ion-exchange), partition and permeation (see next section). The addition of organic solvents can bring into play these secondary effects with a significant contribution to the selectivity of the overall separation. Regrettably, prediction of the effect of addition of organic solvents upon retention is, at this time, not possible; however, Fig. 3.12 shows the effect of including ethanol in the mobile phase on the retention of nucleotides and nucleosides.

## F   Chelation

The retention characteristics of solute ions can be drastically modified by introducing a suitable chelating agent into the mobile phase. For example, carbohydrates may be analysed by IEC by chelating them with the borate ion, and metal ions can have their selectivity altered by incorporating a suitable ligand into the mobile phase. This situation can be operated in reverse by incorporating a metal counterion which can complex with the

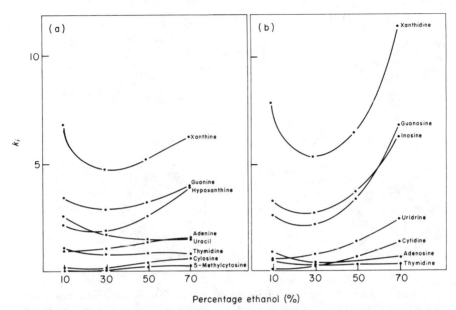

**Fig. 3.12** Effect of ethanol content on the capacity ratio ($k_i$) of some nucleotides (a) and nucleosides (b) on a strong anion-exchanger. [Redrawn with permission from Ekstien, Linssen and Kraak (1978).]

stationary phase and interact selectively with the solute molecules, e.g. $Cu^{2+}$ for the separation of amino acids in cation-exchange chromatography using ammonia as the competing ligand.

## G  The Nature of the Ion-exchanger

A variety of materials has been used to form ion-exchangers, notably synthetic resins of cross-linked polystyrene, cross-linked polydextrans, cellulose and silica. The ion-exchange function is formed on the surface of these beads or particles. With synthetic resins there is a marked swelling of the beads in aqueous buffers which makes them compressible and not suitable for high flow rates which require elevated pressures. This problem can be alleviated by increasing the percentage of cross-linking which can have a benficial effect in selectivity. Unfortunately, this has an adverse effect on pore diameter and mass transfer.

The reduction in pore size restricts access of large ions and hence reduces the effective exchange capacity of the resin with a concomitant reduction in $k$. However, an improvement in capacity and mass transfer characteristics may be obtained by decreasing the bead size to 5 $\mu$m and increasing the pore size of the resin (so-called macroporous resins) or by

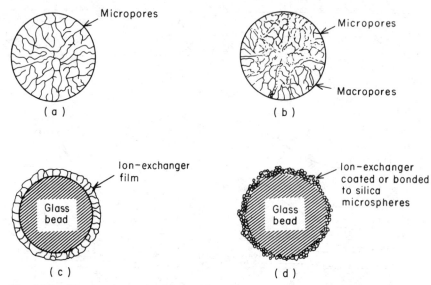

**Fig. 3.13** Structural types of ion-exchange resins: (a) microreticular resin; (b) macroreticular resin; (c) pellicular resin; (d) superficially porous resin. [Redrawn with permission from Leich and De Stephano (1973).]

coating an impervious sphere with a thin layer of resin (so-called pellicular beads). The former improves the mass transfer processes while maintaining an effective capacity while the latter gives good kinetics at the expense of capacity. Similar remarks are also applicable to the superficially porous exchangers (see Fig. 3.13).

In order to overcome the disadvantages of compressibility or poor capacity, ion-exchangers have been prepared by chemically bonding the functional group to silica particles. These products allow high flow rates with good kinetics and hence high speed separations. However, the stability of silica-based exchangers is not high at pH >8 because of dissolution of the silica matrix. Even at pH <8 appropriate reagents can rupture the surface bonds and thus, although useful, silica based ion-exchangers are not satisfactory for routine use.

## VII ION-PAIR CHROMATOGRAPHY

Ion-pair extraction, the forerunner of the chromatographic technique, was developed by Schill and co-workers (see Schill, 1974) and basically involves the extraction by an organic solvent of an aqueous solution of an ion which is paired with a suitable counterion. The process is described by the equilibrium

$$(X^+)_{aq} + (Y^-)_{aq} \xrightleftharpoons{K_{XY}} (X^+Y^-)_{org}$$

in which $X^+$ is the solute ion and $Y^-$ the counterion, or vice versa. In a similar manner to that described earlier, the distribution coefficient is given by the expression

$$D_X = [X^+Y^-]_{org}/[X^+]_{aq} = K_{XY}[Y^-]_{aq}.$$

Which shows that, in the simplest case, $D_X$ depends upon the counterion concentration and the equilibrium constant, which itself is a function of the nature of the organic phase, the nature of the counterion and the temperature. Obviously, it is necessary that the solute should be in a suitably charged state, and thus pH also affects the distribution coefficient.

As in conventional IEC, other factors can play a part in the selectivity observed, namely adsorption and/or partition of uncharged species (which may be present under certain pH conditions), dissociation or dimerization of ion pairs and so on. Thus a complete description of the processes involved is complex.

However, the practice of ion-pair chromatography is relatively simple. It may be carried out in both normal and reversed phase modes. In the normal mode, the aqueous phase is used as the stationary phase, while the organic medium is the stationary phase in the reversed phase mode.

## A   Normal Phase Ion-pair Partition Chromatography

In this mode, the aqueous phase is immobilized on the support (silica) by equilibrating the column with the aqueous phase containing the counterion previously equilibrated in contact with the organic phase. The organic phase – also equilibrated at the same temperature – is passed through the column until a stable system is obtained. Basically, the system is similar to normal partition chromatography described earlier, except for the presence of the counterion and pH control. Figure 3.14 illustrates the separation of some carboxylic acids and demonstrates that the selectivity of the system can be changed together with the order of elution by changing the composition of mobile (organic) phase, and this can be ascribed to the greater solution of the ion pairs by the increased level of butanol present.

A particularly interesting application of "normal" ion-pair chromatography is the enhancement of the detection of solutes which normally do not adsorb strongly in the ultraviolet by using an ion-pairing reagent with strong ultraviolet absorbing properties. An example of this is shown in Fig. 3.15.

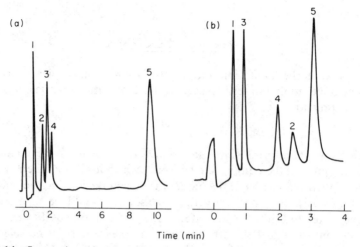

**Fig. 3.14** Separation of carboxylic acids by means of normal phase ion-pair partition chromatography. Stationary phase, 0.1 M tetrabutylammonium hydrogen sulphate + phosphate buffer. Solutes: 1, toluene; 2, VMA; 3, IAA; 4, HVA; 5, HIAA. (a) Mobile phase is butanol–dichloromethane (4 : 96); (b) mobile phase is butanol–dichloromethane–hexane (20 : 30 : 50). [Redrawn with permission from Persson and Karger (1974).]

## B  Reversed Phase Ion-pair Chromatography

This is the most useful form of ion-pair chromatography practised at the present time and compares with the corresponding partition and bonded-phase mode. However, the situation is more complicated in that there are at least two distinct mechanisms which can operate (Hearn, 1980):

(1) The pairing ion remains principally in solution and the mechanism is basically simple partition.

(2) The pairing ion is adsorbed on the surface of the reversed phase ($C_2$, $C_8$ or $C_{18}$), and effectively an ion-exchanger is being produced *in situ*.

From (1), it is clear that a hydrophobic support is a fundamental requirement and the reverse situation to that outlined above for normal ion-pair systems is followed, i.e. the column is equilibrated first with the organic phase containing the counterion and then with the aqueous mobile phase (again after pre-equilibration with the organic phase and at the same temperature). Figure 3.16 shows the separation of some organic acids by reversed phase ion-pair chromatography using pentanol as the organic stationary phase adsorbed on an alkyl-modified silica (RP-2). The mobile

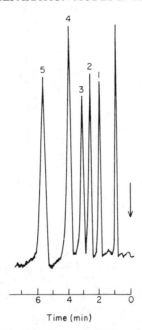

Time (min)

**Fig. 3.15** Separation of dipeptides by normal phase ion-pair partition chromatography. Stationary phase, 0.1 M naphthalene-2-sulphonate; pH 2.3 Mobile phase, chloroform–pentanol (9 : 1). Solutes: 1, Leu-Leu; 2, Phe-Val; 2, Val-Phe; 4, Leu-Val; 5, Met-Val. [Redrawn with permission from Crommen, Fransson and Schill (1977).]

phase is water saturated with pentanol and containing the tetrabutylammonium ion, $(C_4H_9)_4N^+$, as the pairing ion at pH 7.4.

The second mode of undertaking ion-pair chromatography (ii) has been demonstrated by Knox and Laird (1976). They proposed the technique known as "soap chromatography" and employed long chain anions and cations such as sodium dodecyl sulphate and cetrimide (cetyltrimethylammonium bromide) as the ion-pairing reagent, using a reversed phase $C_8$ or $C_{18}$ as the support matrix. Scott and Kucera (1979a) demonstrated that these materials were strongly adsorbed on the surface of the reversed phase, so that effectively the technique produces an ion-exchanger *in situ*. Using this technique, Knox and his collaborators demonstrated some elegant separations of vat dyes (Knox and Laird, 1976) and catacholamines (Knox and Jurand, 1976) (Figs 3.17 and 3.18).

A study of the literature reveals that the use of conventional analytical ion-exchange chromatography is on the wane, its place being taken by the ion-pairing technique. This stems from the convenience of the technique,

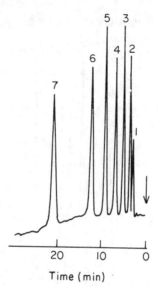

**Fig. 3.16** Reversed phase ion-pair partition chromatography of organic acids. Stationary phase: pentan-1-ol on RP-2. Mobile phase: 0.03 M tetrabutylammonium, pH 7.4. Peaks: 1, 4-aminobenzoic acid; 2, 3-aminobenzoic acid; 3, 4-hydroxybenzoic acid; 4, 3-hydroxybenzoic acid; 5, benzenesulphonic acid; 6, benzoic acid; 7, toluene-4-sulphonic acid. [Redrawn with permission from Fransson, Wahlund, Johansson and Schill (1976).]

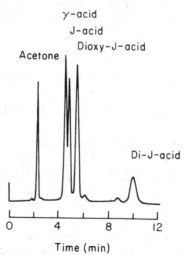

**Fig. 3.17** Separation of dye intermediates by soap chromatography. Column, 120 mm × 5 mm; packing, 7 μm SAS silica; eluent, water–propanol (5 : 2) containing 1% m/v cetrimide. Peak elution order: acetone followed by 1,5, 1,6, 1,7 and 1,8 isomers of napthylaminesulphonic acid. [Redrawn with permission from Knox and Laird (1976).]

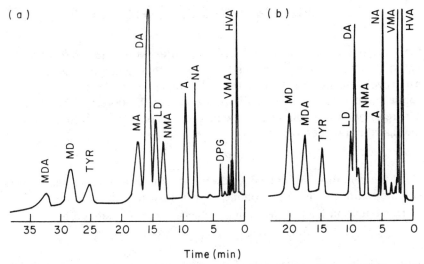

**Fig. 3.18** Soap chromatography of catecholamines and derivatives. Packing, ODS/TMS silica; detection, 280 nm, 0.02 AUFS; eluent, water–methanol–sodium dodecyl sulphate (72.5 : 27.5 : 0.02, v/v/m) with added sulphuric acid: (a) 0.04%; (b) 0.15% (v/v). [Redrawn from Knox and Jurand (1976).]

in so far as the column consists of a single reversed phase column which can be modified by the addition of a suitable counterion, although care has got to be exercised to ensure complete removal of one counterion before a second is used. Furthermore, it depends upon the analyst selecting the appropriate level of counterion, which thus gives him considerable flexibility in his choice of operating conditions. Table 3.4 lists the ion-pairing reagents currently available.

**Table 3.4**   Ion-pairing reagents

| Anionic | Cationic |
|---|---|
| Hexadecyltrimethylammonium bromide | 1-Heptane sulphonic acid (Na salt) |
| Tetrabutylammonium bromide | 1-Hexane sulphonic acid (Na salt) |
| Tetradecyltrimethylammonium bromide | 1-Octane sulphonic acid (Na salt) |
| Tetraethylammonium bromide | 1-Pentane sulphonic acid (Na salt) |
| Tetramethylammonium bromide | Sodium lauryl sulphate |
| Tetrapropylammonium bromide | |

## VIII   GEL PERMEATION CHROMATOGRAPHY

Gel permeation chromatography (GPC) differs from the other modes of liquid chromatography described above in that separation is achieved exclusively (in the ideal case) from differences in molecular size of the solute molecules and their ability to penetrate the pores of the packing. The retention behaviour of a solute in a porous packing is given in terms of the distribution coefficient, $K_{GPC}$, where

$$V_R = V_0 + K_{GPC}V_i.$$

$V_R$ is the retention volume, $V_0$ is the total volume of mobile phase (void volume) and $V_i$ is the total volume of solvent within the packing.

This equation leads to a column capacity factor $k''$ defined by

$$V_R = V_0(1 + k''),$$

and it is important to note that $k''$ is different from $k$, the capacity factor in affinity methods of HPLC.

Many theories have been proposed to account for size separation of solutes; these may conveniently be grouped under two headings – equilibrium models and flow models. It would appear that under normal conditions of flow (i.e. 1 ml min$^{-1}$) practical size separations occur under near equilibrium conditions, and $V_R$ is independent of mobile phase flow rate.

The volume of stationary phase which is accessible to solute molecules of a given size is controlled by $K_{GPC}$, and this is a function of the size of the molecule and the pore size of the packing (Fig. 3.19). Thus very large

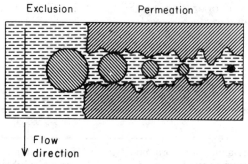

**Fig. 3.19**   Principle of size exclusion chromatography. Smaller molecules penetrate deeper into the available pore volume in the stationary phase: the liquid volume for them is apparently bigger. As the diameter of the sample molecule increases, less and less pore volume can be permeated. (Reproduced by courtesy of Dr I. Molnar.)

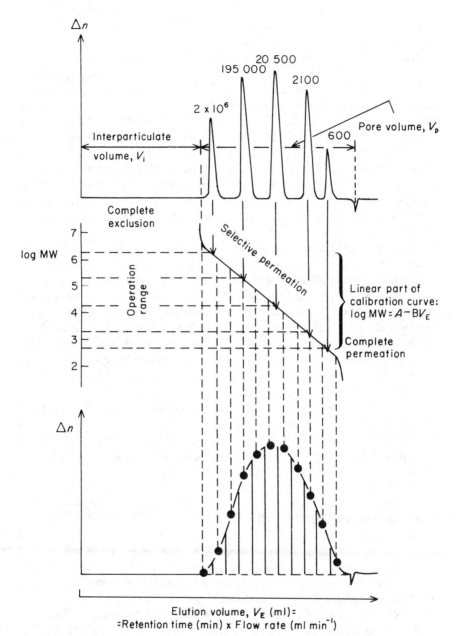

**Fig. 3.20** Calibration and practical evaluation in gel permeation chromatography. First we estimate the elution of standards of known molecular weight (MW) (upper figure) – in this case polystyrenes. The function log MW = $f(V_E)$ is next established (middle figure). Finally, the polydispersity (PD), which is a measure of the molecular weight distribution, can be calculated once $\bar{M}_w$ and $\bar{M}_n$ are known (lower figure). (Reproduced by courtesy of Dr I. Molnar.)

molecules whose sizes in solution prevent them from penetrating the pores of the packing are said to be totally excluded and elute in the dead volume of the column, $V_0$, and here $K_{GPC} = 0$. Small molecules, capable of penetrating all of the available pores, have $K_{GPC} = 1$. Thus, when a sample is placed on a column, and the mobile phase passed, the large totally excluded molecules are eluted first, followed by molecules of decreasing size which can penetrate increasing fractions of the solvent (stationary phase) contained within the pores; finally the small molecules, capable of penetrating all of the pores, are eluted. This is illustrated in Fig. 3.20. The plateau area between total exclusion and total penetration is related to the pore size distribution of the packing and the sizes of the excluded (or included) molecules.

The theoretical curve illustrated in Fig. 3.20 represents the ideal case in which steric exclusion is the only mechanism operating. On occasion, $K_{GPC} > 1$, which clearly indicates that a secondary mechanism is also operating. This may be either the adsorption and/or partition mechanisms and occurs from solute–gel interactions and normally arise from active sites on the gel surface capable of interacting reversibly with the solute molecules.

These effects are discussed in more detail for a number of commercially available packings in Chapter 13. Modern rigid microparticulate materials can be operated at high flow rates, and high pressures, to obtain efficient rapid separations.

An example of a GPC separation of low molecular weight plasticizers is shown in Fig. 3.22. This was obtained by Scott and Kucera (1979b) on $\mu$-Styragel using chloroform as the mobile phase at 2.0 ml min$^{-1}$. It should be noted, however, that it is also possible to separate these components using reversed phase chromatography with appreciably better resolution. Nevertheless, this is a good separation based solely on the exclusion mechanism and illustrates that compounds with quite small differences in molecular size can be differentiated.

A further example of GPC of compounds of low molecular weight is shown in Fig. 3.22. This was obtained by Scott and Kucera (1979b) using a micro-bore column and demonstrates the potential power of the technique.

It will be clear from the foregoing discussion that separation is a function of the mobile phase flow rate, and further, if any meaningful data are to be obtained about the molecular weight distribution of polymers (Billingham, 1976) or, more importantly, an indication of the molecular weight distribution of an eluite, a very precise mobile phase flow is a prerequisite. This situation has become exacerbated with the advent of high efficiency, high speed columns. Fortunately, the development of microprocessor-controlled pumping systems with a positive feedback has alleviated this problem; however, even with these pumping systems an absolute measure

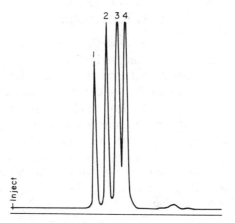

**Fig. 3.21** Separation of alkyl phthalates by GPC using Styragel columns and a chloroform mobile phase. Peaks: 1, dioctylphthalate; 2, dibutylphthalate; 3, diethylphthalate; 4, dimethylphthalate. (Reproduced by courtesy of Waters Associates.)

of a pump's performance is necessary. This may be achieved by connecting the outlet of the detector to a micro-burette and actually measuring the flow rate against a reliable stop-watch. Alternatively, monodisperse markers may be included in the sample as internal standards.

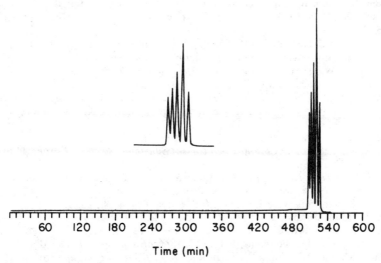

Time (min)

**Fig. 3.22** Exclusion chromatography using a column with an efficiency of 650 000 theoretical plates. Column, 14 m × 1 mm i.d.; packing, 5 $\mu$m Spherisorb; flow rate, 25 $\mu$l min$^{-1}$; sample, 0.5 $\mu$l of a 10% solution of $C_2$ to $C_8$ alkyl benzenes in tetrahydrofuran. [Redrawn with permission from Scott and Kucera, 1979b).]

## IX  CONCLUSION

In an account of this length of the various modes of operation of liquid chromatography, it is obvious that a wealth of information has been omitted. However, it is hoped that the background rationale by which separation is achieved has been made clear; it should also be recognized that there are areas where our knowledge is incomplete.

The interested reader is referred to the many books, review articles, and journals which present a more comprehensive treatment of this widely diversified and fascinating subject; see, for example, Simpson (1982), Snyder and Kirkland (1979), Milton and Hearn (1980), Yau, Kirkland and Bly (1979) and Horváth (1980).

## REFERENCES

Billingham, N. C. (1976). In *Practical HPLC* (C. F. Simpson, ed.), Heyden, London, p. 167.

Crommen, J., Fransson, B. and Schill, G. (1977). *J. Chromat.* **142**, 283.

Davydov, U. Y., Zhurvelev, L. T. and Kiselev, A. V. (1964). *Russ, J. Phys. Chem.* **38**, 1108.

Ekstien, R., Linssen, P. and Kraak, J. C. (1978). *J. Chromat.* **148**, 413.

Fransson, B., Wahlund, K. G., Johansson, J. M. and Schill, G. (1976) *J. Chromat.* **125**, 327.

Halász, I. and Sebastion, I. (1969). Paper presented at the Fifth International Symposium on Advances in Chromatography, Las Vegas.

Hearn, M. T. W. (1980). In *Advances in Chromatography*, Vol. 18 (J. C. Giddings, E. Gruska, J. Cazes and P. R. Brown, eds), Marcel Dekker, New York, p. 59.

Horváth, C. (ed.) (1980). *High Performance Liquid Chromatography – Advances and Perspectives*, Vols I and II, Academic Press, New York.

Huber, J. F. K. (1971). *J. chromatogr. Sci.* **9**, 72.

Huber, J. F. K. and Vodenik, R. (1976). *J. Chromat.* **122**, 331.

Ilher, R. K. (1955). *The Colloid Chemistry of Silica and Silicates*, Cornell University Press, New York.

Ilher, R. K. (1979). *The Chemistry of Silica: Solubility, Polymerization, Colloid and Surface Properties*, John Wiley, New York.

Ilher, R. K. and Pinksney, S. (1947). *Ind. Engng Chem.* **39**, 1379.

Knox, J. H. and Jurand, J. (1976). *J. Chromat.* **125**, 89.

Knox, J. H. and Laird, G. R. (1976). *J. Chromat.* **122**, 17.

Leitch, R. E. and De Stephano, J. J. (1973). *J. chromatogr. Sci.* **11**, 105.

McCann, M., Purnell, H. and Wellington, C. A. (1982). *Proc. Faraday Symp.* **15**, in press.

Majors, R. E. (1980). In *High Performance Liquid Chromatography – Advances and Perspectives*, Vol. I (C. Horváth, ed.) Academic Press, New York, p. 75.

Martin, A. J. P. and Synge, R. L. M. (1941). *Biochem. J.* **35**, 1358.

Melander, W. R. and Horváth, C. (1980). In *High Performance Liquid Chromato-graphy – Advances and Perspectives*, Vol. II (C. Horváth, ed.), Academic Press, New York, p. 113.

Peri, J. B. and Hensley, A. L., Jr (1968). *J. phys. Chem*. **72**, 2226.

Persson, B. A. and Karger, B. L. (1974). *J. chromatogr. Sci*. **12**, 521.

Schill, G. (1974). In *Ion-exchange and Solvent Extraction*, Vol. 6 (J. A. Marinskey and Y. Marcus, eds), Marcel Dekker, New York, Chap. 1.

Scott, R. P. W. (1976). *J. Chromatog*. **122**, 35.

Scott, R. P. W. (1980). *J. chromatogr. Sci*. **18**, 297.

Scott, R. P. W. (1982a). In *Techniques of Liquid Chromatography* (C. F. Simpson, ed.), Wiley–Heyden, Chichester, p. 141.

Scott, R. P. W. (1982b). *Proc. Faraday Symp*., **15**.

Scott, R. P. W. and Kucera, P. (1973a). *J. Chromat.*|**83**, 257.

Scott, R. P. W. and Kucera, P. (1973b). *Analyt. Chem*. **45**, 749.

Scott, R. P. W. and Kucera, P. (1975). *J. Chromat*. **112**, 425.

Scott, R. P. W. and Kucera, P. (1978). *J. Chromat*. **149**, 93.

Scott, R. P. W. and Kucera, P. (1979a). *J. Chromat*. **175**, 51.

Scott, R. P. W. and Kucera, P. (1979b). *J. Chromat.* **169**, 51.

Scott, R. P. W. and Simpson, C. F. (1980). *J. Chromat*. **197**, 11.

Scott, R. P. W. and Traiman, S. (1980). *J. Chromat*. **196**, 193.

Simpson, C. F. (1969). Informal discussion of the Gas Chromatography Discussion Group, Liquid Chromatography Section.

Simpson, C. F. (1974). British Patent no. 1 310 872.

Simpson, C. F. (ed.) (1982). Techniques of Liquid Chromatography, Wiley–Heyden, Chichester.

Snyder, L. R. (1968). *Principles of Adsorption Chromatography*, Marcel Dekker, New York.

Snyder, L. R. (1974). *Analyt. Chem*. **74**, 1384.

Snyder, L. R. and Kirkland, J. J. (1979). *Introduction to Modern Liquid Chromatography*, 2nd edn, John Wiley, New York.

Stewart, H. N. M. and Perry, S. G. (1968). *J. Chromatog*. **37**, 97.

Unger, K. H. (1979). *Porous Silica: Its Properties and Use as Support in Column Liquid Chromatography*, Elsevier, Amsterdam.

Vermont, J., Delevil, M., De Vries, A. J. and Guillemin, C. L. (1975). *Analyt. Chem*. **47**, 1329.

Yau, W. W., Kirkland, J. J. and Bly, D. D. (1979). *Practice of Gel Permeation and Gel Filtration*, John Wiley, New York.

# 4 Data Handling and Automation†

## C. R. LOSCOMBE
Laboratory of the Government Chemist, London, U.K.

## I INTRODUCTION

In recent years HPLC has become one of the most important growth areas in analytical instrumentation. It is particularly useful in fields where previously the analysis of large numbers of samples containing non-volatile or thermally unstable species has required the application of tedious and often imprecise procedures. In addition to improvements in both reproducibility and sensitivity, the savings achieved by the greater sample through-put have been considerable. Further increases in the number of analyses performed, together with significant reductions in labour costs, can be made by automating the chromatographic procedure (Erni, Krummen and Pellet, 1979). The first step that is usually taken in this direction is to use some form of data processsing. This can lead to an increase in both accuracy and precision. It should always be remembered, however, that HPLC is a separation procedure which utilizes a stream of flowing liquid and thus is an ideal technique for *total* automation. A fully automated liquid chromatograph can be used to perform highly reproducible quality control-type analyses but, in addition, can be used in the optimization of semi-developed HPLC methods (Parris, 1978).

Every individual step in the chromatographic process, from injection of sample (usually in the form of a liquid), to detection and computation of results, can be automated. Indeed, the current generation of micro-processor-controlled liquid chromatographs offers considerable facilities for such automation. It is, however, perfectly possible to perform modi-fications to existing equipment, as described by Mills, Mackenzie and Dolphin (1979) and Inman and Schifreen (1980), in order to provide these same abilities. Technicon, well established as a manufacturer of automated systems, has used HPLC as an integral part of an analytical package which they term FAST-LC.

The application of data-processing devices to HPLC analyses is currently the most widespread example of automation. Because of the close resem-blance between HPLC peaks and those generated by gas chromatographs it has been possible to extend the scope of existing GC data techniques to include both forms of chromatography. Various types of integrating device are available and they differ widely in terms of both cost and sophistica-tion, ranging from the now almost obsolete ball and disc mechanical system to the in-house mainframe computer network. Once again the ever present silicon chip has led to a dramatic increase in the use and availability of free-standing integrators. These offer versatile computing abilities, often coupled with the potential for interactive control with the main instrument.

## II  THE HPLC SYSTEM

Manufacturers have not been reticent in producing equipment necessary for the automation of the basic chromatographic processes. As with manu-ally operated apparatus both complete packages and modular systems are available. Historically the modular approach has been favoured in the UK, whilst the total systems package has been more common in the USA. In order to cover both markets, manufacturers will often offer modular units which can be linked by a single keyboard controller (with override facilities) to form an integrated package. A stand-alone microprocessor-

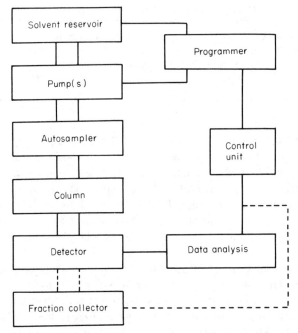

**Fig. 4.1**  Basic components for automated HPLC.

based controller which it is claimed can be interfaced with virtually all modular HPLC equipment has been described (Brenner, 1979). The various components which form the basis of an automated liquid chromatograph are shown in Fig. 4.1.

## A  Pumps and Programmers

Various types of pump design have been utilized. (For additional details see Chapter 2.) The dual head reciprocating system is probably the most popular, although single head systems are currently showing a revival, no doubt due in part to price considerations. Many pumps feature automatic feedback control of flow rate which ensures more reproducible chromatographic results. Some form of electronic pulse damping is sometimes provided.

Associated with the pump module is the solvent programmer. This unit enables the relative ratios of the various components of the mobile phase to be varied during a chromatographic run. This provides for a faster relative elution of compounds than would normally be the case in simple isocratic chromatography. Detection limits for later eluting components

**Table 4.1** Desirable characteristics of a gradient elution system

(1) Reproducible and accurate gradients over a wide range
(2) Good mixing of mobile phase before the column
(3) Minimum solvent volume between mixer and column
(4) Good choice of binary gradient shapes with possible extention to ternary systems

are also generally improved. The detailed theory and practice of gradient elution has recently been reviewed by Snyder (1980). Gradient elution is normally carried out at a constant flow rate but instruments with flow programming options are available. These can be used in order to establish equilibrium rapidly at the original solvent conditions, after a gradient has been run. Solvent programmers can utilize either low or high pressure mixing, but any system should ideally possess the characteristics listed in Table 4.1.

Low pressure mixing systems involve an electronically controlled valve switching between two or more solvent reservoirs prior to the mobile phase reaching the pump. High pressure mixing, on the other hand, is carried out using two pumps and a low volume mixing chamber just before the injector (see Fig. 4.2).† The complexity of the solvent program is only limited by the ingenuity of the instrument designer, although in many cases simple linear, concave or convex ramps will suffice. Peg-boards, keyboards and switch systems are amongst the devices which have been used to preset the controller. This controller will override commands previously set on the pumps. Figure 4.3 shows some of the complex programs available on a ternary low pressure solvent system. The use of ternary solvent mixtures has been shown to be of particular importance in controlling systematically the retention of compounds containing polar functional groups (Roggendorf and Spatz, 1981).

## B Injection Systems

Automation of this module is quite common, with designs often based on those used for gas chromatography. The use of "silent hours" running greatly increases sample through-put. Automatic injectors have reproducibilities comparable with those obtained by manual valve techniques. Samples are preloaded, as a solution, into vials contained in a rack or carousel, which can be rotated mechanically or fed by a gravity mechanism into the path of the injector. The sample is transported from the vial to the

†A more detailed discussion of the relative merits of these two gradient systems will be found in Chapter 2.

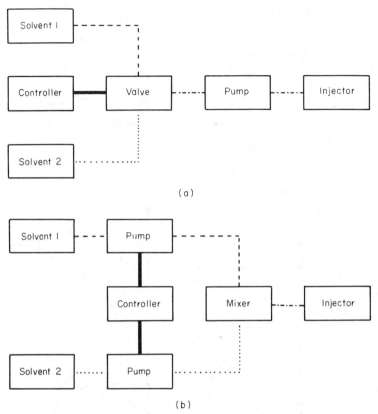

**Fig. 4.2**  Gradient programming systems: (a) low pressure mixing; (b) high pressure mixing.

injector (usually a valve) using suction, gas pressure or some form of displacement system. With valve-based systems sample volume is altered by changing the sample loop manually. Specifications for some commercial autosamplers are given in Table 4.2.

Provision for dealing with solid samples, without a prior manual solution step, is limited at the moment to the Technicon FAST-LC system (see later) and the Chem Research model 1560 sample processor (Meakin and Allington, 1980). A number of auto-injectors are available which do, however, allow for the sample volume and the number of injections per sample to be varied. An example is provided by the Waters Model 710A WISP system which features microprocessor control with each of the sample positions being individually programmable using a keyboard entry system. The time interval between injections is usually fixed by the controls of the

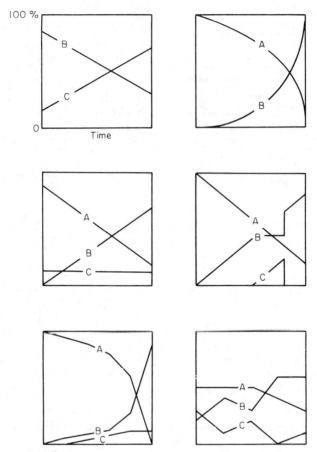

**Fig. 4.3** Some gradient patterns possible using a ternary proportioning valve under microprocessor control. (Reproduced by courtesy of Spectra-Physics Ltd.)

various autosamplers but there is often some provision for external over-ride and thus some feedback from the detector is possible. Some models also feature a cut-out which operates when the solvent reservoir is exhausted.

## C   The Column

Automated column-switching and back-flushing are only available on a limited commercial basis (e.g. from Kontron Instruments Ltd). Systems can, however, be readily constructed using motorized low-dead volume valves and the switching options provided by many integrators and data-

**Table 4.2** Typical HPLC autosampler specifications

| Model | No. of samples | Vial feed mechanism | Sample transport | Max no. inject per sample | Vial size (ml) | Analysis time (min) |
|---|---|---|---|---|---|---|
| Kipp Analytica 9209† | 25 | Gravity | Pressurized air | 2 | 2 | 1–99 or 10–990 |
| Kontron MSI 660 | 30, 60, 120 | Turntable | Suction | 9 | 5, 2, 0.5 | |
| Magnus Scientific M7100 | 60 | Turntable | Suction | 3 | 1 | 0–70 |
| Micrometrics 725 | 64 | Turntable | Positive displacement | 3 | | 0–99 |
| Perkin-Elmer 420B | 42 | Turntable | Pressurized air | 2 | 2 (microvials available) | 1–99 |
| Spectra Physics SP 8010‡ | 210 | Horizontal conveyor track | Suction | 9 | 5 | |
| Spectra Physics SP 8110§ | 4 × 20 | Four turntables | Pressurized air + syringe | — | 2 (microvials available) | 0–9999.9 |
| Waters 710A (WISP) | 48 | Turntable | Syringe | 9 | 4 or 0.3 | 1–540 |
| Varian 8050¶ | 60 | Turntable | Pressurized air | 3 | 2 (microvials available) | 1–99 or 2–198 |

†Similar models available from Altex, ACS, Kratos and Pye Unicam.
‡Only for use with Spectra Physics Model SP 8000 Liquid Chromatograph. (Similar situation with Hewlett-Packard Systems.)
§Only for use with Spectra Physics Model SP8100 Liquid Chromatograph.
¶Similar model available from LDC.

handling devices. Column-switching can both improve and quicken analytical separations whilst back-flushing is particularly useful in preparative HPLC. Erni, Keller, Morin and Schmitt (1981), have recently described a flexible column-switching system employing two six-port valves which can be used for front-cut, heart-cut, end-cut and back-flushing.

Another interesting form of column-switching called "boxcar" chromatography has recently been reported by Snyder, Doland and Van der Wal (1981). In this the partial separation of one or a few compounds of interest is achieved on the first column, with diversion of the resulting fraction to a second column. This second column is filled with several samples at any given time and this allows both high efficiency separations and high sample through-put.

## D Detectors

The most common detection principle used in HPLC is ultraviolet (UV) absorption (see also Chapter 2). Such detectors are available with a preset change of wavelength which occurs on a time basis throughout the chromatogram. This enables the maximum detection limits to be obtained for the various components of a mixture. Stop-flow scanning of peaks to give a UV visible spectrum can also be carried out. Such spectral data can be used to determine both peak purity and identity (Poile and Conlon, 1981). As an alternative to using dual cell detectors, systems have been developed which involve the use of microprocessor memories in order to store background corrections due to absorption by the mobile phase. The Perkin-Elmer model LC-75 is an example of such a detector.

Most UV detectors for HPLC are based on conventional optical systems in which monochromatic light is passed through the sample cell. An alternative approach is the use of a vidicon or scanning diode array device (Haas, Perks and Osten, 1977). Here the sample is illuminated by a wide-band source and only subsequently is the resulting light split into a spectrum by a grating or prism. This spectrum covers the diode array and the required wavelength is selected electronically according to a particular segment on the array (Fig. 4.4). Many wavelengths can be observed simultaneously, the system can also scan very rapidly and in consequence spectra can be recorded without the mobile phase being stopped. This detector does, however, require storage facilities for the large amount of data that is produced. An additional disadvantage with current systems is the generally poor performance at low wavelengths (Titus, 1977).

Perhaps the most widespread application of automation in the field of HPLC detectors is found with pre- and post-column reactions. In post-column derivatization the eluent from the column passes continuously into

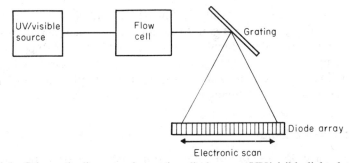

**Fig. 4.4** Schematic diagram of scanning diode array UV/visible light detector.

a reaction system and the reaction product is then measured by a suitable detection system. A simple classification of such reactions is given in Table 4.3. Some dilution is inevitable in any such post-column detection system, together with peak-broadening due to the finite size of the reaction tubing. This can lead to a reduction in the limit of detection but it is compensated by the selective nature of any such detection system.

In pre-column derivatization the reaction is carried out first and the product is then injected on to the column. Both techniques have been well reviewed by Lawrence and Frei (1976). By its very nature post-column derivatization is an automated process. Ross (1977) has reviewed the vari-

**Table 4.3** Classification of post-column reactions for HPLC detection

| Type | Basis of detection | Examples | | Reference |
|------|--------------------|----------|--|-----------|
| | | Reaction | Detector | |
| (1) Addition | Products | Ninhydrin + amino acids | UV | Benson (1975) |
| | | Fluorescamine + amines | Fluorescence | Coppola, Christie and Hanna (1975) |
| (2) Redox | Changed reactant/ solute | Cerium (IV) to cerium (III) | Fluorescence | Katz and Pitt (1972) |
| (3) Complex formation | Loss of reactants | Copper binding | Ion-selective electrode | Loscombe, Cox and Dalziel (1978) |
| (4) pH change | Acidic or basic form | Ochratoxin + ammonia | Fluorescence | Hunt, Philp and Crosby (1979) |

ous pre-column reactions available for HPLC, but they are not always automated. The problems arising from automating such procedures have been described (Gfeller, Huen and Thevenin, 1978). Both types of reaction are important parts of the Technicon FAST-LC system (see later). A self-contained post-column reaction system has recently been announced by Kratos/Schoeffel Instruments.

### E  Fraction Collector

Fraction collectors are not normally used in analytical HPLC but they are of considerable importance in preparative chromatography. They usually involve a moving tray containing sample bottles which are placed under the eluent stream in a preset sequence. Linkage between the collector and the detector unit is desirable in order to limit the number of blank samples collected. For obvious reasons a universal detector such as refractive index is usually employed in preparative systems.

### F  Keyboard Controller

With most integrated systems complete control of the various components can be achieved from a common keyboard. This facility can sometimes be added when modular units are supplied by one manufacturer. A unit which is claimed to be capable of automating most discrete HPLC components has been described by Brenner (1979). Storage of preset programs, using tape or floppy disc for complex ones, is commonly available on control units. Visual display can be provided by printer, LEDs or a VDU. The keyboard controller is often combined with the data-handling device, thus minimizing the number of controls (see later). Indeed a computing integrator can be used for the automated control of an HPLC instrument (Inman and Schifreen, 1980).

## III  THE TECHNICON SYSTEM

In the Technicon approach HPLC forms only one of the units in a continuous-flow automated analytical system. Their Fully Automated Sample Treatment Liquid Chromatography or FAST-LC equipment has a capability ranging from automating straightforward existing UV detection HPLC assay (in this aspect being similar to other manufacturers) to the handling of more complex samples using pretreatment and post-column derivatization. Among the automated sample treatments available are filtering, disintegration, evaporation to dryness, dilution, solvent extraction, dialysis and pre-column derivatization; all of which are based on existing Technicon "Auto-Analyzer" units. Automated post-column derivatization is also available, giving the potential for greatly increased sensitivity and specificity over simple UV detection. Fluorimetry, colorimetry and ion-specific electrodes can be used as alternative detection techniques. The Technicon system is based on the well known air-segmentation principle, originally pioneered by Skeggs (1957), which allows for quite complex operations to be carried out without mixing of samples. FAST-LC systems have already been specifically developed for drugs (such as theophylline) in biological fluids, and fat-soluble vitamins in tablets. The flow diagram for the vitamin analyser is shown in Fig 4.5 and a typical chromatogram, with automated attenuation, in Fig. 4.6. The use of FAST-LC equipment has been reviewed extensively by Burns (1977, 1978). It is also possible to couple Technicon sample treatment modules to many types of existing HPLC equipment.

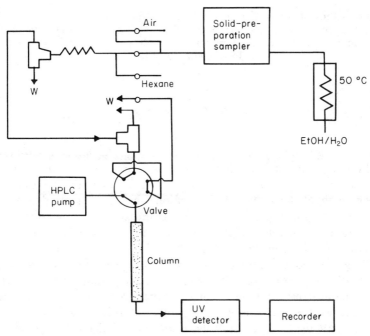

**Fig. 4.5**  Flow diagram for FAST-LC Vitamin Analyzer. (Reproduced by courtesy of Technicon Instruments Co. Ltd.)

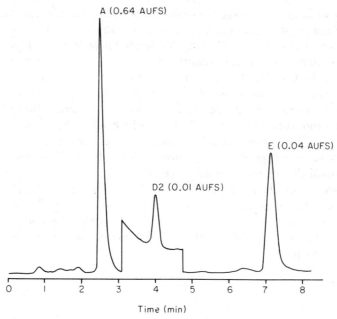

**Fig. 4.6** Chromatogram of vitamin standard mixture, showing effect of automated attenuation on base-line. Column 15 cm × 4.6 mm i.d.; solvent, water–methanol at 1 ml min⁻¹; detector, UV at 254 nm. (Reproduced by courtesy of Technicon Instruments Co. Ltd.)

## IV   DATA HANDLING

Various forms of data processing can be linked to the detector output, with the humble chart recorded forming the most basic approach. The chromatogram so produced allows for qualitative identification of components by manual measurement of retention times. For quantitative analysis either peak height or peak area data is required. Manual measurement of peak height is relatively straightforward but for peak area several methods are available. These include triangulation, the use of a planimeter and cut and weigh methods. Unfortunately all these procedures have the disadvantage of being both time-consuming and prone to human error. The use of a mechanical disc integrator coupled to the recorder can considerably reduce operator time but it is generally only slightly more accurate than the manual methods. Consequently the utilization of digital electronic integrators, in which an ADC converts the detector output into a series of pulses which are then summed, is becoming increasingly important. In addition to the reduction in labour time, electronic integrators offer an increase in both accuracy and precision, good sensitivity, a wide dynamic range and automatic base-line correction. Most machines can measure either peak heights or peak areas, offering a variety of methods for dealing with unresolved peaks, peaks on tails and for rejecting unwanted peaks. Various methods can be used to convert the peak measurements into quantitative results, as described in Chapter 1. Results are given on built-in printers, sometimes combined with plotters which give the actual chromatogram and dispense with the need for a recorder. Retention times of peaks will also be given.

Another useful facility found on many integrators is the ability to carry out timed switching operations; this allows for some degree of automation of the complete HPLC system. In the most common situation these external event controls are used to link the detection system to the autosampler. Increasing use is also being made, however, of timed column-switching and back-flushing techniques, as illustrated by Apffel, Alfredson and Majors (1981). With more sophisticated systems a BASIC (Beginners All-Purpose Symbolic Instruction Code) programmable option is provided in order to increase versatility in both calculation and automated control modes (Montoya and Henselman, 1981). For example, the efficiency of a column can be continuously monitored by repeatedly measuring theoretical plates for a particular compound in the mixture being analysed. When this value falls below a preset minimum either a warning can be sounded or the system can be shut down. A VDU (visual display unit) option will allow BASIC programs to be written and edited without employing the time-consuming and expensive printer/plotter. Generally these more sophisticated computing integrators have facilities for both program and data storage on magnetic tape or floppy disc.

Integrators presently on the market offer various degrees of sophistication. Some details of current models are provided in Table 4.4. Guidelines to enable the potential user to choose the most suitable instrument have recently been published by Bishop (1980). Multi-channel data systems, such as the Spectra Physics SP4000 with up to 16 stations, are also available. It should be noted that it is possible to use a mainframe computer and a data network to handle a large number of instruments simultaneously (Foreman and Stockwell, 1975). This approach does, however, have the disadvantage that a machine "crash" can effectively disable all the chromatographic instrumentation. A safer method is to link individual integrators to the mainframe computer to provide for bulk data storage and increased computational facilities.

**Table 4.4** Typical integrator specifications

| Integrator model | No. of input channels | Input range | Dynamic range | Methods for quantification† | Alphanumeric keyboard | Printer and plotter | External events | Basic option | VDU |
|---|---|---|---|---|---|---|---|---|---|
| Hewlett-Packard 3388A | 2 | −10 mV to +1 V | $10^6$ | 1–5 | Yes | Yes | Yes | Yes | No |
| Hewlett-Packard 3390A | 1 | −10 mV to +1 V | $10^6$ | 1, 4, 5 | No | Yes | Yes (as option) | No | No |
| Infotronics LDC 301 | 1–2 | −5 mV to +1 V | $2 \times 10^6$ | 1–5 | Yes | Yes | Yes | Yes | Yes |
| Infotronics LDC 304/50 | 2 | 0 to +1 V or 0 to +10 V | $10^6$ | 1, 2, 4, 5 | No | Printer only | Yes | No | No |
| Infotronics LDC 308 | 1 | −5 mV to +1 V or −50 mV to +10 V | — | 1, 5 | No | Printer only | No (auto sampler control) | No | No |
| Kemtronix Supergrater 1A | 1 | −10 mV to 1 V or −100 mV to 10 V | — | 1 | No | Printer only | No | No | No (LED option) |
| Kemtronix Supergrater 3A | 1 | −10 mV to 1 V or −100 mV to 10 V | — | 1, 2, 4, 5 | No | Printer only | Yes | No | No (LED option) |

| Instrument | | | | | | | | | |
|---|---|---|---|---|---|---|---|---|---|
| Perkin Elmer Sigma 10 | 1–4 | −5 mV to +10 V | $10^6$ | 1–5 | Yes | Yes | Yes | Yes | As option |
| Pye Unican CDP 1 | 1 | 0 to +10 V | $10^6$ | 1, 3, 5 | No | Printer only | Yes | No | LED display |
| Shimadzu Chromatopac C-E1B | 3 | −5 mV to 1 V | $10^6$ | 1–5 | No | Printer only | As option | No | No |
| Shimadzu Chromatopac C-R1A | 1 | −5 mV to 1 V | $10^6$ | 1–5 | Yes | Yes | Yes | No | No |
| Spectra Physics SP 4-100 | 1 | −1 mV to 10 V | — | 1–5 | Yes | Yes | Yes | Yes | LED display |
| TriVector Trilab 11/111 | 1–4 | 0 to 10 V | $10^6$ | 1–5 | Yes | Yes | Yes | Yes | Yes |
| Varian CDS-111 | 1 | — | — | 1–5 | No | Printer only | Yes | No | No |
| Varian Vista 40 | 1–4 | −400 mV to +1 V | — | 1–5 | Yes | Yes | Yes | No | Yes |
| Waters Model 730 Data Module | 1 | −10 mV to 2 V | $10^6$ | 4, 5 | No (use System Controller 720) | Yes | As option | No | No (use System Controller 720) |

†1, Percentage area; 2, corrected area normalization; 3, scale factor; 4, external standard; 5, internal standard.

# V ADVANTAGES AND DISADVANTAGES OF AUTOMATED HPLC

With the acceptance of HPLC as an analytical technique, automation is becoming increasingly common. Among recent publications have been automated HPLC procedures for the determination of saccharides (Linden and Lawhead, 1975), amino acids (Rose and Schwartz, 1980), food dye intermediates (Singh and Adams, 1979) and organic pollutants in water (Euston and Baker, 1979). Many automated procedures are developed from published manual methods and thus are not recorded specifically in the literature. It is obvious that automation of HPLC is advantageous for routine assays. There is a considerable reduction in staff costs, through-put is generally higher (especially if silent hours running is adopted) and precision is of a high order. To ensure that the maximum advantage is obtained from an automated system it is necessary to use sophisticated data analysis and thus be able to take account of non-routine events. In particular an automated system should shut down when things go wrong, for example column efficiency drops, the solvent reservoir empties or air enters the system. More sophisticated designs allow some of these problems to be overcome and reduce the likelihood of others happening.

The automated system does have the disadvantage of being wasteful for single injections. In addition, it is not always the ideal tool for method development. Although it is true that multiple runs can be carried out with different solvent ratios using a programmer, most chromatographers would suggest that it is still necessary to get the system fairly close to optimum before beginning such a process. Such preliminary work often involves a certain amount of intuition. Although the automation of HPLC has taken great strides in recent years, no manufacturer has yet manged to build intuitive ability into his apparatus.

## REFERENCES

Appffel, J. A., Alfredson, T. V. and Majors, R. E. (1981). *J. Chromat.* **206**, 43.
Benson, J. R. (1975). In *Instrumentation in Amino Acid Sequence Analysis* (R. N. Perham, ed.), Academic Press. New York and London, p. 1.
Bishop, P. (1980). *Lab. Equip. Dig.* **18**, 59.
Brenner, M. (1979). *Am. Lab.* **11**, 41.
Burns, D. A. (1977). In *Advances in Automated Analysis*, Vol. 2 (E.C. Barton, ed.) Technicon Instruments Co., Basingstoke, p. 332.
Burns, D. A. (ed.) (1978). *The Total Automation of HPLC*, Technicon Instruments Co., Basingstoke.
Coppola, E. D., Christie, S. N. and Hanna, J. G. (1975). *J. Ass. off. analyt. Chem.*, **58**, 58.

Erni, F., Krummen, K. and Pellet, A. (1979). *Chromatographia* **12**, 399.

Erni, F., Keller, H. P., Morin, C. and Schmitt, M. (1981). *J. Chromat.* **204**, 65.

Euston, C. B. and Baker, D. R. (1979). *Am. Lab.* **11**, 91.

Foreman, J. K. and Stockwell, P. B. (1975). *Automatic Chemical Analysis*, Ellis Horwood, Chichester.

Gfeller, J. C., Huen, J. M. and Thevenin, J. P. (1978). *J. Chromat.* **166**, 133.

Haas, J. A., Perks, L. J. and Osten, D. E. (1977). *Ind. Res.* **19**, 67.

Hunt, D. C., Philp, L. A. and Crosby, N. T. (1979). *Analyst, Lond.* **104**, 1171.

Inman, S. R. and Schifreen, R. S. (1980). *Am. Lab.* **12**, 67.

Katz, S. and Pitt, J. W. W. (1972). *Analyt. Letters* **5**, 177.

Lawrence, J. F. and Frei, R. W. (1976). *Chemical Derivatization in Liquid Chromatography*, Elsevier, Amsterdam.

Linden, J. C. and Lawhead, C. L. (1975). *J. Chromat.* **105**, 125.

Loscombe, C. R., Cox, G. B. and Dalziel. J. A. W. (1978). *J. Chromat.* **166**, 403.

Meakin, G. and Allington, R. (1980). *Am. Lab.* **12**, 65.

Mills, A. D., Mackenzie, I. and Dolphin, R. J. (1979). *J. autom. Chem.* **1**, 134.

Montoya, E. F. and Henselman, J. (1981). *Int. Lab.* **11**, 99.

Parris, N. A. (1978). *Am. Lab.* **10**, 124.

Poile, A. F. and Conlon, R. D. (1981). *J. Chromat.* **204**, 149.

Roggendorf, E. and Spatz, R. (1981). *J. Chromat.* **204**, 263.

Rose, S. M. and Schwartz, B. D. (1980). *Analyt. Biochem.* **107**, 206.

Ross, M. S. F. (1977). *J. Chromat.* **141**, 107.

Singh, M. and Adams, G. (1979). *J. Ass. off. analyt. Chem.* **62**, 1342.

Skeggs, L. T., Jr (1957). *Am. J. clin. Path.* **28**, 311.

Snyder, L. R. (1980). In *High Performance Liquid Chromatography—Advances and Perspectives*, Vol. 1 (C. Horváth, ed.) Academic Press, New York and London, p. 208.

Snyder, L. R., Doland, J. W. and Van der Wal, Sj. (1981). *J. Cromat.* **203**, 3.

Titus, C. A. (1977). *Dissert. Abstr. Int. B* **38**, 172.

# 5  Introduction to Applications of HPLC to Food Analysis

**R. MACRAE**

Department of Food Science, University of Reading, U.K.

## I  INTRODUCTION

The increasing interest in human nutrition has lead to an increased demand for the analysis of a wide range of food components. These analyses may vary from the comparatively simple determination of macronutrients, such as proteins and lipids, to the more complex assays of micronutrients such as vitamins. The advent of chromatographic methods has increased the ease with which many of these components can be determined, and has even allowed the determination of certain compounds that was not previously possible. It is hoped that many animal assays, and to a lesser extent microbiological assays for micronutrients, will be replaced by more precise chromatographic methods. The improved resolution of these methods also allows determination of the components of complex nutrients, which, as the components may vary in biological activity, will provide more significant data than those chemical methods based on total nutrient determination.

The proportion of processed foods in an "average Western diet" is increasing and this had lead to concern over the effects of such processing on the nutrients in convenience foods. Rapid analytical methods are therefore required for determination of the effect of each stage of production, so that nutrient destruction can be minimized. It is not envisaged that such analytical methods would be used "on-line" for altering processing conditions, but the data are necessary for setting-up the production system. Coupled with the increasing interest in nutrition there is also an increasing concern over food safety. This may be divided into the determination of "naturally" occurring toxic factors, such as mycotoxins, and the determination of synthetic food additives. Clearly there are certain toxic compounds that should not be present in any food system even at extremely low levels, such as aflatoxins, nitrosamines or pesticide residues, and their detection and quantification at levels below 1 $\mu$g kg$^{-1}$ is desirable. Also, with the more widespread use of synthetic food additives, it is important that guidelines for their use in food systems should be adhered to and hence that accurate methods are available for their quantification in foods. The effect of processing on these additives, and indeed the possibilities of interactions between additives, are also important and analytical methods of high resolving power are required to study the complex products that may be formed.

The food processing industry is always endeavouring to improve both the quality and range of products that are made available to the public. This desire for improvement and consistency of quality has led to the need for the introduction of improved quality control methods. Many such quality control checks are in fact organoleptic, but in certain instances there are considerable attractions in being able to use more objective instrumental methods. Here the range of compounds to be analysed, for any one product, is probably small and it is the ability to automate chromatographic methods to handle large numbers of samples that makes the techniques so important (see Chapter 4). Quality control checks can obviously apply to raw materials as well as to the final processed product.

Another important area in the food industry is in the development of new food products, where composition must be checked at each stage of the process. A more complete understanding of classical processing techniques such as smoking and pickling is also highly desirable both from a safety point of view and also so that the processes may be controlled more carefully.

In the USA nutritional labelling of processed foods is now very common (Richardson, 1981) and it is only a matter of time before similar conventions are adopted in Europe. The debate as to the value of such labelling will continue but its instigation will necessitate the use of rapid analytical methods, as the nutrient levels quoted must pertain to the actual sample sold and not to expected values. A major consequence of making such a scheme a legal requirement will be to increase the use of rapid instrumental methods of analysis, particularly those that can be automated, such as HPLC.

## II FEATURES OF HPLC RELEVANT TO FOOD ANALYSIS

All forms of chromatographic analysis have been increasingly applied to the determination of organic compounds in many scientific areas, including food analysis, over the last few decades. HPLC is simply the most recent technique to join this group of chromatographic methods. As discussed in Chapter 1, instrumental methods for high resolution liquid column chromatography suffered a prolonged adolescence as compared with gas chromatography, but it is now generally accepted that there are several areas of analysis where HPLC is a more suitable technique than its gas-phase counterpart. The major advantage of HPLC is its ability to handle compounds of limited thermal stability or volatility. Despite the fact that volatility of many compounds can be increased by derivatization, liquid chromatographic methods, which avoid this stage in the analysis, are generally favoured. A simple example of this attribute would be in the analysis of sugars which can be carried out directly by HPLC but which require prior silylation before GC (Sweeley, Bently, Makita and Wells, 1963).

HPLC is capable of producing good resolution between chemically similar components in mixtures to be analysed but to date resolution comparable to capillary GC has only been achieved in a very limited number of applications using specialized equipment (Scott and Kucera, 1979). Nonetheless, the resolving power of HPLC is more than adequate to handle many food components, even to the extent of separating very closely related vitamins (e.g. tocopherols) or amino acids. However, in practice with food systems it is not the resolving power *per se* that is inadequate but rather the elaborate extraction procedures that are often required to isolate the compound of interest in a sufficiently pure state with high recovery. HPLC undoubtedly has much to offer as a chromatographic method with high resolving power but in several cases thin layer chromatography (TLC), when carried out carefully with modern plates, can produce comparable resolution. TLC plate manufacturers may be out to impress the scientific community with colourful separations of synthetic dyes but they do show the high resolution which is possible with carefully controlled TLC. Thus for many *qualitative* analyses TLC should not be ignored as a powerful alternative to HPLC.

Separation times in HPLC are usually short, often in the region of 5–10 min, with certain simpler analyses being carried out in less than 5 min. This feature of HPLC is often quoted as one of its major advantages over other chromatographic techniques, but it must be remembered that this is the analysis time for one sample, such that the analysis of 20 samples will take 20 times as long. On the other hand a neatly spotted TLC plate (10 cm × 10 cm) can take up to 20 samples. Thus, although the speed of HPLC is attractive for many analyses, the mass screening of samples to give *qualitative* results is often more efficiently carried out by TLC. Examples of such a use would be in the screening of vast numbers of urine samples for drug residues (Kaistha and Tadrus, 1975) or grain samples for mycotoxins (Stoloff *et al.*, 1971)

Modern HPLC is a quantitative technique and in this feature HPLC scores heavily over other liquid chromatographic techniques. TLC can be made a quantitative technique by densitometry, or by removal of the separated component from the layer and subsequent analysis (Kirchner, 1978) but neither of these techniques approaches the precision which can be obtained routinely by HPLC. The coefficient of variation for the chromatographic stage in a particular laboratory may be as low as 1%, depending on the actual analysis involved. This will be influenced by the peak size, resolution and general stability of the chromatographic system. The coefficient of variation for the analysis of food samples is usually much higher, reflecting the difficulty in ensuring efficient extraction and clean-up without loss prior to chromatography. However, some samples require minimal preparation where losses should be negligible, e.g. the determination of sugars in soft drinks (Anon., 1979). There appears to have been very little work carried out in the form of collaborative trials in HPLC methods, which is unfortunate as it is the present author's opinion that such trials would be exceptionally illuminating in terms of the strengths and the weaknesses of the technique.

In addition to its quantitative precision HPLC is a very sensitive technique. This is a major attribute of the method but it does not apply to all analyses. Thus while certain workers may be quoting a detection limit of 750 fg for aflatoxins (Diebold, Karni, Zare and Zeitz, 1979) most routine systems using refractive index detection for sugar analysis require on-column loadings of some 20 $\mu$g for accurate quantification (Johncock and Wagstaffe, 1980), which is considerably more than would be required for colorimetric methods or GC. There is still no sensitive universal detector for HPLC and this is the major chink in the armour of the practising chromatographer. In certain areas of analysis sensitivity may not be a problem: thus in the analysis of glucose syrups there is no problem in obtaining extracts of sufficient concentration for the limited sensitivity of the detection systems available, whereas in other areas, such as pesticide residue analysis, sensitivity remains a serious limitation. The advent of an efficient and reliable interface to allow HPLC to be coupled to mass spectrometry will do much to overcome this problem, but significant development work is required.

HPLC can also be used as an isolation technique rather than a method for quantitative analysis. Thus compounds can be readily recovered from the column eluate and then studied further by additional techniques, chromatographic or otherwise. Unless the columns are scaled up the techniques that can be subsequently used are limited to those that require only milligram quantities of the pure components.

The cost of modern HPLC equipment is horrendous, but when this is compared with the cost of other scientific instruments of similar capabilities it does not appear totally out of line. A simple isocratic system with limited detection facilities can be purchased for between £3000 and £4000, whilst a sophisticated microprocessor-controlled gradient system with multiple detectors may well cost in excess of £20 000. However, it is not the capital costs alone which must be considered but also the running costs, which can be high. Commercial HPLC columns are expensive, although most laboratories with a significant investment in HPLC will almost certainly be packing their own. The myth perpetuated by column manufacturers that packing columns is difficult has now largely been dispelled and most stationary phases are available as loose material. Solvent costs may be high but it is possible to recycle solvents with purification, and in some cases without purification, although the latter technique can lead to complications. Routine maintenance and instrumental spares will also contribute significantly to running costs.

## III  FOOD ANALYSIS

The vast range of column materials and chromatographic solvents that can be used in HPLC means that potentially the technique can be applied to nearly all food components. It is interesting to note that the applications of HPLC to food analysis have trailed well behind those in other biochemical fields, such as pharmaceutical chemistry or forensic toxicology. This may in part be due to the very complex matrices in many food systems and the very low levels of many of the compounds of interest. However, there is a valid comparison between the determination of drug residues and hormones in body fluids and the determination of micronutrients in foods. A further reason may be the traditionalist attitude of many analysts in the food area.

The situation over the last 5 years has changed rapidly and the number of published papers on applications of HPLC to food analysis has increased exponentially. A number of reviews have attempted to scan this mass of literature, e.g. Saxby (1978), Tweeten and Euston (1980) and Macrae (1980, 1981). There is no evidence that the rate of publication is subsiding but, now that the technique is well established in the food area, there should be more emphasis on the analysis of food samples, rather than the chromatographic separation of food components. The latter in most cases is comparatively easy and it is the other stages of the analysis which require attention.

On account of the wide range of different classes of chemical compounds and also the wide range of matrices encountered in food systems, it is not possible to consider a typical analytical scheme. However, irrespective of the actual analysis in question there are potential problems at various stages of the analysis, and these should be considered (see Table 5.1).

The first stage in any analysis is that of sampling, where care must be taken to ensure that the analytical samples are representative of the bulk sample. This is particularly important where there are reasons to suspect heterogeneity, for example in determining mycotoxins formed by fungal contamination. Comparisons between samples are valueless unless they are reported on the same basis. Thus if the water content of a food is likely to be variable some attempt must be made to allow for this, either by drying the sample prior to analysis or by determination of the water content. In most cases the next stage involves extraction of the component of interest, although for some commodities, such as soft drinks and fats, a simple dilution may be all that is required. It is at this stage that severe errors can be introduced by incomplete extraction and some attempt must be made to ascertain the percentage extraction, either by re-extraction and analysis or by a recovery study on added amounts of the component of interest. However, neither of these methods will overcome the problem of "bound" material and it may be necessary to introduce an acid or base hydrolysis (e.g. Association of Official Analytical Chemists, 1970) or even an enzymic hydrolysis (e.g. Ang and Moseley, 1980). In many published methods this

**Table 5.1**   Stages of typical HPLC
food analysis

| | |
|---|---|
| (1) | Sampling |
| (2) | Water determination or drying |
| (3) | Extraction (where applicable) |
| (4) | Clean-up |
| | (a)   Solvent extraction |
| | (b)   Precipitation |
| | (c)   Chromatography |
| (5) | HPLC |
| | (a)   Peak identity |
| | (b)   Peak purity |
| | (c)   Quantification |

stage of the analysis, which is essential, appears to be omitted. The extract may then be in a suitable condition for direct HPLC, but more commonly some form of clean-up will be required. This may involve a simple solvent extraction, precipitation or complex chromatographic step. Irrespective of the procedure adopted it is again necessary to determine the recovery at this stage. It is quite possible that a simple protein precipitation will also remove 50% or more of the compound of interest.

The cleaned extract can then be chromatographed on the HPLC system and the results handled as discussed in Chapter 1. In particular attention must be paid to confirmation of peak identity, peak purity and linearity of the detection system. The method of quantification adopted will depend on the actual determination and whether an internal standard has been added to the food sample earlier in the analysis to compensate for extraction losses. Sufficient replication of the analysis, and not just the chromatographic stage, must be carried out to ensure good precision and ideally a coefficient of variation for the method should be quoted.

The remaining chapters in the book are concerned with the wide range of applications to food analysis. In each area of application the strengths and weaknesses of HPLC are highlighted. HPLC is a powerful analytical technique but it is not a panacea for all food analyses and, as with all analytical instrumental methods, is only as good as the operator. A good understanding of the problems involved in the application of HPLC to food analysis is essential for the production of reliable and accurate data.

## REFERENCES

Ang. C. Y. W. Moseley, F. A. (1980). *J. agric. Fd Chem.* **28**, 483.

Anon. (1979). *Soft Drinks Trade J.* **33**, 18.

Association of Official Analytical Chemists (1970). In *Official Methods of Analysis* (W. Horwitz, ed.) Association of Official Analytical Chemists, Washington, D.C., p. 771.

Diebold, G. J., Karni, N., Zare, R. N. and Zeitz, L. M. (1979). *J. Ass. off. analyt. Chem.* **62**, 564.

Johncock, S. I. M. and Wagstaffe, P. J. (1980). *Analyst, Lond.* **105**, 581.

Kaistha, K. K. and Tadrus, R. (1975). *J. Chromat.* **109**, 149.

Kirchner, J. G. (1978). *Techniques of Chemistry*, Vol. 14, *Thin Layer Chromatography*, John Wiley, New York, pp. 292–334.

Macrae, R. (1980). *J. Fd Technol.* **15**, 93.

Macrae, R. (1981). *J. Fd Technol.* **16**, 1.

Richardson, D. P. (1981). *Proc. Inst. Fd Sci. Technol.* **14**, 87.

Saxby, M. J. (1978). In *Developments in Food Analysis Techniques*, Vol. 1 (R. D. King, ed.), Applied Science Publishers, London, p. 125.

Scott, R. P. W. and Kucera, P. (1979). *J. Chromat.* **169**, 51.

Stoloff, L., Nesheim, S., Yin, L., Rodricks, J. V., Stack, M. and Compbell, A. D. (1971). *J. Ass offic. analyt. Chem.* **54**, 91.

Sweeley, C. C., Bently, R., Makita, M. and Wells, W. W. (1963). *J. Am. chem. Soc.* **85**, 2479.

Tweeten, T. N. and Euston, C. B. (1980). *Fd Technol. December issue*, 29.

# 6  Determination of Carbohydrates

## D. J. FOLKES and P. W. TAYLOR

The Lord Rank Research Centre, High Wycombe, U.K.

## I  INTRODUCTION

The past 20 years has seen rapid progress in the application of chromatographic methods to the analysis of carbohydrates, although in both the application of high resolution gas and liquid chromatographies progress was somewhat slower than in other areas of analysis. Paper and thin layer chromatography are now generally regarded as inferior methods for carbohydrates owing to their poor separation efficiencies, long analysis times and difficulties with quantification. However, it is interesting to note that the development of high performance thin layer chromatography (using media primarily designed for high performance column chromatography) has recently been used by Lee, Nurok and Zlatkis (1979) for the analysis of glucose, fructose and sucrose in molasses. Ion-exchange chromatography can provide excellent separations of carbohydrates but also requires comparatively long analysis times with elevated column temperatures. Gel filtration chromatography is particularly useful for the separation of carbohydrate polymers, with the added advantage that the entire molecular weight range present can be eluted, albeit with limited resolution. Current developments in gel filtration column technology indicate that the use of this technique will probably increase considerably in the near future. Size exclusion chromatography in general may become widely used as a preliminary analysis step for HPLC, replacing current clean-up and sample treatment procedures. Since it can be applied to both aqueous and organic extracts of foods its potential is considerable.

Without doubt the most widely utilized techniques of high resolution chromatography for the qualitative and quantitative analysis of individual sugars are gas–liquid chromatography (GLC) and high performance liquid chromatography (HPLC). The first exhaustive investigation into the application of GLC to a wide range of sugars was reported by Sweeley, Bentley, Makita and Wells (1963) which demonstrated the potential of such a technique in carbohydrate analysis. Their study included both pentoses and hexoses and the analysis of monosaccharides through to tetrasaccharides. Since that time many applications have been published, all of which require the preparation of volatile derivatives, the most frequently used being trimethylsilyl ethers. HPLC has been applied to carbohydrate analysis for several years now, but only in the last two or three years has it become the method of choice, replacing GLC for many analyses. Typical applications have been reported by Havel, Tweeten, Seib, Wetzer and Liang (1977), Hunt, Jackson, Mortlock and Kirk (1977) and Hurst, Martin and Zoumas (1979), amongst others. The most frequently studied sugars have been glucose, fructose, sucrose, maltose and lactose. The main development responsible for this change is the availability of new stable column packing materials at a reasonable cost.

A comparison of the application of HPLC and GLC for carbohydrate analysis can conveniently be made on the basis of analysis time and quantitative aspects and this clearly indicates their relative advantages and disadvantages. Table 6.1 shows typical times for analyses by the two techniques, with HPLC particularly advantageous in terms of speed and simplicity of sample preparation (see Section IV.A for details of sample prepara-

**Table 6.1** Time comparison of GLC and HPLC for sugar analysis

|  | HPLC | GLC |
|---|---|---|
| Sample preparation (including extraction) | 20 min per sample | (a) Foods, 2 h per sample<br>(b) Syrups, etc., 30 min per sample |
| Typical run time per analysis | 15 min | 20–30 min |

**Table 6.2** Quantitative comparison of GLC and HPLC for sugar analysis

|  | HPLC | GLC |
|---|---|---|
| Typical column load | 100–1000 $\mu$g | 2–20 $\mu$g |
| Detector linear range | 10 $\mu$g–4 mg (RI) | 0.1–20 $\mu$g (FID) |
| Minimum detectable amount | 10 $\mu$g | 0.1 $\mu$g |
| Detection limit in sample | 0.5% | 0.05% |

tion). Examination of the quantitative aspects of HPLC and GLC, which are compared in Table 6.2, shows advantages for both techniques. However, practical considerations are strongly in favour of HPLC, since in most cases component sugars of foods are present in adequate concentrations. The limited sensitivity of the refractive index detector in routine use indicates the main area of carbohydrate analysis where GLC can be used with advantage. (The use of alternate means of detection for HPLC is discussed in Section III.) In those cases where only a very limited amount of sample is available (less than 1 mg), GLC can be readily employed. Eklund, Josefsson and Roos (1977) demonstrated the analysis of picogram amounts of monosaccharides using trifluoroacetyl derivatives and electron capture detection, utilizing the advantages of GLC to their extreme. It should be emphasized that such applications of carbohydrate analysis are not typical and are more likely to arise from biochemical and phytochemical studies than from food analysis.

The development of carbohydrate analysis by HPLC using modified silica with bonded polar phases was initially constrained by the high cost of commercial columns (between £200 and £300 in 1976) and the instability of the bonded phase. This latter factor consequently gave rather short column life. Column packings have now been much improved and, although the most frequently used bonded amino columns suffer from a slow but unavoidable stripping of amino groups from the silica surface, this change in column characteristics can often be countered by adjustment of the solvent composition. In the case of the usual combination of acetonitrile and water, the acetonitrile content is increased as the amino groups decrease and thus retention times and resolution can be approximately maintained.

The combination of stable columns and refractive index detection with adequate sensitivity has thus provided the basic requirements for separation and detection for quantitative analysis of carbohydrates by HPLC. Simple modification of the method renders it applicable to mono-, di- and oligosaccharides.

## II   DEVELOPMENT OF COLUMN TECHNOLOGY

During the period preceding 1970, most carbohydrate analyses using liquid chromatography were carried out using low resolution columns of charcoal or Sephadex. Although charcoal columns are now obsolete the faster Sephadex columns are still used for some analyses of high molecular weight carbohydrates. In both cases the analyses were extremely slow, requiring at least 12 h for the analysis of oligo- and polysaccharides. A typical application is the separation of the oligo- and polysaccharides of chicory root, shown in Fig. 6.1.

The introduction of Bio-Gel gel filtration columns for carbohydrate separations offered considerably faster analyses, enabling the analysis above to be achieved in approximately 6 h. Fig. 6.2 illustrates the improvements obtained, providing a much better chromatogram for quantification. Further improvements in Bio-Gel materials have been reported since the analysis in Fig. 6.2 was obtained.

A considerable advance in column technology was reported by Hobbs and Lawrence (1972) who demonstrated the properties of strong cation-exchange columns. They were able to resolve mixtures containing mono- and disaccharides in 1–4 h depending upon the component sugars present. Such columns, however, require operation at elevated temperatures, which is a practical disadvantage. Although more recent developments have enabled considerably faster analyses to be obtained, these columns have proved to be less versatile than the bonded-phase columns developed since 1976.

It is undoubtedly the development of polar bonded-phase materials pre-

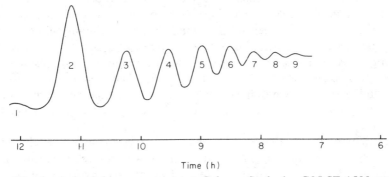

Time (h)

**Fig. 6.1**   Analysis of chicory root extract. Column, Sephadex G25 SF, 1500 mm × 25 mm; solvent, water, 0.57 ml min$^{-1}$; colorimetric detection. (Numbers refer to degree of polymerization.)

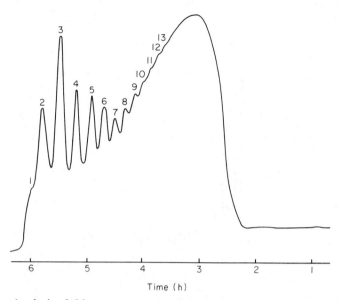

**Fig. 6.2**   Analysis of chicory root extract. Column, Bio-Gel P4, 1000 mm × 4 mm, 50 °C; solvent: water, 2 ml h$^{-1}$; refractive index detection. (Numbers refer to degree of polymerization.)

pared from 5 μm and 10 μm silica which has lead to the full advantages of HPLC becoming applicable to the determination of carbohydrates. Although both cyano and amino bonded phases have been utilized for carbohydrate analyses, it is the latter which has been most extensively studied and subsequently used for routine analysis of sugars in foods.

Schwarzenbach (1976) reported the first evaluation of the preparation and properties of an amino bonded stationary phase. This was prepared from silica and 3-aminopropyltriethoxysilane and gave columns with good resolution of low molecular weight carbohydrates. Subsequently, Jones, Burns, Selling and Cox (1977) published a detailed report on the preparation and application of amino bonded columns to food sugars analysis. In the present authors' hands, the columns produced using the above methods, although providing good resolution, always had the disadvantage of a short lifetime. In the past 3 years various amino bonded-phase materials have become commercially available. Columns of these materials give good resolution, a stable base-line and have a lifetime of many analyses. The great improvement in resolution is shown in Fig. 6.3, which illustrates analysis of the same mono- through to polysaccharide sample as shown in Figs 6.1 and 6.2.

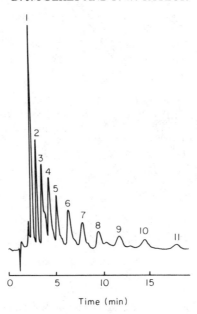

**Fig. 6.3**   Analysis of chicory root extract. Column, Spherisorb S5 NH$_2$, 250 mm × 4.6 mm; solvent, 60% acetonitrile, 40% water (v/v), 2 ml min$^{-1}$; refractive index detection. (Numbers refer to degree of polymerization.)

The versatility of such columns is exemplified in Fig. 6.4 which now shows complete separation of the common food sugars: fructose, glucose, sucrose, maltose and lactose. It should be noted that in Fig. 6.3 the mobile phase is acetonitrile–water in the ratio 60 : 40 (v/v), whereas the ratio is 75 : 25 (v/v) in Fig. 6.4. Thus, selection of the mobile phase composition allows a choice between resolution and molecular weight range eluted. These examples demonstrate that columns of amino bonded-phase materials provide an excellent medium for analysis of carbohydrates by HPLC.

Finally, an unusual but interesting separation of saccharides was published by Wells and Lester (1979) utilizing reversed phase chromatography of the peracetylated saccharides. Gradient elution with acetonitrile–water gave good resolution of linear glucose polymers from one to 35 saccharide units. The acetylation procedure may readily be carried out so that this mode of analysis could be of interest to those involved in the determination of polysaccharides.

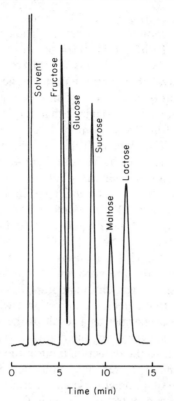

**Fig. 6.4** Analysis of common food sugars. Column, Spherisorb S5 NH$_2$, 250 mm × 4.6 mm; solvent, 75% acetonitrile, 25% water (v/v), 2 ml min$^{-1}$; refractive index detection.

## III DETECTION AND QUANTIFICATION

The detection of sugars separated by liquid chromatography is most commonly carried out by the use of refractive index monitoring. Since analysis for sugars, particularly in foods, is rarely required at trace levels, the limited sensitivity of refractive index detectors is usually adequate. Thus, there has been little requirement to examine other means of detection. Binder (1980) has reported a detailed comparison of ultraviolet and refractive index detection of monosaccharides but with no overall advantage to be gained from the use of the former. Since a detection wavelength below 200 nm must be used, which means that solvent purity is exceptionally critical, ultraviolet detection is probably not practicable for routine use.

Refractive index detectors are available based on various optical systems such as deflection, reflection and interference. The basic principle common to all designs is that of differential refractometry (see Chapter 2). The detector used in our laboratory (Waters R401), which is similar in performance to most others available, gave the performance data shown in Table 6.3. These data are probably typical of those obtainable with most RI detectors, although it must be recognized that detector noise levels (which are dependent on the overall system) may differ, enabling lower detection limits to be attained. The data in Table 6.3 show that the peak area response is linear for injections of fructose from 40 to 4000 $\mu$g. The peak height response is not linear over the whole of this range since 4000 $\mu$g fructose overloads the column causing considerable peak-broadening. Scobell, Brobst and Steele (1977) demonstrated that the peak area response factors for glucose and its polymers (maltose, maltotriose and maltohexaose) are almost identical. Therefore, the RI detector can be calibrated with glucose for HPLC analysis for glucose polymers and thus other high purity saccharides are not required for calibation. In contrast, oligosaccharide analysis by gas chromatography requires calibration for each sugar. This is due to the decrease in response with increasing molecular weight of the oligosaccharides as a result of thermal degradation.

The responses obtained for 40 $\mu$g injections of sugars onto the column indicate that this is near to the detection limit of the system used as shown by the deviation in some of the response data quoted in Table 6.3, e.g. the data from the injection of 40 $\mu$g glucose and sucrose. In routine analysis of foods, individual sugar concentrations down to 0.5% can be measured and smaller concentrations can be estimated. This is adequate in most cases of food analysis. If the measurement of lower sugar concentrations is required then a more sensitive refractive index detector (e.g. Optilab 902) must be

**Table 6.3**  Response of refractive index detector for sugar analysis

| Amount injected ($\mu$g) | Peak area/amount injected | | |
|---|---|---|---|
| | Fructose | Glucose | Sucrose |
| 40 | 789 | 942 | 795 |
| 80 | 782 | 820 | 856 |
| 200 | 797 | 805 | 827 |
| 400 | 797 | 789 | 825 |
| 1000 | 802 | | |
| 2000 | 797 | | |
| 4000 | 792 | | |

used, although it is necessary to thermostat the entire HPLC system to get the best performance.

A further major disadvantage of RI detection is that it is not possible to carry out gradient elution. Thus in those cases where a food extract contains a number of simple sugars and higher molecular weight species it is usually necessary to carry out more than one chromatographic run, altering the solvent composition. Limited gradients can be performed with UV detection but even with high grade solvents a marked shift in baseline is obtained. A recently introduced detector, the mass detector, overcomes this problem as the solvent is removed by evaporation prior to detection, see Chapter 2. To date only one publication has appeared on the application of this detector to the determination of carbohydrates (Macrae and Dick, 1981) but in addition to allowing gradient elution it appears more sensitive than most RI detectors.

An alternative approach is to form sugar derivatives with a strong UV absorbance which can then be detected with an ultraviolet detector. Thompson (1978) has reported the analysis of the benzyloxime–perbenzoyl derivatives of mono- and disaccharides with picomole sensitivity. Although a single quantifiable derivative of each sugar is easily obtained, there are problems of separation with some mixtures. Application of this technique to complex foods may also reveal other interferences with UV detection. In practice, analysis by GLC is probably the best technique if very high sensitivity is required.

Examples of the analytical data obtainable from HPLC of sugars with RI detection are given in Tables 6.4 and 6.5. Table 6.4 shows results obtained from the analysis of some model sugar solutions, confirming the validity of the method and Table 6.5 shows some results from food samples. There is good agreement with results achieved by enzymic methods and the replication, in general, is good. The poor replication observed in some cases is probably due, at least in part, to sample inhomogeneity.

**Table 6.4** Analysis of model sugar solutions by HPLC

| | Sugar concentrations (g l$^{-1}$) | | | | | |
|---|---|---|---|---|---|---|
| | Fructose | | Glucose | | Sucrose | |
| Sample | Actual | Measured | Actual | Measured | Actual | Measured |
| A | 30 | 30.7 | 10 | 9.4 | 10 | 9.9 |
| B | 10 | 9.9 | 20 | 19.7 | 20 | 19.9 |
| C | 10 | 9.8 | 10 | 10.1 | 30 | 29.8 |
| D | 5 | 5.2 | 30 | 29.2 | 15 | 15.2 |

**Table 6.5** Analysis for commonly occurring food sugars

| Sample | Fructose (%) | | Glucose (%) | | Sucrose (%) | | Maltose (%) | | Lactose (%) | |
|---|---|---|---|---|---|---|---|---|---|---|
| | Replicates | Mean | Replicates | Mean | Replicates | Mean | Replicates | Mean | Replicates | Mean |
| Custard mix | <0.3<br><0.3<br><0.3 | <0.3 | <0.3<br><0.3<br><0.3 | <0.3 | 62.0<br>62.2<br>62.0 | 62.1 | <0.3<br><0.3<br><0.3 | <0.3 | <0.3<br><0.3<br><0.3 | <0.3 |
| Sponge mix | <0.3<br><0.3 | <0.3 | 1.2<br>1.2 | 1.2 | 35.3<br>35.8 | 35.6 | 0.4<br>0.3 | 0.4 | <0.3<br><0.3 | <0.3 |
| Marzipan | 7.4<br>7.0 | 7.2 | 8.6<br>9.1 | 8.9 | 46.7<br>47.3 | 47.0 | 1.1<br>1.2 | 1.2 | <0.5<br><0.5 | <0.5 |
| Crunch bar | 1.4<br>1.5 | 1.5 | 1.2<br>1.1 | 1.2 | 25.3<br>25.3 | 25.3 | <0.5<br><0.5 | <0.5 | 1.1<br>1.1 | 1.1 |
| Bath buns | 1.1<br>1.7 | 1.4 | 11.0<br>12.0 | 11.5 | <0.5<br><0.5 | <0.5 | 1.9<br>1.9 | 1.9 | <0.5<br><0.5 | <0.5 |

## IV APPLICATIONS

### A Low Molecular Weight Sugars

Analysis for low molecular weight sugars can be arbitrarily defined as the determination of individual mono- to tetrasaccharides. A frequent request for the food analyst is for the determination of some or all of the sugars: fructose, glucose, sucrose, maltose and lactose. Such analyses are required for comparison and control of ingredients and finished products and monitoring of production processes. Analysis for fructose, glucose and sucrose provides information on the extent of inversion or use of invert sugar in a product. Sucrose is the most common ingredient sugar and lactose can be used as an index of non-fat milk powder content. These examples are typical of the use of HPLC analysis for common food sugars. A rather more specialized application is the determination of raffinose and stachyose, which is of interest due to the development of products based on soya, in which these tri- and tetrasaccharides are major carbohydrates.

The advantage of HPLC analysis in terms of sample preparation is an important factor in favour of its use where many and varied samples have to be analysed. A simple and rapid sample preparation used in our laboratory requires only aqueous extraction at 65 °C, dilution with acetonitrile and filtration prior to injection on to the HPLC column. This method has proved to be satisfactory for various food samples without the need for a de-proteinizing step. The common sugars are well separated by HPLC in 15 min or less, as shown in Fig. 6.4. The chromatogram from a typical application, the analysis of a breakfast cereal, is shown in Fig. 6.5. Sucrose is the major sugar constituent with fructose, glucose and maltose also being detected. The peak eluted immediately before fructose is due to pentoses.

Basically similar HPLC methods have been reported in the literature by many analysts and the references that follow provide a brief review of typical applications. Timbie and Keeney (1977) reported the analysis of confectionery products for fructose, glucose, sucrose, maltose and lactose and also discussed the necessary protection of the analytical column by sample preparation procedure and use of a pre-column. Hurst and Martin (1977) investigated the determination of carbohydrates in milk chocolate, again paying particular attention to sample preparation. A HPLC method for sucrose, raffinose and stachyose determination in textured soya has been described by Havel et al. (1977). The same analysis of soybeans and soybean products was reported by Black and Bagley (1978), with quantification achieved by an internal standard method employing $\beta$-cyclodextrin as the standard. The method published by Hunt et al. (1977) for the determination of the common food sugars employed ribose as the internal standard

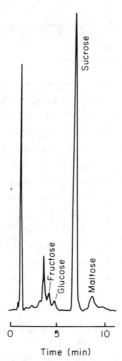

Time (min)

**Fig. 6.5**   Analysis of breakfast cereal. Column, Spherisorb S5 $NH_2$, 250 mm ×
4.6 mm; solvent, 75% acetonitrile, 25% water (v/v), 2.5 ml $min^{-1}$; refractive index
detection.

and also compared results with those from enzymic analyses. The wide
range of application of the HPLC determination of food sugars was clearly
demonstrated by Hurst *et al*. (1979) who reported analytical data for fruit
juices, soft drinks, fruits, chocolate, dairy products and snack foods.
Wartheson and Kramer (1979) reported the analysis of sugars in milk and
ice cream. Milk products were also analysed by Euber and Brummer (1979)
with quantification by an internal standard method using xylose. Changes
in the concentrations of different sugars during fermentation and baking
were studied by Varo, Westermarck-Rosendahl, Hyvönen and Koivistoinen (1979) using HPLC for the analyses. Gotz (1979) has reported
various carbohydrate analyses in foods demonstrating that mobile phase
composition must be selected for different carbohydrates. De Vries, Heroff
and Egberg (1979) evaluated a HPLC method for determination of the
common food sugars for reproducibility, recoveries and comparison with
chemical methods.

Most published work reports direct injection of a small volume (typically 10 μl) of the aqueous extract on to the HPLC column. The method employed in our laboratory involves dilution of the aqueous extract with 1.5 volumes of acetonitrile. Injection of 100 μl of filtered diluted extract maintains peak shape and resolution, whereas injection of aqueous extracts (of similar column load) was found to cause unacceptable peak broadening. The addition of acetonitrile also causes precipitation of material which would otherwise deposit on the column and cause a deterioration in performance over a period of time.

HPLC analysis of sugars also has application in basic food research. The analysis of wheat germ carbohydrates is illustrated in Fig. 6.6. Fructose, glucose, galactose, sucrose and raffinose can be identified but there are several minor components eluted between sucrose and raffinose which remain unidentified. The poor separation between glucose and galactose should be noted, since this is one particular separation where column technology has so far failed to find a solution. Isocratic conditions are essential

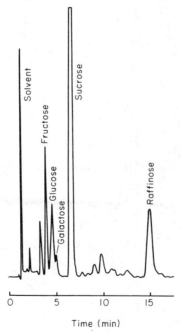

Time (min)

**Fig. 6.6** Analysis of wheat germ sugars. Column, Spherisorb S5 NH₂, 250 mm × 4.6 mm; solvent, 75% acetonitrile, 25% water (v/v), 3 ml min⁻¹; refractive index detection.

162            D. J. FOLKES AND P. W. TAYLOR

for stable operation of a refractive index detector. However, the use of an alternative detection system, such as the mass detector, allowing gradient elution to be employed can overcome this separation problem (Macrae and Dick, 1981). HPLC has also been successfully applied to the analysis of carbohydrates in barley, rye and triticale germs, indicative of the many potential applications in biochemistry and plant science.

The use of an ultraviolet detector in series with a refractive index detec-

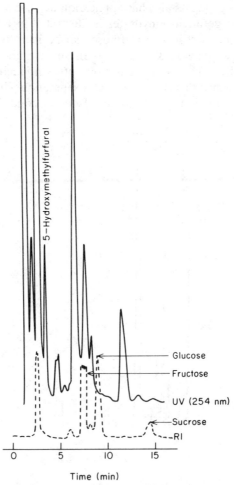

**Fig. 6.7** Analysis of hydrolysed chicory root extract. Column, Spherisorb S5 NH$_2$, 250 × 4.6 mm; solvent, 65% acetonitrile, 35% water (v/v), 2 ml min$^{-1}$; refractive index and UV detection.

tor can provide additional analytical information by monitoring other sample components which give little or no refractive index response. With an UV detector operating at 254 nm which does not respond to sugars (fructose gives a very small response), UV absorbing compounds can be detected. An example from dual detector HPLC is shown in Fig. 6.7 which illustrates the analysis of a crude chicory root hydrolysate and enables 5-hydroxymethylfurfural to be detected on the UV channel.

## B Oligosaccharides

Although the majority of HPLC analyses for carbohydrates require the determination of individual sugars, there is also a considerable need for molecular weight profile analysis. This is required for examination of syrups containing glucose polymers produced by acidic or enzymic hydrolysis of wheat or corn starch. Determination of the oligosaccharides enables the extent and nature of the depolymerization to be followed throughout production. Another application of oligosaccharide analysis is the study of plant storage carbohydrates and their depolymerization, of which the depolymerization of inulin (fructose polymer) in chicory roots is a typical example. The chromatogram in Fig. 6.3 illustrates such an analysis.

HPLC analysis is frequently used in our laboratory to determine the molecular weight profile of wheat starch syrups containing glucose polymers. Samples are rapidly prepared by dissolution in water, addition of an equal volume of acetonitrile and filtration of the solution prior to injection. Ideally, individual component concentrations should be in the range 1–5 mg ml$^{-1}$ in the final solution. Since the solubility of carbohydrates in acetonitrile decreases with increasing molecular weight, the proportion of water in the mobile phase must be increased to decrease retention times and hence the overall time of analysis. The optimum mobile phase composition depends upon the age and condition of the column but would normally be in the range 35–40% (v/v) water. This enables mono- to decasaccharides to be eluted within 15 min with complete resolution of the monosaccharide and disaccharide. Increasing the water content in the mobile phase to 45% (v/v) enables polysaccharides up 15 monomer units to be eluted. The chromatography of sugars on bonded amino columns is normally carried out at room temperature but operation at elevated temperature may be advantageous in this application. Elution of polysaccharides of up to 20 or more monomer units may then be possible. However, the sample preparation described above could change the composition by precipitation of higher saccharides and totally aqueous solutions may be required for injection on to the column. Although adjustment of the HPLC operating conditions enables a greater range of polysaccharides to be ana-lysed, this can only be achieved at the expense of the separation between lower molecular weight saccharides. Thus, for a full analysis it may be necessary to carry out chromatography with more than one set of operating conditions. Here the use of a gradient system, with suitable detection, would be advantageous.

However, it should be emphasized that it is impossible to chromatograph very high molecular weight saccharides using a bonded amino column and acetonitrile–water eluent. To elute the entire molecular weight range (e.g. glucose to starch) it is necessary to use aqueous size exclusion chromato-graphy employing columns packed with either cation resin or hydrophobic gels such as TSK.

The analysis of a low DE (dextrose equivalent) syrup is illustrated by Fig. 6.8. This shows that glucose polymers of up to 10 monosaccharide units can be sufficiently resolved for accurate integration and quantifi-cation. Increasing the water content in the mobile phase gives sharper peaks

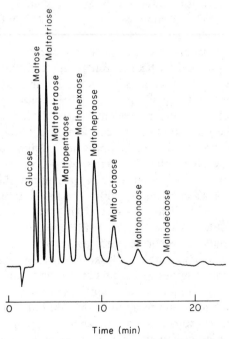

Time (min)

**Fig. 6.8** Analysis of corn syrup (low DE). Column, Spherisorb S5 NH$_2$, 250 mm × 4.6 mm; solvent, 62% acetonitrile, 38% water (v/v), 2 ml min$^{-1}$; refractive index detection.

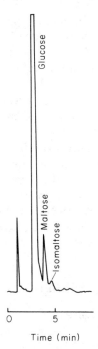

**Fig. 6.9** Analysis of corn syrup (high DE). Column, Spherisorb S5 NH$_2$, 250 × 4.6 mm; solvent, 65% acetonitrile, 35% water (v/v), 2 ml min$^{-1}$; refractive index detection.

(and shorter retention times) which gives better sensitivity and possibly greater accuracy in quantification of low levels of polysaccharides.

Depending upon the conditions used, hydrolysis of starch can produce oligosaccharides with both 1 : 4 and 1 : 6 glucose linkages, e.g. maltose and isomaltose. The conditions for oligo- and polysaccharide analysis will not resolve such sugar pairs. A mobile phase of 30–35% (v/v) water enables maltose and isomaltose to be separated as shown in Fig. 6.9. Although this is an analysis of a high DE syrup, oligosaccharides are present but in much lower concentrations. The presence of isomaltose in a food product is indicative of the use of wheat or corn syrup in its formulation.

## V FUTURE DEVELOPMENTS

The foregoing sections have described the development and applications of HPLC analysis for carbohydrates. Considerable progress has been made such that HPLC is now routinely and reliably used for quantitative analysis of food sugars and should rapidly become the standard method without further development. Although a few separations remain to be improved, it seems unlikely that any major development in column technology, specific to carbohydrate analysis, will occur. It is possible that bonded phases with different organic loadings, and also prepared from different amino alkyl silanes, may be of interest and become commercially available. Column development is of less interest than the current development of the mass detector as an alternative to refractive index monitoring. If this detector can provide performance comparable to or better than most current refractive index detectors then its use with gradient elution chromatography offers much scope in improving the resolution of carbohydrates (Macrae and Dick, 1981). Reduction of the noise level of refractive index detectors would also be of considerable benefit, allowing greater sensitivity to be achieved.

## REFERENCES

Binder, H. (1980). *J. Chromat.* **189**, 414.
Black, L. T. and Bagley, E. B. (1978). *J. Am. Oil. Chem. Soc.* **55**, 228.
De Vries, J. W., Heroff, J. C. and Egberg, D. C. (1979). *J. Ass. off. analyt. Chem.* **62**, 1292.
Eklund, G., Josefsson, B. and Roos, C. (1977). *J. Chromat.* **142**, 575.
Euber, J. R. and Brummer, J. R. (1979). *J. Dairy Sci.* **62**, 685.
Gotz, H. (1979). Hewlett Packard Application Note AN 232–14.
Havel, E., Tweeten, T. N., Seib, P. A., Wetzer, D. L. and Liang, Y. T. (1977). *J. Fd Sci.* **42**, 666.
Hobbs, J. S. and Lawrence, J. G. (1972). *J. Sci. Fd Agric.* **23**, 45.
Hunt, D. C., Jackson, P. A., Mortlock, R. E. and Kirk, R. S. (1977). *Analyst, Lond.* **102**, 917.
Hurst, W. J. and Martin, R. A. (1977). *J. Ass. off. analyt. Chem.* **60**, 1180.
Hurst, W. J., Martin, R. A., Jr and Zoumass, B. L. (1979). *J. Fd Sci.* **44**, 892.
Jones, A. D., Burns, I. W., Sellings, S. G. and Cox, J. A. (1977). *J. Chromat.* **144**, 169.
Lee, K. W., Nurok, D. and Zlatkis, A. (1979). *J. Chromat.* **174**, 187.
Macrae, R. and Dick, J. (1981). *J. Chromat.* **210**, 138.
Schwarzenbach, R. (1976). *J. Chromat.* **117**, 206.
Scobell, H. D., Brobst, K. M. and Steele, E. M. (1977). *Cereal Chem.* **54**, 905.
Sweeley, C. C., Bentley, R., Makita, M. and Wells, W. W. (1963). *J. Am. chem. Soc.* **85**, 2497.
Thompson, R. (1978). *J. Chromat.* **166**, 201.
Timbie, D. J. and Keeney, P. G. (1977). *J. Fd Sci.* **42**, 1590.
Varo, P., Westermarck-Rosendahl, C., Hyvönen, L. and Koivistoinen, P. (1979). *Lebensmittel-Wiss. Technol.* **12**, 153.
Warthesen, J. J. and Kramer, P. L. (1979). *J. Fd Sci.* **44**, 626.
Wells, G. B. and Lester, R. L. (1979). *Analyt. Biochem.* **97**, 184.

# 7 Determination of Lipids

## E. W. HAMMOND
Unilever Research, Sharnbrook, Bedford, U.K.

## I INTRODUCTION

Aitzetmüller (1975) is quoted as saying in his excellent review entitled "The liquid chromatography of lipids" that "HPLC has advanced very rapidly during the past few years". To repeat that comment now would be something of an understatement. The improvements in instrumentation and chromatographic media have been immense. The applications of HPLC are very diverse and this review will concern itself only with those applications to lipid analysis. Hitchcock and Hammond (1980) partially reviewed the application of HPLC to food lipids and lipid soluble vitamins.

I have attempted to divide this chapter into four main sections: neutral lipids and free fatty acids (FFAs) (see Section II); polar lipids (see Section III); argentation HPLC (see Section IV) and detection and detectors (see Section V). It will be found that both adsorption and reversed phase (RP) partition are used for separation in Sections II, III and IV.

Many excellent separating schemes have been published. However, re-

167

producing some of these separations can be very difficult, as the author has found in his own laboratory. The reasons for this are not necessarily that the schemes do not work, but are more likely to be due to differences in equipment. Three different makes of solvent pumping equipment are in use in our laboratory and a particular elution scheme has to be modified to suit each piece of equipment. This can also be required even when using different pumps from the same manufacturer. A point which manufacturers might like to record.

Thin layer chromatography (TLC) has traditionally tended to use silica adsorption as the main separating mode for lipids. This has been much extended by HPLC methodology. Also, where TLC was qualitative or at best semi-qualitative, HPLC is certainly a quantitative analytical tool. However, the most widely used mode of HPLC for lipid analysis is now RP, where a hydrophobic stationary phase competes with a hydrophilic mobile phase for the lipid component.

## II NEUTRAL LIPIDS AND FREE FATTY ACIDS

Silica adsorption HPLC is perhaps the most convenient system to use where groups of lipids of largely differing polarities are to be separated. For instance, repeated and rapid separations of triglyceride (TG), FFA, 1 : 3 and 1 : 2 diglyceride (DG) and monoglyceride (MG) fractions are readily achieved as shown in Fig. 7.1. This sort of system might be ideal for industrial quality control, where rapid analytical turnaround is necessary, with long column life (c. 6 months full time use). This can be particularly important in the commercial production of emulsifiers for food use. In this particular area, Sudraud, Coustard and Retho (1981) described the separation of a number of distilled monoglycerides, lactylated monoglycerides and tartaric acid esters of partial glycerides. Aitzetmüller and his colleagues used a similar system for the separation of sebum lipids (Fig. 7.2; Aitzetmüller and Koch, 1978) polyglycerols and polyols (Aitzetmüller, Bohrs and Arzberger, 1979).

The partial separations of a class of lipids (e.g. TG) into its individual species may be achieved by using RP partition systems. However, some partial separations of chain length can be achieved on straight silica, although the efficiency is not as good. Plattner and Payne-Wahl (1979) compared the two systems in the separation of TGs. The silica separation was less efficient than $\mu$-Bondapak $C_{18}$ and also the elution order was reversed (Fig. 7.3).

A number of papers have appeared in the last few years describing

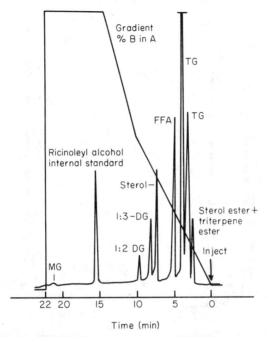

Time (min)

**Fig. 7.1** HPLC of netural lipids. Column, 5 $\mu$m LiChrosorb, 15 cm × 4 mm i.d.; solvent A, toluene–hexane (1 : 1 v/v); solvent B, toluene–ethyl acetate (3 : 1 v/v); flow rate, 1.5 ml min$^{-1}$. [Redrawn with permission from Hammond (1981).]

RP-HPLC separations of neutral lipids and FFAs. In the RP-HPLC analysis of FFAs, the carboxylic acid is normally derivatized to an ester. However, Bush, Russell and Young (1979) have demonstrated a useful separation of volatile FFAs, from acetic to isovaleric, without derivatization. This was achieved on $\mu$-Bondapak C$_{18}$ using ultraviolet (UV) detection at 210 nm.

The derivatization of FFAs can be used as a means of improving their chromatography, by reducing or increasing their polarity. More particularly in HPLC, it is used as a means of producing a UV-absorbing chromaphore for improving detection sensitivity. The most commonly reported FFA derivative is that first used by Borch (1975), the phenacyl ester. Pei, Kossa, Ramachandran and Henly (1976), using *p*-bromophenacyl esters, showed an interesting RP separation of *cis* and *trans* unsaturated fatty acids. Specific applications of this derivatization technique and RP-HPLC have been made by Mell and Bussell (1979) to a complex separation of bacterial fatty acids. Also, Bussell, Gross and Miller (1979) showed some excellent separations of fatty acids (Fig. 7.4), in their

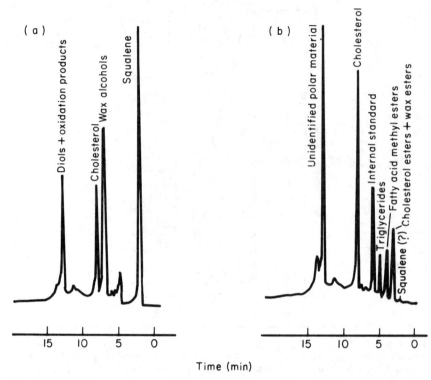

Time (min)

**Fig. 7.2** (a) Chromatogram of unsaponifiable portion of human sebum. (b) Chromatogram of atheroma lipids (the extract was methylated by diazomethane prior to analysis). [Redrawn with permission from Aitzetmüller and Koch (1978).]

study on the relationship between skin surface fatty acids and the degree of microbial contamination.

Most of these RP-HPLC separations suffer [as with gas–liquid chromatography (GLC) to some extent] from the problem of overlapping "critical pairs" of fatty acids. Some attempt at alleviating this problem has been made by the use of $C_{30}$ and $C_{18}$ bonded silicas by Takayama, Jardi and Benson (1980). They concluded that while the two columns complemented each other by providing different separations, the $C_{30}$ column was in general better. Özcimder and Hammers (1980) tackled this problem in a different way for fish oil fatty acid methyl esters (FAMEs). They preferred GLC as the final stage quantitative analytical tool, and overcame the "critical pair" problem by the use of small scale preparative RP-HPLC, or conventional argentation HPLC. The use of the methyl derivative of FFAs has also been reported by Pei, Henly and Ramachandran (1975), who showed some rapid HPLC separations of medium and long chain FAMEs.

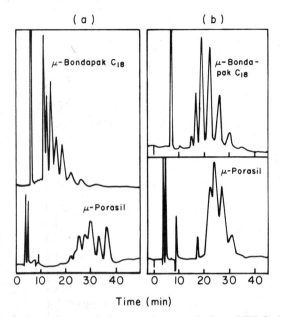

Time (min)

**Fig. 7.3** Comparison of normal phase and reversed phase HPLC chromatograms of linseed oil (a) and soybean oil (b). HPLC conditions: $\mu$-Bondapak $C_{18}$ eluted with acetonitrile–acetone (2 : 1 v/v) at 1.0 ml min$^{-1}$; $\mu$-Porasil eluted with 2,2,4-trimethylpentane–ether–acetic acid (99 : 1 : 1 v/v/v) at 1.0 ml min$^{-1}$. [Redrawn with permission from Plattner and Payne-Wahl (1979).]

Apart from normal fatty acid separations, RP-HPLC has usefully been employed in the separation of prostaglandins by Nagayo and Mizumo (1979), hydroxyl and cyclopropane fatty acids by Bussell and Miller (1979) and unsaturated fatty acid autoxidation products by Chan and Levett (1977) and Neff and Frankel (1980).

The separation and quantification of mixed TGs is of increasing importance to food and confectionery manufacturers. The application of GLC to this problem is increasingly well established, but Hammond (1981) has demonstrated that quantitative problems do exist. These problems are associated with the high temperatures required in the GLC analysis of TGs. The RP-HPLC separation of TGs might be a simpler approach and should yield more information, due to the much higher chromatographic selectivity and efficiency. Since the initial RP-HPLC separation of TGs shown by Pei, Henly and Ramachandran (1975), a number of workers have shown this to be a very promising approach. Plattner, Spencer and Kleiman (1977) published extensive work, followed by a number of others, including Herslof, Podlaha and Toregard (1979) (Fig. 7.5) and Vonach

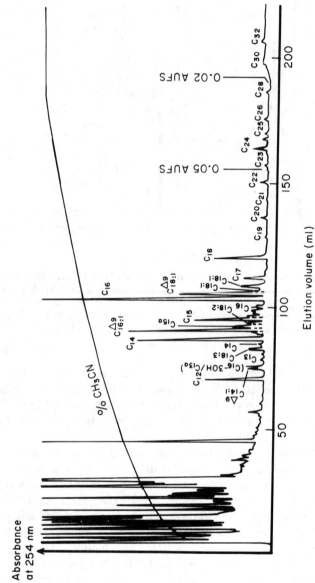

**Fig. 7.4**  Example of HPLC chromatogram of the free fatty acids from the finger-tips. Columns, two 30 cm × 3.9 mm $\mu$-Bondapak $C_{18}$; eluent, acetonitrile–water (40 : 60% to 100 : 0%); running time, 180 min; flow rate, 1.0 ml min⁻¹. [Redrawn with permission from Bussell *et al.* (1979).]

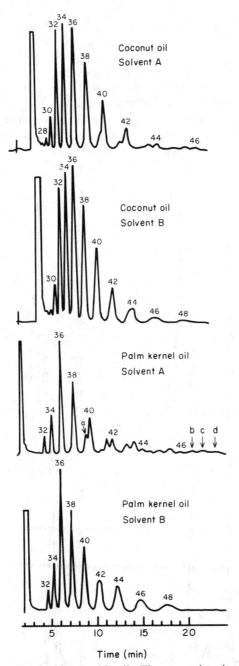

**Fig. 7.5**  Chromatograms of vegetable oils. The separations in solvent A reflect the degree of unsaturation. For example, in palm kernel oil in solvent A, peak a represents monounsaturated triglycerides while peak 40 is saturated (confirmed by fractional GC analysis); peaks b, c, and d correspond in retention time to model compounds, OOO, sn-OOP and sn-PPO respectively. [Redrawn with permission from Herslof *et al*. (1979).]

and Schomburg (1978). The latter demonstrated separations not only by chain length but also by level of unsaturation, separating $C_{48}$ TGs containing oleic and palmitic acids. The separation of a series of estolide TGs, which occur in a number of natural oils containing hydroxy fatty acids (Morris and Hall, 1966), was demonstrated by Payne-Wahl, Plattner, Spencer and Kleiman (1979).

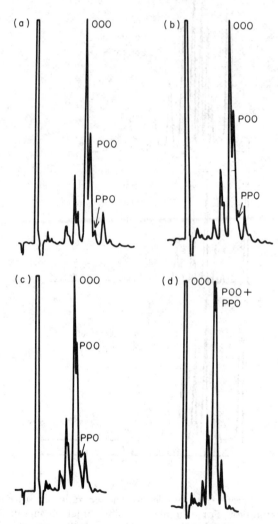

**Fig. 7.6** Separation of triglycerides in olive oil at different temperatures: (a) 14.5 °C; (b) 17.5 °C; (c) 20.5 °C; (d) 25.5 °C. [Redrawn with permission from Jensen (1981).]

El-Hamdy and Perkins (1981) have studied the effect of different variables in the RP-HPLC of TGs. Their conclusions were that $C_{18}$ bonded silica was better than $C_8$; increasing the bonded alkyl content improved separation and reducing the particle size also improved resolution. However, there remains the problems of critical pairs. Bezard and Ouedraogo (1980) approached the problem with a combination of argentation TLC prefractionation, followed by RP-HPLC of the fractions. Jensen (1981), by careful control of solvent composition and column temperature, was able to show partial resolution of some critical pairs (see Fig. 7.6). Complete resolution of the very complex mixture of TGs present in natural oils is still elusive, but perhaps the new 3 $\mu$m bonded phases will help in this respect.

## III PHOSPHOLIPIDS AND RELATED POLAR LIPIDS

Much of the analytical interest in phospholipids is in determining the proportion of the different classes present in a given biological extract. For this reason, most effort has been put into the use of straight silica columns. The amphipathic nature of polar lipids provides a challenging task to the analyst using HPLC.

Jungalwala, Turel, Evans and McCluer (1975) described a method for the separation of phospholipids containing primary amine groups, such as phosphatidylethanolamine. Detection in the UV at 268 nm relied on the formation of biphenylcarbonyl derivatives. The method could also determine plasmalogens separately from diacyl and alkylacyl aminophospholipids. Jungalwala, Evans and McCluer (1976) broadened the application, without using derivatization, relying on the absorption around 200–280 nm. These applications used silica columns. Fager, Shapiro and Litman (1977) showed similar separations for egg yolk phospholipids on a preparative scale. They achieved a partial subfractionation on the basis of chain length and unsaturation on 20–40 $\mu$m silica. Hax and Guerts van Kessel (1977) described a rapid and efficient separation of extracted biological and synthetic phospholipids on a small scale, using 5 $\mu$m and 10 $\mu$m silicas. A useful commercial application of such a system, to the analysis of lecithin in chocolate samples, was reported by Hurst and Martin (1980).

A major problem in the HPLC of phospholipids is the necessity to use very polar solvents, often including water. Regeneration of the column activity between runs becomes difficult. These problems led Kiuchi, Ohta and Ebine (1979) and Hanson, Osborn, Park, Walker and Kival (1980) to

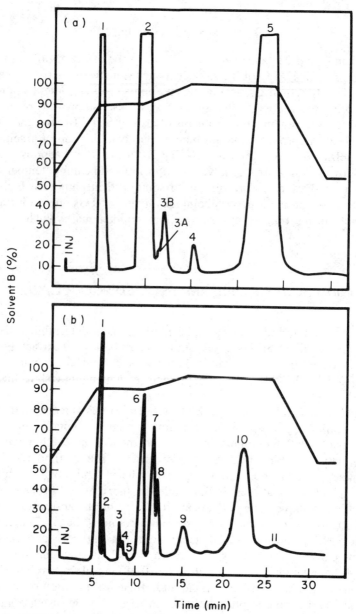

**Fig. 7.7** HPLC separation of methanol-extracted egg yolk phospholipids (a) and a mixture of natural phospholipid standards (b). In (a) there are five species as follows: 1, neutral lipids; 2, phosphatidylcholine; 3A and 3B, sphingomyelin; 4, phosphatidylcholine; 7,8, sphingomyelin; 9, lysophosphatidylcholine; 10, phosphatidylethanolamine; 11, lysophosphatidylethanolamine. Conditions were the same in both parts: column, Ultrasil-NH₂; eluents, solvent A, hexane–propan-2-ol (5.5 : 8.0 v/v), and solvent B, hexane–propan-2-ol–methanol–water (5.5 : 8.0 : 1.0 : 1.5 v/v/v/v); gradient technique used as indicated; flow rate, 0.7 ml min$^{-1}$, increased to 0.85 ml min$^{-1}$ at 3 min; temperature, ambient; detection, UV absorbance at 206 nm. [Redrawn with permission from Hanson *et al*. (1981).]

use amine-modified silicas. The latter authors reported difficulties in eluting phosphatidyl serine and phosphatidyl inositol, but the columns were superior to silica gel in their resistance to water deactivation. Also some separation of the molecular species of sphingomyelin, cerebrosides and phosphatidyl ethanolamine was achieved. Hanson, Park and Kival (1981) have since shown an improved separation of egg yolk phospholipids and natural phospholipids (see Figure 7.7) using Ultrasil-NH$_2$ and UV detection.

An anion-exchange resin was used by Kaitaranta, Geiger and Bessman (1981) in the analysis of phosphorylated metabolic intermediates, such as glycerylphosphoryl choline and glycerylphosphate. They found that column behaviour was quite reproducible.

The RP-HPLC of phospholipids has been more or less restricted to work on lecithin (phosphatidylcholine) fractions, both natural and synthetic. Porter, Wolf and Nixon (1979) showed good separation of 10 different synthetic lecithins using a $\mu$-Bondapak C$_{18}$ column. Compton and Purdy (1980) studied phospholipid separations on Spherisorb S5 ODS (octadecyl bonded, 5 $\mu$m) and LiChrosorb C$_{18}$ (10 $\mu$m) columns. Elevation of column temperatures and the use of mineral acid modifiers or ion-pairing agents were found useful in achieving improved separations of lecithins. They concluded that the use of modifiers in the aqueous methanol mobile phase gave greater chromatographic efficiency.

The RP-HPLC analysis of phospholipids is an area of work deserving more research. Efficient and quantitative analytical methods are much required in industry and also in the study of phospholipid metabolism and interactions in lipid biosynthesis.

## IV  ARGENTATION HPLC

The first papers describing the chromatography of lipids on adsorbents impregnated with silver nitrate were published in 1962 (De Vries, 1962; Barrett, Dallas and Padley, 1962; Morris, 1962). These demonstrated that a given class of compounds can be separated according to the number of isolated double bonds, and to the *cis/trans* geometry of those bonds. Subsequently De Vries and Jurriens (1963) showed that positionally isomeric dienes could be separated. Following this Morris, Wharry and Hammond (1967) showed that a range of positionally isomeric *cis* and *trans* monoenes could also be separated.

This technique of argentation chromatography was widely applied and became an important method in lipid analysis. However, there were major difficulties in obtaining true quantitative data from argentation TLC. HPLC has extended this technique and a number of workers have made different approaches to its application. The addition of silver to an aqueous eluent in RP-HPLC has been reported by Chan and Levett (1978),

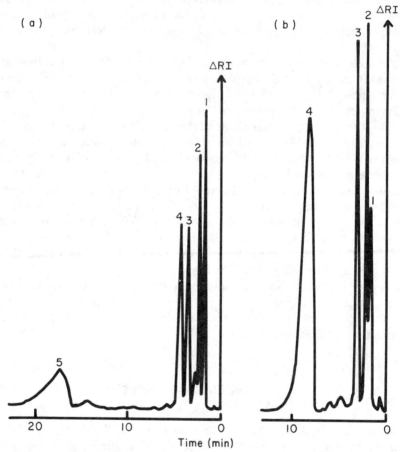

**Fig. 7.8** (a) Separation of a test mixture on 10 μm Nucleosil 10 SA (Ag⁺). (b) Separation of a test mixture on Nucleosil 10 SA (Ag⁺). Five species are seen in (a) as follows: 1, tetradecan-1-ol; 2, *trans*-9-tetradecen-1-ol; 3, *cis*-11-tetradecen-1-ol; 4, *trans*-4,*cis*-7-tridecadien-1-ol acetate; 5, *cis*-4,*trans*-7 *cis*-10-tridecatrien-1-ol acetate. Four species are seen in (b) as follows: 1, unknown; 2, methyl *trans*-9-octadecenoate; 3, methyl *cis*-9-octadecenoate; 4, methyl *cis*-9,*cis*-12-octadecadienoate; 5, methyl *cis*-6,*cis*-9,*cis*-12-octadecatrienoate was retained on the column. Chromatographic conditions were the same for parts (a) and (b): column, 250 mm × 4.6 mm i.d.; solvent, methanol; flow rate, 2 ml min⁻¹; pressure, 50 kg cm⁻²; temperature, 7 °C; attentuator, ×4; sample size, 2 μl of methanolic solution. [Redrawn with permission from Houx and Voerman (1976).]

Health, Tumlinson and Doolittle (1977) and Vonach and Schomburg (1978). However, the separation in these systems may not be generally as good as those separations shown on silver-loaded silica, and the high level of silver required in the eluent makes the system prohibitively expensive as a routine exercise.

The use of custom-prepared silver-loaded ion-exchangers was reported by Houx and Voerman (1976). They showed good and rapid separations of insect pheromones (Fig. 7.8). They also reported very long column life with no silver bleed. However, previous to this Emken and his colleagues described the use of silver-treated macroreticular cation-exchange resins (Emken, Scholfield and Dutton, 1964; Scholfield and Emken, 1966; Emken, Scholfield, Davison and Frenkel, 1967). These early resins gave very slow chromatography. This group of authors has since been more active in an attempt to produce an improved resin with faster chromato-

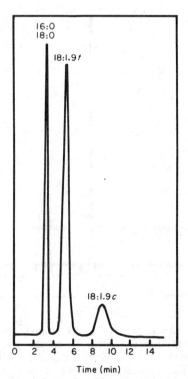

Time (min)

**Fig. 7.9**  Chromatogram of 5 μl of a 10% methyl palmitate, stearate, elaidate, oleate mixture on a 2 mm i.d. × 61 cm column. Flow rate, 0.5 ml min⁻¹; solvent, methanol; refractometer attentuation, ×16. [Redrawn with permission from Scholfield (1980).]

graphic properties (Scholfield and Mounts, 1977; Rakoff and Emken, 1978; Emken, Hartman and Turner, 1978). Although the separations are potentially very good, the chromatography times are still of the order of hours, with very broad peaks. More recently Scholfield (1980) has shown rapid chromatography of methyl esters on a modified silver-loaded resin (Fig. 7.9) and also reviewed this particular area of argentation chromatography (Scholfield, 1979a). However, the production of a generally available, fast, silver-loaded resin is still elusive.

Perhaps the most widely used mode of argentation HPLC (AgHPLC) is silver nitrate-loaded silica. Scholfield (1979b) showed good separation of saturated and *cis* and *trans* monoenoic|fatty methyl esters. Subsequently, Battaglia and Frohlich (1980) demonstrated separation of *cis* and *trans* positional isomers of methyl esters (see Fig. 7.10). Column packing preparation was by suspending 5 µm Spherisorb S 5W in an aqueous silver

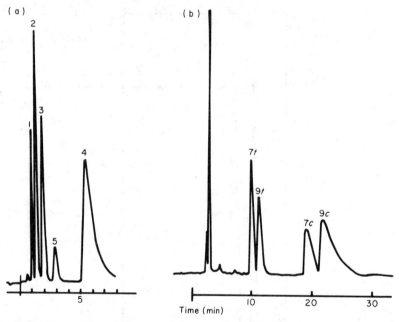

**Fig. 7.10** Two chromatograms of fatty acid methyl ester test mixtures. (a) Column, 240 mm × 3 mm i.d. packed with Partisil 20 impregnated with 2% Ag-$NO_3$. Peaks are as follows: 1, methyl ester of $C_{18:0}$; 2, methyl ester of $C_{18:1}$, $\omega 9$ *trans*; 3, methyl ester of $C_{18:1}$, $\omega 9$ *cis*; 4, methyl ester of $C_{18:2}$, $\omega 6$ *cis*, $\omega 9$ *cis*; 5, an impurity of 4. (b) Column, 240 mm × 5 mm i.d. packed with Spherisorb S5W impregnated with 5% $AgNO_3$. Peaks are as follows: methyl esters of the following $C_{18:1}$ positional isomers: 7t, $\omega 7$ *trans*; 9t,|$\omega 9$ *trans*; 7c, $\omega 7$ *cis*, 9c. $\omega 9$ *cis*. [Redrawn with permission from Battaglia and Frohlich (1980).]

nitrate solution. This was then dried under vacuum and reactivated. Hsieh, Welch and Turcotte (1981) showed that improvements in the resolution of certain FAMEs could be achieved by using truly anhydrous solvents. The separation of saturated and isomeric unsaturated TGs, using AgHPLC, was demonstrated by Smith, Jones and Hammond (1980a). At that time the silica was coated using an acetonitrile–methanol solution of silver nitrate. Hammond (1981) subsequently reported improvements in the silver loading procedure and the effects of loading from an aqueous as opposed to a non-aqueous solvent. In this separation, SOS and SSO isomers of TGs separate, as do SOO and SLS TGs (where S = saturated, O = monoenoic, L = dienoic fatty acids on the 1, 2 or 3 positions of glycerol). To reduce the leaching of silver from the column by solvent, a pre-column of silver nitrate crystals was placed in the high pressure solvent delivery line.

The simplest and cheapest AgHPLC system to apply at the present is silver-loaded silica, although the advent of a good, reliable commercial silver-loaded resin would offer considerable advantages in column life. It is clear, however, from the work being carried out on RP-HPLC separations, particularly of TGs, that RP-HPLC would offer certain advantages over AgHPLC, if the problems of critical pairs could be overcome.

## V  DETECTION SYSTEMS

The problem of detection in the HPLC of lipids may be seen as the rate-limiting step to the development of the technique. The advent of the moving wire detector (Pye Unicam LCM) was perhaps too early and thus its development suffered from a lack of application. Finally it was taken off the market, at a time when the interest in the HPLC analysis of lipids was increasing. Aitzetmüller (1975) stressed the great importance of a non-specific detector for HPLC and has been most active in reiterating this point to the present day. This problem of detection is particularly relevant to lipids, but not exclusively so, since all non-UV absorbing species undergoing gradient elution require a non-specific detection system that is not affected by the solvent gradient.

However, it is notable that certain types of lipid HPLC systems can utilize other detectors. Smith et al. (1980a) used refractive index (RI) detection to good effect in the AgHPLC of triglycerides, but the range of triglyceride species that could be analysed was restricted by the solvent. RI detection was preferred to that of UV at 215 nm when the two were compared in the HPLC of lecithins by Porter et al. (1979). This is mainly because the extinction coefficient for phospholipids in the UV is less than 200 and signal to noise level is poor. In fact triglycerides have been characterized by Karleskind (1977) after iodination; the iodinated species absorb at 265 nm but saturated triglicerides are not detected. Other lipids which do not absorb at longer wavelengths may be detected in the end-absorption region between 195 nm and 215 nm (Hax and Guerts van Kessel, 1977; Guerts van Kessel, Hax, Demel and De Gier, 1977; Risson and Hoffmeyer, 1978). However, solvent selection restricts the separations achieved and the signal to noise ratio is not generally good. There are very few quoted applications of infrared (IR) as a lipid detector in HPLC, although Parris (1978) has shown that it offers some promise for triglycerides.

The major use of UV detection in lipid HPLC is in the analysis of fatty acids as their phenacyl esters (Borch, 1975; Pei et al., 1976; Mell and Bussell, 1979; Bussell et al., 1979). This particular derivative has a high extinction coefficient and gives excellent sensitivity. Other UV absorbing derivatives of FFAs have been used, such as benzyl, p-nitrobenzyl and 2-naphthacyl derivatives.

Light scattering or nephelometric detection in liquid chromatography was reported by Jorgenson, Smith and Novotny (1977) and Smith, Jorgenson and Novotny (1980b). While this appeared useful, insufficient quantitative data were supplied. Another light scattering detector, called a mass detector, has been commercially developed by Applied Chromatography Systems (Luton, Bedfordshire, England). Charlesworth (1978) has studied the variables influencing the response and linearity of such an instrument. The linear range of the earliest instruments appeared poor, and because it is an evaporative system, there were problems of sample loss by volatilization. The detector has been successfully applied to the analysis of carbohydrates, where sample volatization is not a problem (Macrae and Dick, 1981), but its application to lipid analysis requires further study.

The role of spectral selectivity in fluorescence detection for HPLC has been discussed by Ogan, Katz and Porro (1979). Fluorescence labelling has been applied to lipids using phenanthrimidazole derivatives (Lloyd, 1980) and 4-methyl-7-methoxycoumarin derivatives (Zelenski and Huber, 1978). The derivatization procedures are rather involved but sensitivity is high.

## VI  CONCLUSION

HPLC is a very powerful tool in lipid analysis. There are two major areas which offer exciting prospects for the future as more research work develops them, namely phospholipid analysis and reversed phase analysis of triglycerides.

The main problem for the future is the commercial development and production of a non-specific detector that will have a wide range of application to lipid analysis.

## ACKNOWLEDGEMENTS

I would like to thank Unilever Research for providing the stimulating environment in which this review has been prepared. Also thanks to my wife, Nuala, for her patience and careful typing.

## REFERENCES

Aitzetmüller, K. (1975). *J. Chromat.* **113**, 231.

Aitzetmüller, K. and Koch, J. (1978). *J. Chromat.* **145**, 195.

Aitzetmüller, K., Bohrs, M. and Arzberger, E. (1979). *Fette Seifen. AnstrichMittel.* **81**, 436.

Barrett, C. B., Dallas, M. S. J. and Padley, F. B. (1962). *Chemy Ind.* **1962**, 1050.

Battaglia, R. and Frohlich, D. (1980). *Chromatographia* **13**, 428.

Bezard, J. A. and Oeudraogo, M. A. (1980). *J. Chromat.* **196**, 279.

Borch, R. F. (1975). *Analyt. Chem.* **47**, 2347.

Bush, K. J., Russell, R. W. and Young, J. W. (1979). *J. Liquid Chromat.* **2**, 1367.

Bussell, N. E. and Miller, R. A. (1979). *J. Liquid Chromat.* **2**, 697.

Bussell, N. E., Gross, A. and Miller, R. A. (1979). *J. Liquid Chromat.* **2**, 1337.

Chan, H. W. S. and Levett, G. (1977). *Chemy Ind.* **1977**, 692.

Chan, H. W. S. and Levett, G. (1978). *Chemy Ind.* **1978**, 578.

Charlesworth, J. M. (1978). *Analyt. Chem.* **50**, 1414.

Compton, B. J. and Purdy, W. C. (1980). *J. Liquid Chromat.* **3**, 1183.

De Vries, B. (1962). *Chemy Ind.* **1962**, 1049.

De Vries, B. and Jurriens, G. (1963). /*Fette Seifen. AnstrichMittel.* **65**, 725.

El-Hamdy, A. H. and Perkins, E. G. (1981). *J. Am. Oil Chem. Soc.* **58**, 49.

Emken, E. A., Scholfield, C. R. and Dutton, H. J. (1964). *J. Am. Oil Chem. Soc.* **41**, 388.

Emken, E. A., Scholfield, C. R., Davison, V. L. and Frenkel, E. N. (1967). *J. Am. Oil Chem. Soc.* **44**, 373.

Emken, E. A., Hartman, J. C. and Turner, C. R. (1978). *J. Am. Oil Chem. Soc.* **55**, 561.

Fager, R. S., Shapiro, S. and Litman, B. J. (1977). *J. Lipid Res.* **18**, 704.

Geurts van Kessel, W. S. M., Hax, W. M. A., Demel, R. A. and De Gier, J. (1977). *Biochim. biophys. Acta* **486**, 524.

Hammond, E. W. (1981). *J. Chromat.* **203**, 397.

Hanson, V. L., Osborn, T. W., Park, J. Y., Walker, W. V. and Kival, R. M. (1980). *Fedn Proc. Am. Socs exp. Biol.* **39**, 1040.

Hanson, V. L., Park, J. Y. and Kival, R. M. (1981). *J. Chromat.* **205**, 393.

Hax, W. M. A. and Guerts van Kessel, W. S. M. (1977). *J. Chromat.* **142**, 735.

Health, R. R., Tumlinson, J. H. and Doolittle, R. E. (1977). *J. chromatogr. Sci.* **15**, 10.

Herslof, B., Podlaha, O and Toregard, B. (1979). *J. Am. Oil Chem. Soc.* **56**, 864.

Hitchcock, C. and Hammond, E. W. (1980). In *Developments in Food Analysis Techniques*, Vol. 2 (R. D. King, ed.), Applied Science Publishers, London, p. 185.

Houx, N. W. H. and Voerman, S. (1976). *J. Chromat.* **129**, 456.

Hsieh, J. Y. K., Welch, D. K. and Turcotte, J. G. (1981). *Analyt. chim. Acta* **123**, 41.

Hurst, W. J. and Martin, R. A. (1980). *J. Am. Oil Chem. Soc.* **57**, 307.

Jensen, G. W. (1981). *J. Chromat.* **204**, 407.

Jorgenson, J. W., Smith, S. C. and Novotny, M. (1977). *J. Chromat.* **142**, 233.

Jungalwala, F. B., Turel, R. J., Evans, J. E. and McCluer, R. H. (1975). *Biochem. J.* **145**, 517.

Jungalwala, F. B., Evans, J. E. and McCluer, R. H. (1976). *Biochem. J.* **155**, 55.

Kaitaranta, J. K., Geiger, P. J. and Bessman, S. P. (1981). *J. Chromat.* **206**, 327.

Karleskind, A. (1977). *Revue. fr. Corps. Gras.* **24**, 419.

Kiuchi, K., Ohta, T. and Ebine, H. (1979). *J. Chromat.* **133**, 226.

Lloyd, J. B. F. (1980). *J. Chromat.* **189**, 359.

Macrae, R. and Dick, J. (1981). *J. Chromat.* **210**, 138.

Mell, L. P. and Bussell, N. E. (1979). *J. Liquid Chromat.* **2**, 407.

Morris, L. J. (1962). *Chemy Ind.* **1962**, 1238.

Morris, L. J. and Hall, S. W. (1966). *Lipids* **1**, 188.

Morris, L. J., Wharry, D. M. and Hammond, E. W. (1967). *J. Chromat.* **31**, 69.

Nagayo, K. and Mizumo, N. (1979). *J. Chromat.* **178**, 347.

Neff, W. E. and Frankel, E. N. (1980). *Lipids* **15**, 587.

Ogan, K., Katz, E. and Porro, T. J. (1979). *J. chromatogr. Sci.* **17**, 597.

Ozcimder, M. and Hammers, W. E. (1980). *J. Chromat.* **187**, 307.

Parris, N. A. (1978). *J. Chromat.* **149**, 615.

Payne-Wahl, K., Plattner, R. D., Spencer, G. F. and Kleiman, R. (1979). *Lipids* **14**, 601.

Pei, P. T.-S., Henly, R. S. and Ramachandran, S. (1975). *Lipids* **10**, 152.

Pei, P. T.-S., Kossa, W. C., Ramachandran, S. and Henly, R. S. (1976). *Lipids* **11**, 814.

Plattner, R. D. and Payne-Wahl, K. (1979). *Lipids* **14**, 152.

Plattner, R. D., Spencer, G. F. and Kleiman, R. (1977). *J. Am. Oil Chem. Soc.* **54**, 511.

Porter, N. A., Wolf, R. A. and Nixon, J. R. (1979). *Lipids* **14**, 20.

Rakoff, H. and Emken, E. A. (1978). *J. Am. Oil Chem. Soc.* **55**, 564.

Risson, T. and Hoffmeyer, L. (1978). *J. Am. Oil Chem. Soc.* **55**, 649.

Scholfield, C. R. (1979a). In *Geometric and Positional Fatty Acid Isomers* (H. J. Dutton and E. A. Emken, eds), American Oil Chemists' Society, New York, p. 31.

Scholfield, C. R. (1979b). *J. Am. Oil Chem. Soc.* **56**, 510.

Scholfield, C. R. (1980). *J. Am. Oil Chem. Soc.* **57**, 331.

Scholfield, C. R. and Emken, E. A. (1966). *Lipids* **1**, 235.

Scholfield, C. R. and Mounts, T. L. (1977). *J. Am. Oil Chem. Soc.* **54**, 319.

Smith, E. C., Jones, A. D. and Hammond, E. W. (1980a). *J. Chromat.* **188**, 205.
Smith, S. L., Jorgenson, J. W. and Novotny, M. (1980b). *J. Chromat.* **187**, 111.
Sudraud, G., Coustard, J. M. and Retho, C. (1981). *J. Chromat.* **204**, 397.
Takayama, K., Jardi, H. C. and Benson, R. (1980). *J. Liquid. Chromat.* **3**, 61.
Vonach, B. and Schomburg, G. (1978). *J. Chromat.* **149**, 417.
Zelenski, S. G. and Huber, J. W. (1978). *Chromatographia* **11**, 645.

# 8 Determination of Vitamins

## P. J. VAN NIEKERK

National Food Research Institute, Council for Scientific and Industrial Research, Pretoria, South Africa

## I INTRODUCTION

A vitamin has been defined as a biologically active organic compound which acts as a controlling agent for an organism's (human's) normal health and growth. It is not synthesized within the organism but is available in the diet in small amounts, and is carried in the circulatory system in small concentrations to act on target organs or tissues (Kutsky, 1973). The levels of the vitamins present in foods may be as low as a few micrograms per 100 g and they are often accompanied by an excess of compounds with similar physical and chemical properties. Separation of the vitamins from each other and from interfering substances is, therefore, usually necessary when their levels in food have to be determined. Vitamins generally are labile compounds and many are susceptible to oxidation and breakdown when exposed to oxygen, heat or ultraviolet radiation.

HPLC has developed during the past decade as a separating technique with many advantages relevant to vitamin analysis. It can be performed at room temperature. Light and oxygen are easily excluded during chromatography. Exposure to radiation during detection is brief. HPLC will perform many separations in a much shorter time than other separating techniques but speed is not necessarily its main advantage. Separations may be obtained far more consistently than with the older column or thin layer techniques. Because of the superior ability of HPLC to separate the compounds of interest from interfering substances, coupled to the availability of selective detectors, sample preparation is kept to a minimum and it is generally not necessary to prepare derivatives of the vitamins, as is mostly the case where gas chromatography is used. Errors are minimized by keeping down the number of manipulations in a method.

The literature contains numerous examples of the separation of vitamins but not all are applicable to food samples and in this chapter the emphasis will be placed on those methods which have been applied to foods. The vitamins may be divided into a fat-soluble group and a water-soluble group and will be discussed under these two headings.

## II  THE FAT-SOLUBLE VITAMINS

This group of vitamins includes vitamins A, D, E and K and consists of derivatives of partially cyclized isoprenoid polymers. The extensive literature on the use of HPLC for the determination of these vitamins in foods has been covered by various review articles (Parrish, 1977, 1979, 1980; Macrae, 1980; Christie and Wiggins, 1978) so that an exhaustive coverage will not be attempted here. The emphasis will rather be on giving an overview of the now established methods.

### A  Vitamin A

Vitamin A activity is shared by a number of compounds of which retinol, retinyl esters and retinaldehyde are the most important for the food analyst (Parrish, 1977). Certain carotenoids also possess vitamin A activity and of these β-carotene is the most widespread and biologically most active (Green, 1970).

### 1  Retinol and its Derivatives

The use of HPLC for the determination of retinol is a vast improvement on the older methods which were predominantly colorimetric and column chromatographic methods that depended very heavily on the skills and experience of the operator to obtain consistent results. Owing to the extreme sensitivity of retinol to oxidation and to destruction by light, it is important that all manipulations be performed in the absence of oxygen and shielded from direct sunlight and from the light of fluorescent lamps.

(a) *Sample preparation*  Unsupplemented foods mostly contain retinol and mixed retinyl esters. To simplify the analysis, the sample is usually saponified to convert the esters to the free alcohol. The saponification also serves to free the vitamin A from the food matrix and eliminate bulk components such as triglycerides.

Retinyl acetate or palmitate is used to supplement a variety of food products and saponification can also be used to hydrolyse the vitamin A in these samples. The procedure is, however, time-consuming and a number of methods have been developed to eliminate this step (Aitzetmüller, Pilz and Tasche, 1979; Thompson, Hatina and Maxwell, 1980; Widicus and Kirk, 1979). The retinyl acetate or palmitate is then determined directly after solvent extraction. Care is needed with this procedure as simple solvent extraction may not always completely extract the vitamin A from certain sample matrices such as products derived from spray-dried homogenized milk (van Niekerk, unpublished results) or products containing encapsulated vitamin A. Some form of hydrolysis is necessary in these instances.

Various modifications of the saponification procedure are in use (Association of Official Analytical Chemists, 1975; Roels and Mahadevan, 1967; Strohecker and Henning, 1965; Thompson and Maxwell, 1977) and the method should be adapted to the type of sample being analysed. Parrish (1977) considers the use of an atmosphere of nitrogen gas unnecessary during refluxing for hydrolysis, or during evaporation under partial vacuum, because the area above the liquid is filled with solvent vapours, displacing most of the oxygen.

Extraction after saponification may be carried out with diethyl ether, *n*-hexane, pentane or petroleum ether. The hydrocarbon solvents have certain advantages, when compared to diethyl ether, because of their lower solubility in water and a reduced tendency to form emulsions. When the solvents are evaporated to dryness, it is essential to have an antioxidant present (van Niekerk and du Plessis, 1976). The antioxidant may be one naturally present in the sample but, if any doubt exists, it should be added before evaporation, to ensure quantitative recovery of the vitamin A.

Egberg, Heroff and Potter (1977) circumvented the time-consuming extraction after saponification, by acidifying with acetic acid in acetonitrile and filtering off the precipitated fatty acid salts. After dilution with water an aliquot could be injected directly onto a reversed phase column. Unfortunately the advantage of concentrating low potency samples through solvent extraction is lost.

Lipid removal from a sample may also be accomplished by enzymic hydrolysis (Barnett, Frick and Baine, 1980) or gel permeation chromatography (Landen and Eitenmiller, 1979; Landen, 1980; Williams, Schmit and Henry, 1972).

(b) *Chromatography* Vitamin A may be chromatographed on either normal or reversed phase columns. When simple solvent extraction is used for sample preparation, normal phase is generally used because the lipid material that is co-extracted has high solubility in solvents used for this type of chromatography (Aitzetmüller *et al.*, 1979; Thompson *et al.*, 1980; Widicus and Kirk, 1979). Non-aqueous reversed phase chromatography affords the same advantages of high solubility for lipids but aqueous reversed phase chromatography systems are normally used for samples which are low in lipids or where the bulk of the lipids have been removed by saponification or gel permeation chromatography (Bui-Nguyên and Blanc, 1980; Thompson and Maxwell, 1977; Williams *et al.*, 1972).

Egberg *et al.* (1977) demonstrated the presence of 13-*cis*-retinol together with all-*trans*-retinol in a number of food products and could separate the two isomers on either a normal or reversed phase system. The 11-*cis* and 9-*cis* isomers may also be separated from all-*trans*- and 13-*cis*-retinol on a silica gel column. Since the 13-*cis* isomer has approximately 75% (Tsukida, Kodama, Ito, Kawamoto and Takahashi, 1977) of the biological activity of all-*trans*-retinol, reports on the vitamin A content of foods should be more specific and state that the result is, for example, for all-*trans*-retinol and not just retinol (Wiggins, 1979a).

Fluorescence detectors are inherently more selective and often more sensitive than UV detectors. Since vitamin A has a natural fluorescence, this method of detection should be chosen where possible. On the other hand, UV detectors at 313–328 nm will give satisfactory results with most samples. This is borne out in the chromatogram of a margarine sample depicted in Fig. 8.1.

Typical conditions for HPLC analysis of retinol and retinyl esters are summarized in Table 8.1.

## 2  *Provitamin A Carotenoids*

There are nearly 300 naturally occurring carotenoids known at present (Weedon, 1971). Any carotenoid which contains the retinol structure in

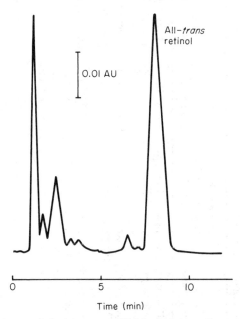

All–*trans*
retinol

0.0I AU

Time (min)

**Fig. 8.1** Chromatogram of the non-saponifiable fraction of margarine. Stainless steel column (25 cm × 0.4 cm i.d.) slurry packed with LiChrosorb SI 60 (5 $\mu$m) with dioxane and $n$-hexane (8 : 92, v/v) as mobile phase (3 ml min$^{-1}$) and UV detection at 328 nm.

the molecule may be converted *in vivo* to retinol (Pitt, 1971). Of these the following are the most important: $\beta$-carotene, $\alpha$-carotene, $\gamma$-carotene, echinenone, cryptoxanthin, torularhodin, $\beta$-apo-8'-carotenal, $\beta$-apo-8'-carotenoic acid ethyl ester, $\beta$-apo-12'-carotenal, $\beta$-zeacarotene and citranaxanthin (Bauernfeind, Brubacher, Kläui and Marusich, 1971). The large number of compounds to be considered complicates the task of determining the total provitamin A content of a sample. The result is that many analysts concentrate on only a few compounds which they consider important in a specific sample. $\beta$-Carotene and, to a lesser extent, $\alpha$-carotene are the provitamin A carotenoids which are determined most often (Zakaria, Simpson, Brown and Krstulovic, 1979; Calabro, Micali and Curro, 1978; van Niekerk and du Plessis, 1976; Reeder and Park, 1975).

(a) *Sample preparation*    Carotenoids may be extracted from dry samples by water-immiscible solvents, whereas for wet samples water-miscible solvents, such as mixtures of acetone and methanol, are used (Liaaen-Jensen, 1971; van Niekerk and du Plessis, 1976). Carotenes may be re-extracted from aqueous acetone extracts into chloroform or hexane in order to

**Table 8.1**  HPLC systems used for retinol and retinyl esters

| Compounds determined | Type of sample | Sample preparation | Stationary phase | Mobile phase (proportions by volume) | Detector | Reference |
|---|---|---|---|---|---|---|
| Retinyl palmitate | Margarine | Heptane extraction | 5 μm LiChrosorb SI 60 (Merck) | Heptane–diisopropyl ether (95 : 5) | UV 325 nm | Aitzetmüller et al. (1979) |
| Retinyl palmitate | Milk, margarine | Hexane extraction | 5 μm LiChrosorb SI 60 (Merck) | Hexane–diethyl ether (2 : 98) | UV 325 nm | Thompson et al. (1980) |
| Retinol | Margarine, fish liver oil, fortified milk powder | Saponify and extract | 5 μm LiChrosorb SI 60 (Merck) | Hexane–dioxane (92 : 8) | UV 328 nm; fluorescence, ex. 330 nm, em. 480 nm | van Niekerk (unpublished) |
| Retinyl palmitate | Breakfast cereals | Extract with chloroform + ethanol | μ-Porasil (Waters) | Hexane–chloroform (85 : 15) | UV 313 nm; fluorescence, ex. 360 nm, em. 415 nm | Widicus and Kirk (1979) |

| Compound | Food | Sample preparation | Column | Mobile phase | Detection | Reference |
|---|---|---|---|---|---|---|
| Retinol | Milk, cheese | Saponify and extract | Nucleosil 10 C$_{18}$ (Machery and Nagel) | Acetonitrile–water (95 : 5) | UV 328 nm | Bui-Nguyên and Blanc (1980) |
| Retinol, retinyl acetate | Milk, soya-based infant formulae | Enzymic hydrolysis and extraction | Zorbax ODS (Dupont) | Methanol–ethyl acetate–acetonitrile gradient | UV 325 nm for retinol; 365 nm for retinyl palmitate | Barnett et al. (1980) |
| Retinyl palmitate | Margarine | Extraction and lipid removed by gel permeation chromatography | μ-Bondapak C$_{18}$ 10 μm (Waters) | Methylene chloride–acetonitrile (30 : 70) | UV 313 nm | Landen and Eitenmiller (1979) |
| Retinol | Infant formula, fortified milk, margarine | Saponify and extract | 10 μm Li-Chrosorb reversed phase (Merck) | Methanol–water (90 : 10) | UV 325 nm | Thompson and Maxwell (1977) |
| All-trans-and 13-cis-retinol | Butter, fish, liver oil, breakfast cereal, cat food | Saponify, add acetic acid, filter, add water, inject | 10 μm Vydac ODS (Separations Group) | Acetonitrile–water (65 : 35) | UV 328 nm; fluorescence, ex. 365 nm, em. 510 nm | Egberg et al. (1977) |

eliminate the water, which is difficult to remove by evaporation. Carotenoids have also been extracted from fruit juices by filtration through magnesium oxide, which will adsorb the pigments (Stewart, 1977). The carotenoids were then removed from the adsorbent by washing with water-immiscible solvents.

Saponification is generally used to remove unwanted lipid material and is usually performed at room temperature with methanolic potassium hydroxide (Zakaria *et al.*, 1979; Reeder and Park, 1975). Saponification at elevated temperatures has also been applied to samples such as margarine

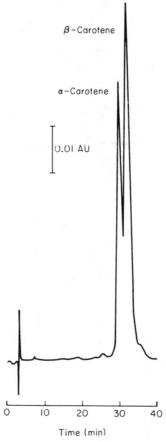

**Fig. 8.2** Chromatogram of an acetone extract of carrots. Stainless steel column (25 cm × 0.4 cm i.d.) slurry packed with Nucleosil 5 $C_{18}$, with chloroform and acetonitrile (4 : 96, v/v) as mobile phase (1 ml min$^{-1}$) and absorption detection at 440 nm.

**Table 8.2**  HPLC systems used for carotenoid analysis

| Compounds determined | Type of sample | Sample preparation | Stationary phase | Mobile phase (proportions by volume) | Detection wavelength (nm) | Reference |
|---|---|---|---|---|---|---|
| α- and β-carotenes, β-cryptoxanthin | Orange juice | Adsorb on magnesia, extract, saponify, extract | Magnesia | n-Hexane–acetone gradient (99 : 1 to 80 : 20) | 440 | Stewart (1977) |
| β-Carotene, lycopene | Tomatoes | Extract with acetone, saponify, extract | Partisil PXS 5/ODS (Whatman) | Chloroform–acetonitrile (8 : 92) | 470 | Zakaria et al. (1979) |
| α- and β-carotenes | Orange juice | Extract, saponify, extract | HC-ODS/Sil X (Perkin Elmer) | Methanol–water (100 : 6) | 450 | Calabro et al. (1978) |
| α- and β-carotenes, β-cryptoxanthin | Orange juice | Extract, saponify, extract | 18–30 μm Basic Alumina (Woelm) | Benzene–hexane (3 : 5) + 0.01% BHT | 440 | Reeder and Park (1975) |
| β-Carotene | Margarine | Saponify, extract | 10 μm LiChrosorb reversed phase (Merck) | Methanol–water (99 : 1) | 453 | Thompson and Maxwell (1977) |
| β-Carotene | Margarine | Dissolve in hexane, wash with 60% ethanol and inject | 5 μm LiChrosorb SI 60 (Merck) | Diethyl ether–hexane (2 : 98) | 453 | Thompson et al. (1980) |
| β-Carotene | Margarine | Dissolve in methylene chloride fractionate on gel permeation column | μ-Bondapak C$_{18}$ 10 μm (Waters) | Methylene chloride–acetonitrile (30 : 70) | 436 | Landen and Eitenmiller (1979) |

(Thompson and Maxwell, 1977), but may be avoided by washing a hexane solution of the margarine with aqueous alcohol (Thompson *et al*., 1980) or by subjecting the sample to high performance gel permeation chromatography prior to HPLC (Landen and Eitenmiller, 1979). Hydroxycarotenoids are often present in samples as their fatty acid esters (Weedon, 1971) and saponification will convert them to the free xanthophylls. Acidic carotenoids will similarly be freed from their esters by saponification.

(b) *Chromatography* Because of their sensitivity to oxidation, carotenoids should be chromatographed immediately after extraction. Traditionally adsorption chromatography on magnesia, zinc carbonate, alumina or silica, has been used (Stewart and Wheaton, 1971; van de Weerdhof, Wiersum and Reissenweber, 1973; Reeder and Park, 1975), but the newer reversed phase packing materials have also been applied (Zakaria *et al*., 1979; Calabro *et al*., 1978; Botey Serra and Garcia Fite, 1975). Figure 8.2 shows a chromatogram of the separation of $\alpha$- and $\beta$-carotene in a carrot extract on a non-aqueous reversed phase system. Reversed phase systems have the advantage, when analysing for carotenes, that the more polar compounds are not retained on the column and thus the need for frequent regeneration of the column is eliminated. A further advantage is that reversed phase systems are better suited to gradient elution which is necessary when compounds with a wide range of polarities are to be separated (Botey Serra and Garcia Fite, 1975). Table 8.2 summarizes typical conditions that are used for the HPLC analysis of carotenoids.

## B  Vitamin D

Vitamin D is the most potent of the fat-soluble vitamins and occurs in foods (mainly animal products) in very small amounts. The fact that it is accompanied by an excess of compounds, such as sterols, vitamin E and vitamin A, with similar physical and chemical properties, poses a formidable problem to the analyst and it is, therefore, no wonder that the tedious and time-consuming biological methods (Kodicek and Lawson, 1967) have only recently been replaced by chemical methods. HPLC has gone a long way in solving some of the analytical problems and has been used routinely in certain areas (Henderson and McLean, 1979), although some development is still necessary in others.

Vitamin $D_3$ (cholecalciferol) is the major form of vitamin D present in animal products but vitamin $D_2$ (ergocalciferol) may also be present in fortified food products. 25-Hydroxyvitamin $D_3$ or $D_2$ is formed in the liver and is present in animal products in small amounts (Koshy and van der Slik, 1977). In solution, vitamin D isomerizes to pre-vitamin D and forms an equilibrium mixture with a composition which is temperature-dependent. In order to secure a valid estimation of the total vitamin D activity, the pre-vitamin D content of the sample must be taken into account (Hofsass, Grant, Alicino and Greenbaum, 1976). This may be done by separating the pre-vitamin D and determining it separately or, alternatively, to pre-isomerize the vitamin D to a constant ratio of vitamin D and pre-vitamin D (Borsje, Craenen, Esser, Mulder and de Vries, 1978). Because the isomerization rates of vitamins $D_2$ and $D_3$ are equal (Hanewald, Mulder and Keuning, 1968) the potential vitamin D (pre-vitamin D plus vitamin D) content of a sample is determined when vitamin $D_2$ is used as an internal standard for vitamin $D_3$ (or vice versa).

## 1 Sample Preparation

Methods generally involve extraction, hydrolysis and sample clean-up before HPLC analysis. Problems which may be encountered in sampling, especially when stabilized forms of vitamin D have been added, have been dealt with in detail by Parrish (1979). It is clear that hydrolysis (saponification with ethanolic potassium hydroxide) of the sample should precede the extraction when the form, in which the vitamin D is added, is unknown. The sample size should also be sufficient for low potency samples to ensure that it is representative of the bulk of the sample. Saponification serves the purpose of freeing the vitamin D from the sample matrix and eliminating the bulk of the lipids, but further purification is usually needed before HPLC is applied. Methods include purification procedures which were used in the older chemical methods. These procedures include precipitation of sterols with digitonin, alumina column chromatography, Florex XXS chromatography, Celite partition chromatography and various other chromatographic steps (Henderson and Wickroski, 1978; Adachi and Kobayashi, 1979; Ali, 1978; Antalick, Debruyne and Faugere, 1977). Purification of samples may also be accomplished by preparative HPLC (Egaas and Lambertsen, 1979; Cohen and Wakeford, 1980; van Niekerk and Smit, 1980) or chromatography on hydroxyalkoxypropyl Sephadex (HAPS) (Thompson, Maxwell and L'Abbé, 1977). Purification by HPLC has the advantage that the resolution is much higher than with older column chromatographic methods and retention volumes are more constant.

## 2 Chromatography

With HPLC it is possible to separate vitamin $D_2$ from vitamin $D_3$ (Williams et al., 1972; Wiggins, 1977), vitamin $D_3$ from its precursors (Hofsass et al.,

1976) and the metabolites of vitamins $D_2$ and $D_3$ (Tanaka, De Luca and Ikekawa, 1980). In applications to food samples both normal and reversed phase chromatography have been used but only reversed phase was able to resolve vitamins $D_2$ and $D_3$, with the result that methods which use adsorption chromatography make no distinction between vitamins $D_2$ and $D_3$ and

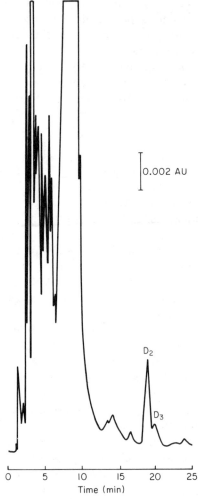

**Fig. 8.3** Chromatogram of the purified extract of fish liver containing 17 International Units of vitamin $D_3$ per gram. Vitamin $D_2$ was added as internal standard. Stainless steel column (25 cm × 0.4 cm i.d.) slurry packed with Nucleosil 5 $C_{18}$ with methanol and water (95 : 5, v/v) as mobile phase (1 ml min$^{-1}$) and UV detection at 264 nm. $D_2$ is vitamin $D_2$ and $D_3$ is vitamin $D_3$.

will report the total amount if both are present. Vitamins $D_2$ and $D_3$ may be separated from their corresponding pre-vitamins on most reversed phase or normal phase systems. Both vitamin $D_2$ and vitamin $D_3$ have absorption maxima at 265 nm (Morton, 1970) which can be used for detection with variable wavelength detectors but, owing to the inherent greater stability of fixed wavelength detectors, lower detection limits (1 ng vitamin D) may be obtained at their operating wavelength of 254 nm (van Niekerk and Smit, 1980) even though this is shifted from the absorption maximum.

HPLC methods have been used for such samples as fortified milk, margarine, butter, cod liver oil and other fish products and an example of a chromatogram obtained from an extract of fish liver containing 17 International Units of vitamin $D_3$ per gram, is given in Fig. 8.3. The vitamin $D_2$ in the sample was added as an internal standard. Table 8.3 summarizes typical HPLC methods for the determination of vitamin D in food.

## C Vitamin E

Vitamin E, as it occurs naturally, consists of eight compounds which belong to two series of methyl-substituted chromanols, with either a saturated (the tocopherols) or unsaturated (the tocotrienols) side chain in the 2-position. Each of these compounds has different vitamin E activities and antioxidant properties, with the result that the concentration of each individual compound must be known before an assessment of the vitamin E and antioxidant activity of a sample can be made. The problem is further complicated by the fact that synthetic isomers and esters, especially of $\alpha$-tocopherol, may be added to foods. The structures and activities of the various compounds with vitamin E activity have been discussed in detail by Parrish (1980).

Numerous methods for the determination of vitamin E have been reported (Bunnell, 1967, 1971; Parrish, 1980). HPLC methods are to be preferred because of their speed, ease of operation and reliability. A minimum of sample preparation is required with the result that losses are kept to a minimum.

### 1 Sample Preparation

Vegetable oils contain tocopherols in the unesterified form (Chow, Draper and Csallany, 1969) and may be injected on to the column directly, without any pretreatment, apart from mixing (van Niekerk, 1973; van Niekerk and du Plessis, 1976) or may be dissolved in the mobile phase or hexane prior to chromatography (Abe, Yuguchi and Katsui, 1975; Eriksson and Toeregaard, 1977; Deldime, Lefebvre, Sadin and Wybauw, 1980). Solid food samples have to be extracted before chromatography (Shaikh, Huang

**Table 8.3** HPLC systems used for vitamin D determinations

| Compounds determined | Type of sample | Sample preparation | Stationary phase | Mobile phase (proportions by volume) | Detection wavelength (nm) | Reference |
|---|---|---|---|---|---|---|
| Vitamins $D_2$ plus $D_3$ | Fortified milk | Saponify at room temperature, extract, HAPS chromatography | 5 $\mu$m LiChrosorb SI 60 (Merck) | Propan-2-ol–hexane (1 : 99) | 265 | Thompson et al. (1977) |
| Vitamin D | Cod liver oil | Saponify at 70 °C, extract, precipitate with digitonin, Florex XXS chromatography | Partisil 10 PXS (Whatman) | Chloroform–hexane –acetic acid (70 : 30 : 1) | 268 | Ali (1978) |
| Vitamins $D_2$ or $D_3$ | Fish products | Saponify with heating, extract, crystallize cholesterol at −18 °C, clean up on HPLC silica column. Vitamin $D_2$ or $D_3$ internal standard | LiChrosorb 10 $RP_{18}$ (Merck) | Methanol–water (95 : 5) | 263 | Egaas and Lambertsen (1979) |

| Compound | Sample | Preparation | Column | Mobile phase | Wavelength (nm) | Reference |
|---|---|---|---|---|---|---|
| Vitamins $D_2$ or $D_3$ | Fortified skim milk and chocolate milk | Saponify at 80 °C, extract with hexane | 10 μm Vydac TP reversed phase $C_{18}$ (Separations Group) | Acetonitrile–methanol (90 : 10) | 265 | Henderson and McLean (1979) |
| Vitamin $D_3$ | Non-fat dried milk | Extract with dichloromethane, Sep-Pak silica cartridge clean-up on HPLC Partisil PAC column | 10 μm LiChrosorb $NH_2$ (Merck) | Dichloromethane–hexane–propan-2-ol (50 : 50 : 0.2) | 264 | Cohen and Wakeford (1980) |
| Vitamins $D_2$ or $D_3$ | Margarine and butter | Saponify with heating, extract, clean-up twice on HPLC silica column. Vitamin $D_2$ or $D_3$ internal standards | 5 μm Nucleosil 5 $C_{18}$ (Machery and Nagel) | Methanol–water (95 : 5) | 264 or 254 | van Niekerk and Smit (1980) |
| 25-Hydroxyvitamin $D_3$ | Egg yolk | Extract with chloroform–methanol. Partition between acetonitrile and hexane, coarse silica column, microparticulate silica column, Celite column | 5 μm Zorbax ODS (Dupont) | Acetonitrile–methanol–water (94 : 3 : 3) | 254 | Koshy and van der Slik (1979) |

and Zielinski, 1977; Hung, Cho and Slinger, 1980; Widicus and Kirk, 1979). Thompson and Hatina (1979) extracted food samples such as spinach, beef and cereals with boiling propan-2-ol. The digest was homogenized, acetone was added and the mixture was filtered. The residue was re-extracted with acetone, and hexane was added to the combined filtrate. Water was added and the phases separated. The aqueous phase was extracted twice with hexane and the pooled hexane extracts were washed twice with water before evaporating under reduced pressure. Extraction efficiency was better than 97%.

Tocopheryl esters present in a sample must be hydrolysed if fluorescence is to be used for detection and this is usually accomplished by saponification. Saponification is also used to free vitamin E for extraction from the sample matrix when it has been added in an encapsulated form, or when the samples are based on homogenized spray-dried milk powder (van Niekerk, unpublished).

The tocopherols are very sensitive to oxygen in the presence of alkali (Bunnell, 1967) and care should be exercised to avoid destruction during saponification. Air may be excluded from the saponification flask by flushing with nitrogen or, better still, by boiling the mixture of the sample plus solvent (ethanol most commonly) under reflux for a few minutes to displace the air with solvent vapours, before adding the potassium hydroxide through the condenser. The mixture is boiled for another 5–10 min before being cooled in ice. Antioxidants such as ascorbic acid or pyrogallol are usually added to the saponification mixture as a further precaution against oxidation losses. The nonsaponifiable fraction may be extracted with diethyl ether, petroleum ether or hexane, but the hydrocarbon solvents have the advantage of having a low solubility for water. When evaporating the solvent it should never be allowed to go to complete dryness (Thompson and Hatina, 1979) and an antioxidant such as BHT may be added.

## 2   Chromatography

The four tocopherols were initially separated by HPLC on a 1.5 m column of pellicular (37–50 $\mu$m) silica gel (van Niekerk, 1973). Cavins and Inglett (1974) improved on this by separating all the tocopherols and tocotrienols in 80 min on a 2 m column of pellicular silica gel. The analysis time was later reduced to less than 10 min by using a column of 10 $\mu$m microparticulate silica gel (van Niekerk, 1975; van Niekerk and du Plessis, 1976) and this is illustrated by the chromatogram in Fig. 8.4. A variety of HPLC methods for vitamin E analysis in foods have since been published and typical conditions are summarized in Table 8.4. Reversed phase systems have been used for vitamin E analyses, especially when other vitamins are

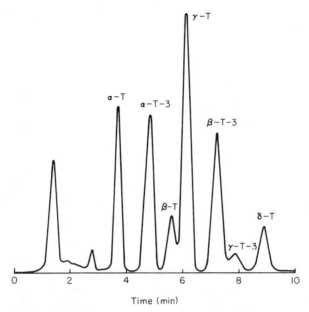

Fig. 8.4  Chromatogram obtained from a mixture of maize, wheat bran and soya bean oil. Stainless steel column (64 cm × 0.4 cm i.d.) slurry packed with Merckosorb SI 60 (10 $\mu$m) with dioxane and *n*-hexane (4 : 96, v/v) as mobile phase (5.4 ml min$^{-1}$) and fluorescence detection (ex. 295 nm, em. 330 nm). $\alpha$-T is $\alpha$-tocopherol, $\beta$-T is $\beta$-tocopherol, $\gamma$-T is $\gamma$-tocopherol, $\delta$-T is $\delta$-tocopherol, $\alpha$-T-3 is $\alpha$-tocotrienol, $\beta$-T-3 is $\beta$-tocotrienol, $\gamma$-T-3 is $\gamma$-tocotrienol.

determined simultaneously (Barnett *et al.*, 1980; Mankel, 1979), but only normal phase systems are able to separate the eight natural forms of vitamin E. To ensure consistent retention times with adsorption columns it is important to control the water content of the mobile phase.

The tocopherols exhibit strong fluorescence and, because of the inherent sensitivity and selectivity of a fluorescence detector, it is the detector of choice for vitamin E determinations. Tocopheryl esters do not fluoresce and an ultraviolet absorption detector must be used if they are not hydrolysed before chromatography. Absorption detectors have been used for the detection of the tocopherols, but it has been reported (Abe *et al.*, 1975; Thompson and Hatina, 1979) that vegetable oils and food extracts contain a number of compounds which interfere with the tocopherol peaks if this type of detector is used. A fluorescence detector does not detect these compounds and, so far, the only substances in lipid extracts which have been confirmed to produce peaks near those of the tocopherols are plastochromanol-8 and butylated hydroxyanisole (Thompson and Hatina,

**Table 8.4** HPLC systems used for vitamin E determinations

| Compounds determined | Type of sample | Sample preparation | Stationary phase | Mobile phase (proportions by volume) | Detector | Reference |
|---|---|---|---|---|---|---|
| $\alpha$-, $\beta$-, $\gamma$-, $\delta$-Tocopherols | Vegetable oils | Dissolve in hexane | Jascopack-wc-03-500 (normal phase partition) | Diisopropyl ether–hexane (2 : 98) | Fluorescence, ex. 298 nm, em. 325 nm; UV 295 nm | Abe et al. (1975) |
| $\alpha$-, $\beta$-, $\gamma$-, $\delta$-Tocopherols, $\alpha$-, $\beta$-, $\gamma$-tocotrienols | Cereals, vegetable oils, milk | Extract with propan-2-ol and acetone, add hexane and water, use epiphase for HPLC. Saponify for $\alpha$-tocopheryl acetate | 5 $\mu$m LiChrosorb SI 60 (Merck) | Diethyl ether–hexane (5 : 95) or propan-2-ol–hexane (0.2 : 99.8) | Fluorescence, ex. 290 nm, em. 330 nm; UV 295 nm | Thompson and Hatina (1979) |
| Tocopherols and tocotrienols | Vegetable oils | Dissolve in mobile phase | 10 $\mu$m Rsil (RSL) | Ethyl acetate–hexane (3 : 97) | Fluorescence, ex. 303 nm, em. 328 nm | Deldime et al. (1980) |
| $\alpha$-Tocopheryl acetate | Cereal products | Extract with chloroform, ethanol plus water | $\mu$-Porasil (Waters) | Chloroform–hexane (15 : 85) | UV 280 nm | Widicus and Kirk (1979) |
| $\alpha$-Tocopherol, $\alpha$-tocopheryl acetate | Milk and soya-based infant formulae | Enzymic hydrolysis of lipids, extract with pentane | Zorbax ODS (Dupont) | Methanol–ethyl acetate–acetonitrile gradient | UV 265 nm | Barnett et al. (1980) |
| Tocopherols | Foods | Saponify with KOH and neutralize with acetic acid | LiChrosorb RP-8 (E.M.) | Methanol–water (95 : 5), pH 4.0, with acetic acid | Fluorescence, ex. 295 nm, em. 330 nm; UV 308 nm | Devries, Egberg and Heroff (1979) |
| $\alpha$-, $\beta$-, $\gamma$-, $\delta$-Tocopherols, $\alpha$-, $\beta$-, $\gamma$-tocotrienols | Vegetable oils | Inject directly | 10 $\mu$m Merckosorb SI 60 (Merck) | Dioxane–hexane (4 : 96) | Fluorescence, ex. 295 nm, em. 330 nm | van Niekerk and du Plessis (1976) |

1979; van Niekerk and du Plessis, 1980). When a number of solvents were tested, the tocopherols fluoresced most strongly in dioxane and diethyl ether (Thompson, Erdody and Maxwell, 1972; van Niekerk, 1975) and the inclusion of these solvents in the mobile phase increases the sensitivity of the detector measurably and a detection limit of 4 ng for $\alpha$-tocopherol has been obtained (Thompson and Hatina, 1979).

## D   Vitamin K

There are a number of compounds which exhibit vitamin K activity. Vitamin $K_1$ or phylloquinone (2-methyl-3-phytyl-1,4-naphthoquinone) is widely distributed in photosynthetic plants. The vitamin $K_2$ group or menaquinones are substituted naphthoquinones with unsaturated side chains of varying lengths and are found in many photosynthetic as well as non-photosynthetic micro-organisms (Green, 1970).

Very few papers on the application of HPLC to the analysis of vitamin K in foods have been published. Thompson, Hatina and Maxwell (1979) analysed a variety of foods, such as peas, cabbage, spinach and milk, for vitamin K, by HPLC. The lipids from foods were extracted with hot propan-2-ol and acetone and the low polarity lipids were obtained by the addition of hexane and water. The extracts were purified by chromatography on one or two hydroxyalkoxypropyl Sephadex columns and further cleaned by preparative HPLC on a silica gel column. The final chromatography was on a LiChrosorb reversed phase column using a water–methanol gradient, and various phylloquinones and menaquinones could be separated. UV detection was used at 262 nm.

Barnett et al. (1980) determined vitamin $K_1$ together with vitamins A, D and E in infant formulae and dairy products. The lipid components of the sample were hydrolysed enzymically and extracted with pentane after precipitation of the fatty acids as metal salts. Separation was achieved with a methanol–ethyl acetate–acetonitrile gradient on two Zorbax ODS columns in series. Detection was accomplished using a microprocessor-controlled variable wavelength UV detector. Vitamin $K_1$ was detected at 265 nm.

## III   THE WATER-SOLUBLE VITAMINS

The water-soluble vitamins (Kutsky, 1973) consist, in general, of derivatives or substituted derivatives of sugars (vitamin C), pyridine (niacin, vitamin $B_6$), purines and pyrimidines (folic acid, vitamin $B_2$, vitamin $B_1$), amino acid–organic acid complexes (folic acid, biotin, pantothenic acid) and a porphyrin–nucleotide complex (vitamin $B_{12}$).

The application of HPLC to the determination of water-soluble vitamins in foods has received far less attention than is the case with the fat-soluble vitamins. Consequently, for certain vitamins, no HPLC methods for food samples have to date been reported and, therefore, only vitamins $B_1$, $B_2$, $B_6$, C, niacin and folic acid will be discussed in this section.

Despite the smaller interest, HPLC has made a definite impact on the analysis of water soluble vitamins and for some vitamins ($B_1$ and $B_2$ especially) HPLC methods are applied routinely in a number of laboratories.

## A  Vitamin $B_1$

The principle form of thiamine (vitamin $B_1$) in animal products is the pyrophosphate ester, but in plant materials it is largely present in the free form (Pearson, 1967a). Other phosphate esters may also be present in small amounts in food samples.

### 1  Sample Preparation

Thiamine is, to some extent, protein-bound and a mild acid hydrolysis (thiamine is heat labile in alkaline medium) is normally used to free it. This is done by autoclaving the sample in $c$. 0.1 M sulphuric or hydrochloric acid for approximately 30 min. If phosphate esters are present they have to be converted to the free form by treatment with an enzyme containing phosphatase activity such as takadiastase at a pH between 4 and 4.6. In the case of plant materials the takadiastase has the additional function of hydrolysing the starch present in the sample and in this way speeding up the filtration step. For the extraction of meat and meat products, papain has been used in addition to takadiastase (Ang and Mosely, 1980; van de Weerdhof et al., 1973).

### 2  Chromatography

Thiamine has been chromatographed on silica gel columns (van de Weerdhof et al., 1973) or on reversed phase columns with an ion-pairing reagent in the mobile phase (Henshall, 1979; Toma and Tabekhia, 1979; Kamman, Labuza and Warthesen, 1980). Ion-pairing reagents that have been used are sodium salts of pentane-, hexane- or heptane-sulphonic acid, at concentrations of $c$. 0.005 M. The position of the thiamine peak, relative to the other peaks in the chromatogram, is determined by the properties of the ion-pairing reagent. One of the ion-pairing reagents, or a combination of them, is usually chosen to ensure that the thiamine is adequately separated from interfering substances. Interfering substances may also be eliminated by size exclusion chromatography on LiChrosphere SI 100 columns connected in series with the analytical column (Henshall, 1979). Chromatographic conditions for the analysis of vitamin $B_1$ in foods are summarized in Table 8.5.

A UV detector at 254 nm may be used for samples that contain sufficient amounts of thiamine but Ang and Moseley (1980) found it inadequate for the low levels (0.062–0.486 mg per 100 g) that are present in meat or meat products. They converted thiamine to thiochrome prior to chromatography and used a fluorescence detector which has the necessary sensitivity and selectivity for samples that are low in thiamine. Thiamine may also be converted to thiochrome, after chromatography, by continuously mixing the mobile phase with an alkaline solution of potassium ferricyanide (van de Weerdhof et al., 1973; Gubler and Hemming, 1979; Kimura, Fujita, Nishida and Itokawa, 1980). In this way the advantages of the fluorescence detector are obtained, while the additional labour and variability of the manual conversion to thiochrome are eliminated.

## B  Vitamin B₂

The principal forms of riboflavin (vitamin $B_2$) found in nature are riboflavin 5'-phosphate and flavin adenine dinucleotide, both of which are protein bound. Free riboflavin occurs rarely in nature (Pearson, 1967b). Vitamin $B_2$ is easily destroyed at an alkaline pH and is sensitive to light in the blue and violet regions. Special precautions should, therefore, be taken to avoid these conditions.

### 1  Sample Preparation

Extraction with hot dilute acids (c. 0.1 M sulphuric or hydrochloric acid) suffices to split vitamin $B_2$–protein compounds but the phosphoric acid esters of riboflavin can only be hydrolysed completely by means of an enzyme (diastase) (Strohecker and Henning, 1965). The enzyme has the additional function of hydrolysing the starch present in the sample and so ensuring complete extraction of vitamin $B_2$ and faster filtration. Wiggins (1979b) has found that several commercially available forms of takadiastase failed to convert riboflavin 5'-phosphate quantitatively to riboflavin and for that reason acid phosphatase is now used for this purpose.

### 2  Chromatography

Although ion-exchangers (Floridi, Fini, Palmerini and Rossi, 1976; Williams, Baker and Schmit, 1973) and silica gel (van de Weerdhof et al.,

**Table 8.5** HPLC systems used for thiamine analysis

| Compounds determined | Type of sample | Sample preparation | Stationary phase | Mobile phase (proportions by volume) | Detector | Reference |
|---|---|---|---|---|---|---|
| Thiamine | Meat, potatoes | Extract with 0.25M $H_2SO_4$, heat at 120 °C, incubate with takadiastase, incubate with papain, add tri-chloroacetic acid, heat and filter | 20–30 $\mu$m silica gel | 0.1M phosphate buffer (pH 6.8)–ethanol (90 : 10) | Post-column oxidation with potassium ferri-cyanide; fluo-rescence, ex. 362 nm, em. 464 nm | Van de Weerdhof et al. (1973) |
| Thiamine | Fortified meat | Extract with 0.1M $H_2SO_4$, autoclave, pH adjusted to 4–6. In-line clean-up with LiChrosorb SI 100 columns | 10 $\mu$m LiChrosorb RP 8 (Merck) | $5 \times 10^{-3}$M sodium hexane sulphonate –5% acetic acid–tetrahydrofuran (16 : 79 : 5) | UV 254 nm | Henshall (1979) |

| Compound | Sample | Extraction/preparation | Column | Mobile phase | Detection | Reference |
|---|---|---|---|---|---|---|
| Thiochrome (from oxidized thiamine) | Meat and meat products | Extract with 0.1 M HCl autoclave. Incubate with taka-diastase and papain at pH 4–4.5, precipitate with tri-chloroacetic acid, oxidize with potassium ferricyanide, extract with butanol | 20 $\mu$m Spherisorb silica gel | Chloroform–methanol (90 : 10) | Fluorescence, ex. 367 nm, em. filter KV 418 (Schoeffel) | Ang and Moseley (1980) |
| Thiamine | Rice and rice products | Extract with 0.05 M $H_2SO_4$, autoclave. Incubate with taka-diastase and papain | $\mu$-Bondapak $C_{18}$ (Waters) | Methanol–acetic acid–water (39 : 1 : 60) plus 25 ml each of PIC A and PIC 7 (Waters) | UV 254 nm | Toma and Tabekhia (1979) |
| Thiamine | Enriched cereal products | Extract with 0.1 M HCl, autoclave, filter | $\mu$-Bondapak $C_{18}$ (Waters) | Acetonitrile–0.01 M phosphate buffer, pH 7 (12.5 : 87.5) containing $5 \times 10^{-3}$ M sodium heptane sulphonate | UV 254 nm | Kamman et al. (1980) |

210 P. J. VAN NIEKERK

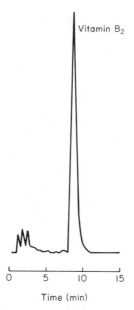

**Fig. 8.5** Chromatogram of an extract of Mopani worms (an African delicacy) containing 4 mg riboflavin per 100 g. Stainless steel Micro Pak MCH 10 column (30 cm × 0.4 cm i.d.), with 1% acetic acid and methanol (70 : 30, v/v) as mobile phase (1.5 ml min$^{-1}$) and fluorescence detection (ex. 449 nm, em. 520 nm).

1973; Richardson, Favell, Gidley and Jones, 1978) have been used for the chromatography of riboflavin, reversed phase columns are to be preferred (Fig. 8.5). Polar compounds, which may bind irreversibly to silica gel, are not held back by reversed phase materials and are eluted before the riboflavin. Any non-polar compounds which may remain on the reversed phase column are easily removed by flushing with methanol or acetonitrile once a day. Rouseff (1979) separated riboflavin and riboflavin 5'-phosphate in orange juice with an ionic strength gradient on a Partisil-10 PAC column. Conditions for the chromatography of riboflavin are summarized in Table 8.6.

Riboflavin has a strong inherent fluorescence and this property allows it to be detected very specifically and with high sensitivity. Although it has its maximum intensity of fluoresence at a pH between 6 and 7, it is better to measure it in the pH range 3–5, where the intensity–pH curve is horizontal (Strohecker and Henning, 1965). For this reason the mobile phase should be buffered at a pH in this range. UV detection at 254 nm has been used for riboflavin (Toma and Tabekhia, 1979; Kamman *et al.*, 1980) but is subject to interference and only suitable for samples that are sufficiently

**Table 8.6** Conditions for HPLC of vitamin $B_2$

| Compounds determined | Type of sample | Sample preparation | Stationary phase | Mobile phase (proportions by volume) | Detector | Reference |
|---|---|---|---|---|---|---|
| Riboflavin, riboflavin 5'-phosphate | Citrus juice | Add metaphosphoric acid and centrifuge | Partisil 10 PAC (Whatman) | Ionic strength gradient with propan-2-ol–acetate buffer, pH 4 | Fluorescence, ex. 450 nm, em. 520 nm | Rouseff (1979) |
| Lumiflavin (from irradiated riboflavin) | Meat and meat products | Extract with 0.1M HCl, autoclave, incubate with taka-diastase + papain (pH 4–4.5), precipitate with trichloro-acetic acid, irradiate with UV light in alkaline solution, extract with chloroform | 20 $\mu$m Spherisorb silica | Chloroform–methanol (90 : 10) | Fluorescence, ex. 450 nm, em. filter KV 418 (Schoeffel) | Ang and Moseley (1980) |
| Riboflavin | Components of meals | Extract with 0.125M $H_2SO_4$, autoclave, incubate with taka-diastase (pH 4.6), filter | 10 $\mu$m silica gel | 0.1M Sodium acetate buffer, pH 4.6 | Fluorescence, ex. 457 nm, em. 510 nm | Richardson et al. (1978) |

**Table 8.6** Continued

| Compounds determined | Type of sample | Sample preparation | Stationary phase | Mobile phase (proportions by volume) | Detector | Reference |
|---|---|---|---|---|---|---|
| Riboflavin | Rice and rice products | Extract with 0.05M $H_2SO_4$, autoclave, incubate with taka-diastase + papain (pH 4.5), filter | $\mu$-Bondapak $C_{18}$ (Waters) | Methanol–acetic acid–water (39 : 1 : 60) plus 25 ml each of PIC 5 and PIC 7 (Waters) | UV 254 nm | Toma and Tabekhia (1979) |
| Riboflavin | Cereals and other foods | Extract with acid, incubate with acid phosphatase (pH 5.6) | 5 $\mu$m Spherisorb ODS | Methanol–citrate buffer, pH 5.8 (30 : 70) | Fluorescence, ex. 449 nm, em. 520 nm | Wiggins (1979b) |
| Riboflavin | Milk | Add trichloroacetic acid, centrifuge | ODS Sil-X-1 (Perkin Elmer) | Acetonitrile–water (6 : 94) | Fluorescence, ex. 453 nm, em. 520 nm | Williams and Slavin (1977) |
| Riboflavin | Fortified cereals | Extract with 0.1M HCl, autoclave, centrifuge, filter | $\mu$-Bondapak $C_{18}$ (Waters) | Acetonitrile–0.01M phosphate buffer, pH 7 with $5 \times 10^{-3}$M sodium heptane sulphonate (12.5 : 87.5) | UV 254 nm | Kamman et al. (1980) |

high in vitamin $B_2$. Ang and Moseley (1980) converted riboflavin to lumi-flavin by irradiating with ultraviolet light in alkaline medium before chromatography, followed by fluorescence detection.

## C Vitamin B₆

Vitamin $B_6$ occurs in foods mainly as pyridoxine (pyridoxol, PN), pyridoxal (PL), pyridoxamine (PM) and their corresponding phosphates. The physiological forms in animals are pyridoxal-5-phosphate (PLP) and pyridoxamine phosphate (PMP). In plants the physiological forms are pyridoxine-5-phosphate (PNP), pyridoxal-5-phosphate and pyridoxamine phosphate (Kutsky, 1973). Vitamin $B_6$ is sensitive to light in alkaline medium.

### 1 *Sample Preparation*

Vitamin $B_6$ may be extracted from foods by heating (autoclaving) the sample in dilute (0.055–0.4 M) hydrochloric acid (Sauberlich, 1967; Yasumoto, Tadera, Tsuji and Mitsuda, 1975). The phosphates of the vitamin $B_6$ are then converted to the free forms by incubating the sample with enzymes such as acid phosphatase and diastase (Wong, 1978; Gregory and Kirk, 1978). In this way the number of compounds that have to be separated and determined are reduced. Extracts may be further purified by ion-exchange chromatography (Wong, 1978; Yasumoto *et al.*, 1975), with a Sep-Pak $C_{18}$ cartridge (Lim, Young and Driskell, 1980) or two size exclusion columns placed in series with the analytical column (Henshall, 1979).

Vanderslice, Maire, Doherty and Beecher (1980) extracted vitamin $B_6$ with sulphosalicylic acid. This reagent preserves the phosphate forms while still releasing the vitamins bound to other food components. The extracts were further purified on an ion-exchange column before being analysed for the complete range of vitamin $B_6$ compounds present in foods.

### 2 *Chromatography*

Pyridoxine, pyridoxal and pyridoxamine have been separated on ion-exchange columns (Yasumoto *et al.*, 1975; Wong, 1978; Floridi *et al.*, 1976; Williams and Cole, 1975) and reversed phase columns (Lim *et al.*, 1980; Gregory and Kirk, 1978). Vanderslice, Stewart and Yarmas (1979) separated pyridoxine, pyridoxal, pyridoxamine and their corresponding phosphates on an ion-exchange column with a two-step gradient or on two ion-exchange columns and a single mobile phase with column switching (Vanderslice and Maire, 1980; Vanderslice *et al.*, 1980). Table 8.7 summarizes conditions for the chromatography of vitamin $B_6$.

**Table 8.7** HPLC conditions for vitamin $B_6$

| Compounds determined | Type of sample | Sample preparation | Stationary phase | Mobile phase (proportions by volume) | Detector | Reference |
|---|---|---|---|---|---|---|
| PL, PN, PM | Rice bran, wheat flour, rat liver | Extract with 0.4M HCl, autoclave, centrifuge, concentrate, purify on ion-exchange column | Aminex A-5 (Bio-Rad) | Step gradient with phosphate buffer–propanol mixture (98 : 2) | Post-column reaction with the diazide of 5-chloroaniline 2,4-disulphonyl chloride, monitor at 440 nm | Yasumoto et al. (1975) |
| PL, PN, PM | Carrots, enriched extracts of apples, pears, spinach | Extract with ethanol, add 0.1M HCl and heat. Incubate with diastase and papain, purify on cation-exchanger | Zipax SCX (Dupont) | 0.1M phosphate buffer, pH 4.35 | UV 210 nm | Wong (1978) |

| Compounds | Sample | Preparation | Column | Mobile phase | Detection | Reference |
|---|---|---|---|---|---|---|
| PL, PN, PM | Skim milk | Milk extract cleaned by Sep-Pak $C_{18}$ cartridge | 10 $\mu$m Spherisorb ODS | 0.033M phosphate buffer, pH 2.2–acetonitrile (99 : 1) | UV 280 nm | Lim et al. (1980) |
| PL, PN, PM | Dehydrated model foods, fortified breakfast cereals | Extract with 0.2M potassium acetate, pH 4.5, sonicate, centrifuge, incubate with acid phosphatase, add trichloroacetic acid, centrifuge | $\mu$-Bondapak $C_{18}$ (Waters) | 0.033M potassium phosphate buffer, pH 2.2 | Fluorescence, ex. 295 nm, em. 370 nm | Gregory (1980) |
| PL, PN, PM, PLP, PNP, PMP | Breakfast cereals, non-fat dry milk, carp, pork, hamburger | Extract with sulpho-salicylic acid in omni-mixer, add methylene chloride, mix, centrifuge, filter water layer, clean up on ion-exchanger | Bio-Rad A-25 (column switching) | 0.04M NaCl, 0.01M glycine, 0.005M semi-carbazide, to pH 10 with NaOH | Fluorescence, ex. 310 nm, em. 380 nm for PMP, PM, PNP, PN; ex. 280 nm, em. 487 nm, for PLP and PL | Vanderslice, et al. (1980) |

UV detectors have been used for the detection of vitamin $B_6$ compounds at wavelengths that vary from 210 to 280 nm (Wong, 1978; Henshall, 1979; Lim et al., 1980). Fluorescence detectors, however, are reported to be more specific and sensitive with detection limits down to 0.1 ng ml$^{-1}$ (Gregory, 1980; Vanderslice et al., 1979, 1980). Yasumoto et al. (1975) detected vitamin $B_6$ compounds with the diazide of 5-chloroaniline 2,4-disulphonyl chloride as a colorimetric reagent in a post-column reaction detector.

## D  Niacin

Niacin (nicotinic acid) is present in foods as free niacin and niacinamide (nicotinamide) or as niacinamide bound to nucleotides. It is one of the most stable of the water-soluble vitamins and is resistant to heat, light, oxidation, acids and alkalis.

### 1  Sample Preparation

Niacinamide and its naturally occurring derivatives can be hydrolysed to free nicotinic acid either by acids or alkali. Alkaline hydrolysis is usually complete within a much shorter time than acid hydrolysis (Strohecker and Henning, 1965). Acid hydrolysis is, therefore, normally supplemented by incubating the sample with takadiastase and papain (Osborne and Voogt, 1978; Toma and Tabekhia, 1979). Acid extracts have been purified by two LiChrosorb SI 100 columns connected in series with an analytical column (Henshall, 1979).

Tyler and Shrago (1980) extracted niacin from cereals with a calcium hydroxide solution in an autoclave and purified the extracts by chromatography on an anion-exchange column, by oxidizing with potassium permanganate, and by passing them through a Sep-Pak $C_{18}$ cartridge.

### 2  Chromatography

Niacin has been chromatographed on a silica gel column with an acetate buffer as mobile phase (Osborne and Voogt, 1978) but, more often, reversed phase columns are used with ion-pairing reagents in the mobile phase (Wills, Shaw and Day, 1977; Henshall, 1979; Toma and Tabekhia, 1979; Tyler and Shrago, 1980). Since niacin has an acid as well as a basic group in the molecule, the ion-pairing reagent may either be an alkyl sulphonate, or a quaternary ammonium compound. The chromatographic conditions which are used for the determination of niacin in foods are summarized in Table 8.8.

UV detection at 254 nm is mostly used for niacin but Osborne and

**Table 8.8** HPLC conditions for niacin

| Compounds determined | Type of sample | Sample preparation | Stationary phase | Mobile phase | Detector | Reference |
|---|---|---|---|---|---|---|
| Niacin | Foods generally | Extract with 0.125M $H_2SO_4$, autoclave, incubate with taka-diastase, treat with papain, add tri-chloroacetic acid, centrifuge | 10 $\mu$m Merckosorb SI 60 (Merck) | Acetate buffer containing 2.72% (m/v) sodium acetate and 1.2% (m/v) acetic acid | React with cyanogen bromide and $p$-aminoaceto-phenone; fluorescence detection, ex. 435 nm, em. >500 nm | Osborne and Voogt (1978) |
| Niacin | Rice | Extract with 0.05M $H_2SO_4$, autoclave, incubate with taka-diastase + papain, filter | $\mu$-Bondapak $C_{18}$ (Waters) | Methanol–acetic acid–water (39 : 1 : 60, v/v/v) plus 25 ml each PIC 5 and PIC 7 (Waters) | UV 254 nm | Toma and Tabekhia (1979) |
| Niacin | Cereals | Extract with calcium hydroxide solution, autoclave, purify on anion-exchanger, oxidize with potassium permanganate, clean with Sep-Pak cartridge | $\mu$-Bondapak $C_{18}$ (Waters) | Methanol–water (5 : 95, v/v) with PIC A (Waters) | UV 254 nm | Tyler and Shrago (1980) |

Voogt (1978) employed a post-column reaction detector. The niacin is detected by fluorescence after reaction with cyanogen bromide and *p*-aminoacetophenone.

## E  Folates

The parent compound is pteroylglutamic acid (folic acid) but the naturally occurring folates are tetrahydrofolates which usually exist as conjugates with up to six molecules of glutamic acid, linked at the gamma carbon, incorporated in the structure. Most of the folates carry a one-carbon unit (e.g. formyl, hydroxymethyl, methyl and formimino) attached to the $N^5$ or $N^{10}$ positions or linked across both positions. The folates are sensitive to heat, strong acids, oxidation and light. They are, therefore, difficult to extract from foods without some oxidation or deconjugation taking place.

Clifford and Clifford (1977) determined tetrahydrofolic acid, $N^5$-methyltetrahydrofolic acid, dihydrofolic acid and folic acid in fruit juices and nuts by HPLC. Samples were extracted with a potassium phosphate buffer (pH 7.5), containing mercaptoethanol as a reducing agent, by heating for 3 min in a boiling water bath followed by rapid cooling on ice. The centrifuged extracts were chromatographed on a pellicular strong anion-exchange resin (Pellionex SAX) with a potassium chloride gradient in phosphate buffer at pH 7.5. They used UV detection at 280 nm.

Reed and Archer (1976) chromatographed folic acid derivatives on a pellicular weak anion-exchange column (AL-Pellionex WAX) with a phosphate buffer (pH 4.8) as mobile phase. Allen and Newman (1980) and Reingold, Picciano and Perkins (1980) separated folates on reversed phase columns with an ion-pairing agent (tetrabutylammonium phosphate) in the mobile phase. Rouseff (1979) used HPLC with dual detectors (fluorescence and UV) to demonstrate that folic acid, contrary to published reports, does not fluoresce. All fluorescence appeared to be associated with impurities present in standards.

## F  Vitamin C

The vitamin C activity of foods may be derived from both L-ascorbic acid and its oxidation product L-dehydroascorbic acid. Dehydroascorbic acid is unstable and readily converts to diketogulonic acid which does not possess vitamin C acitivity. As opposed to diketogulonic acid, dehydroascorbic acid is easily reduced back to ascorbic acid by reducing agents such as hydrogen sulphide and homocysteine (Hughes, 1956). D-Ascorbic acid (isoascorbic acid, erythorbic acid) has a biological activity which is about

20 times less than that of L-ascorbic acid. It does not occur in natural products but is sometimes added to foods as an antioxidant.

Vitamin C is sensitive to heat, alkali, oxygen and light.

## 1  Sample Preparation

Dilute solutions (3–6%) of oxalic or metaphosphoric acid are generally used for the extraction of vitamin C from foods. These two acids are chosen because of their superior ability to prevent catalysis of the oxidation of ascorbic acid by cupric or ferric ions. Although the two acids are about equal in this respect, metaphosphoric acid is the more widely applicable reagent because of its ability to precipitate proteins and inactivate ascorbic acid oxidase (Roe, 1967). The sample is normally homogenized with the extracting solution and then centrifuged or filtered before chromatography. Pachla and Kissinger (1976) diluted the extract with 0.05 M cold perchloric acid in order to reduce the ionic strength of the solution used for injection.

To determine the total vitamin C content of a sample, the dehydro-ascorbic acid may be reduced to ascorbic acid by homocysteine, at pH 6.8, before chromatography (van Niekerk, Smit and Strydom, 1981). The difference in the concentration of ascorbic acid before and after reduction is an indication of the dehydroascorbic acid content of the sample. This method gives results that agree well with results obtained by the fluorometric method of Deutsch and Weeks (1965) for ascorbic plus dehydroascorbic acid.

## 2  Chromatography

Ascorbic acid may be chromatographed on pellicular (Pachla and Kissinger, 1976) or microparticulate (Rouseff, 1979) strong anion-exchange columns, but reversed phase systems (Augustin, Beck and Marousek, 1981; Pachla and Kissinger, 1979; Sood, Sartori, Wittmer and Haney, 1976) are often preferred. Ion-pairing reagents such as quaternary ammonium compounds and tertiary amines are added to the mobile phase in order to retain ascorbic acid on reversed phase columns. The properties of the ion-pair reagent determines the relative retention of ascorbic acid and other components in the sample, and, by proper selection of the ion-pair reagent, the chromatographic separation of ascorbic acid from interfering compounds may be optimized.

Vitamin C is stable in acid solutions only. The mobile phase should, therefore, be at a pH of 5 or lower. Traces of metals (especially iron and copper) should also be removed from the mobile phase since they can catalyse the oxidation of ascorbic acid during chromatography. By degas-

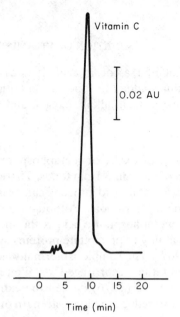

**Fig. 8.6** Chromatogram of an orange juice sample. Stainless steel column (25 cm × 0.4 cm i.d.) slurry packed with Nucleosil 5 $C_{18}$, with a mobile phase consisting of 0.005% tetrahexylammonium hydrogen sulphate and 0.006% acetic acid, adjusted to pH 5 with sodium hydroxide (1.5 ml min$^{-1}$) and UV detection at 264 nm.

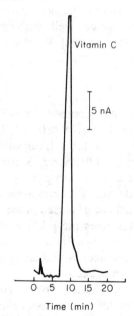

**Fig. 8.7** The same chromatogram as in Fig. 8.6 but using an electrochemical detector with a glassy carbon electrode at 0.8 V versus the Ag/AgCl electrode.

**Table 8.9** HPLC conditions for vitamin C analysis

| Compounds determined | Type of sample | Sample preparation | Stationary phase | Mobile phase | Detector | Reference |
|---|---|---|---|---|---|---|
| Ascorbic acid | Fruits, fruit juices, infant formulae | Extract with 6% metaphosphoric acid, homogenize, filter | μ-Bondapak C$_{18}$ (Waters) | Methanol–water (50 : 50, v/v) with $10^{-3}$M tridecylammonium formate (pH 5) | UV 254 nm | Sood et al. (1976) |
| Ascorbic acid | Fruits, infant foods, milk | Extract with 3% metaphosphoric acid + 8% acetic acid, centrifuge, dilute with $5 \times 10^{-2}$M perchloric acid | LiChrosorb RP 18 (Merck) | Methanol–$8 \times 10^{-2}$M acetate buffer + $10^{-3}$M tridecylamine (15 : 85, v/v), pH 4.5 | Electrochemical with carbon paste electrode at 800 mV versus Ag/AgCl electrode | Pachla and Kissinger (1979) |
| Ascorbic acid | Orange juice | Add metaphosphoric acid, centrifuge | 10 μm strong anion-exchange (Dupont) | Gradient with boric acid buffer (pH 8) | UV 254 nm | Rouseff (1979) |
| Ascorbic acid | Potatoes and potato products | Extract with 6% metaphosphoric acid, homogenize, filter | μ-Bondapak C$_{18}$ (Waters) | $5 \times 10^{-4}$M tridecylammonium formate, pH 4.5–methanol (40 : 60, v/v) | UV 254 nm | Augustin et al. (1981) |
| Ascorbic acid plus dehydroascorbic acid | Tomatoes, green pepper, orange juice | Extract with 3% metaphosphoric + 8% acetic acid, homogenize, filter, reduce with homocysteine (pH 6.8), add metaphosphoric acid, filter | Nucleosil 5 C$_{18}$ (Machery and Nagel) | $5 \times 10^{-3}$% (m/v) tetrahexylammonium hydrogen sulphate plus $6 \times 10^{-3}$% (m/v) acetic acid, to pH 5 with sodium hydroxide | UV 264 nm or electrochemical with glassy carbon electrode at 800 mV versus Ag/AgCl electrode | van Niekerk et al. (1981) |

sing the mobile phase and preventing the re-entry of oxygen, through sparging with an inert gas, the destruction of ascorbic acid may further be prevented. The destruction of ascorbic acid during chromatography is usually evidenced by a calibration curve that does not pass through zero.

L-Ascorbic acid and D-ascorbic acid have been separated on a 10 $\mu$m LiChrosorb $NH_2$ column with a mobile phase consisting of 75% acetonitrile in a $5 \times 10^{-3}$ M phosphate buffer at a pH between 4.4 and 4.7 (Bui-Nguyên, 1980). Tweeten (1979) used the same stationary phase to separate ascorbic acid from dehydroascorbic acid with a gradient from 80 to 40% acetonitrile in $5 \times 10^{-3}$ M phosphate buffer at pH 4. The application of this separation to food samples should allow the determination of total vitamin C, an aspect which is neglected by most HPLC methods.

A UV detector set at 254 or 264 nm is usually used, but it has been reported that an electrochemical detector affords better sensitivity and selectivity for ascorbic acid (Pachla and Kissinger, 1979). The use of UV and electrochemical detectors is illustrated by the chromatograms in Figs. 8.6 and 8.7. A sample of orange juice was chromatographed on a reversed phase column using a UV and electrochemical detector in series. Figure 8.6 shows the trace of the UV detector and Fig. 8.7 the trace of the electrochemical detector. For this type of sample either of the two detectors may be used, but when samples are analysed that are much lower in vitamin C, the electrochemical detector gives a more stable base-line.

Chromatographic conditions that can be used for vitamin C analyses are summarized in Table 8.9.

## REFERENCES

Abe, K., Yuguchi, Y. and Katsui, G. (1975). *J. Nutr. Sci. Vitaminol.* 21, 183.

Adachi, A. and Kobayashi, T. (1979). *J. Nutr. Sci. Vitaminol.* 25, 67.

Aitzetmüller, K., Pilz, J. and Tasche, R. (1979). *Fette Seifen Anstrich mittel.* 81, 40.

Ali, S. L. (1978). *Fresenius Z. analyt. Chem.* 293, 131.

Allen, B. A. and Newman, R. A. (1980). *J. Chromat.* 190, 241.

Ang, C. Y. W. and Moseley, F. A. (1980). *J. agric. Fd. Chem.* 28, 483.

Antalick, J. P., Debruyne, H. and Faugere, J. G. (1977). *Annls Falsif. Expert. chim.* 70, 497.

Association of Official Analytical Chemists (1975). In *Official Methods of Analysis* (W. Horwitz, ed.), Association of Official Analytical Chemists, Washington, D.C., p. 816.

Augustin, J., Beck, C. and Marousek, G. I. (1981). *J. Fd. Sci.* 46, 312.

Barnett, S. A., Frick, L. W. and Baine, H. M. (1980). *Analyt. Chem.* 52, 610.

Bauernfeind, J. C., Brubacher, G. B., Kläui, H. M. and Marusich, W. L. (1971). In *Carotenoids* (O. Isler, ed.), Birkhäuser Verlag, Basle and Stuttgart, p. 744.

Borsje, B., Craenen, H. A. H., Esser, R. J. E., Mulder, F. J. and de Vries, E. J. (1978). *J. Ass. off. analyt. Chem.* 61, 122.

Botey Serra, J. and Garcia Fite, D. (1975). *Afinidad.* **32**, 249.
Bui-Nguyên, M. H. (1980). *J. Chromat.* **196**, 163.
Bui-Nguyên, M. H. and Blanc, B. (1980). *Experientia* **36**, 374.
Bunnell, R. H. (1967). In *The Vitamins*, 2nd edn, Vol. VI (P. György and W. N. Pearson, eds), Academic Press, New York and London, p. 261.
Bunnel, R. H. (1971). *Lipids* **6**, 245
Calabro, G., Micali, G. and Curro, P. (1978). *Atti-Conv. naz. Olii Essenz. Deriv. Agrum.* **7**, 171.
Cavins, J. F. and Inglett, G. E. (1974). *Cereal Chem.* **51**, 605.
Chow, C. K., Draper, H. H. and Csallany, A. S. (1969). *Analyt. Biochem.* **32**, 81.
Christie, A. A. and Wiggins, R. A. (1978). In *Developments in Food Analysis Techniques*, Vol. 1 (R. D. King, ed.), Applied Science Publishers, London, p. 1.
Clifford, C. K. and Clifford, A. J. (1977). *J. Ass. off. analyt. Chem.* **60**, 1248.
Cohen, H. and Wakeford, B. (1980). *J. Ass. off. analyt. Chem.* **63**, 1163.
Deldime, P., Lefebvre, G., Sadin, Y. and Wybauw, M. (1980). *Revue fr. Corps Gras* **27**, 279.
Deutsch. M. J. and Weeks, C. E. (1965). *J. Ass. off. agric. Chem.* **48**, 1248.
Devries, J. W., Egberg, D. C. and Heroff, J. C. (1979). In *Liquid Chromatographic Analysis of Food and Beverages*, Vol. 2 (G. Charalambous, ed.), Academic Press, New York and London, p. 477.
Egaas, E. and Lambertsen, G. (1979). *Int. J. Vitam. Nutr. Res.* **49**, 35.
Egberg, D. C., Heroff, J. C. and Potter, R. H. (1977). *J. agric. Fd Chem.* **25**, 1127.
Eriksson, A. and Toeregaard, B. (1977). *Conference: Scand. Symp. Lipids (Proc.)* **9**, 45.
Floridi, A., Fini, C., Palmerini, C. A. and Rossi, A. (1976). *Riv. Sci. Technol. Alimenti Nutr. Um.* **6**, 197.
Green, J. (1970). In *Fat-soluble Vitamins* (R. A. Morton, ed.), Pergamon Press, New York, p. 71.
Gregory, J. F. (1980). *J. agric. Fd Chem.* **28**, 486.
Gregory, J. F. and Kirk, J. R. (1978). *J. Fd Sci.* **43**, 1801.
Gubler, C. J. and Hemming, B. C. (1979). *Methods Enzymol.* **62**, 63.
Hanewald, K. H., Mulder, F. J. and Keuning, K. J. (1968). *J. Pharm. Sci.* **57**, 1308.
Henderson, S. K. and McLean, L. A. (1979). *J. Ass. off. analyt. Chem.* **62**, 1358.
Henderson, S. K. and Wickroski, A. F. (1978). *J. Ass. off. analyt. Chem.* **61**, 1130.
Henshall, A. (1979). In *Liquid Chromatographic Analysis of Food and Beverages*, Vol. 1 (G. Charalambous, ed.), Academic Press, New York and London, p. 31.
Hofsass, H., Grant, A., Alicino, N. J. and Greenbaum, S. B. (1976). *J. Ass. off. analyt. Chem.* **59**, 251.
Hughes, R. E. (1956). *Biochem. J.* **64**, 203.
Hung, S. S. O., Cho, Y. C. and Slinger, S. J. (1980). *J. Ass. off. analyt. Chem.* **63**, 889.
Kamman, J. F., Labuza, T. P. and Warthesen, J. J. (1980). *J. Fd Sci.* **45**, 1497.
Kimura, M., Fujita, T., Nishida, S. and Itokawa, Y. (1980). *J. Chromat.* **188**, 417.
Kodicek, E. and Lawson, D. E. M. (1967). In *The Vitamins*, 2nd edn, Vol. VI (P. György and W. N. Pearson, eds), Academic Press, New York and London, p. 211.
Koshy, K. T. and van der Slik, A. L. (1977). *J. agric. Fd Chem.* **25**, 1246.
Koshy, K. T. and van der Slik, A. L. (1979). *J. agric. Fd Chem.* **27**, 180.
Kutsky, R. J. (1973). *Handbook of Vitamins and Hormones*, Van Nostrand Reinhold, New York.

Landen, W. O. (1980). *J. Ass. off. analyt. Chem.* **63**, 131.
Landen, W. O. and Eitenmuller, R. R. (1979). *J. Ass. off. analyt. Chem.* **62**, 283.
Liaaen-Jensen, S. (1971). In *Carotenoids* (O. Isler, ed.), Birkhäuser Verlag, Basle and Stuttgart, p. 63.
Lim, K. L., Young, R. W. and Driskell, J. A. (1980). *J. Chromat.* **188**, 285.
Macrae, R. (1980). *J. Fd Technol.* **15**, 93.
Mankel, A. (1979). *Dt. Lebensmitt Rdsch.* **75**, 77.
Morton, R. A. (1970). In *Fat-soluble Vitamins* (R. A. Morton, ed.), Pergamon Press, New York, p. 27.
Osborne, D. R. and Voogt, P. (1978). *The Analysis of Nutrients in Food*, Academic Press, London, New York and San Francisco.
Pachla, L. A. and Kissinger, P. T. (1976). *Analyt. Chem.* **48**, 364.
Pachla, L. A. and Kissinger, P. T. (1979). *Methods Enzymol.* **62**, 15.
Parrish, D. B. (1977). *CRC crit. Rev. Fd Sci. Nutr.* **9**, 375.
Parrish, D. B. (1979). *CRC crit. Rev. Fd Sci. Nutr.* **12**, 29.
Parrish, D. B. (1980). *CRC crit. Rev. Fd Sci. Nutr.*, **13**, 161.
Pearson, W. N. (1967a). In *The Vitamins*, 2nd edn, Vol. VII (P. György and W. N. Pearson, eds), Academic Press, New York and London, p. 53.
Pearson, W. N. (1967b). In *The Vitamins*, 2nd edn, Vol. VII (P. György and W. N. Pearson, eds), Academic Press, New York and London, p. 99.
Pitt, G. A. J. (1971). In *Cartotenoids* (O. Isler, ed.), Birkhäuser Verlag, Basle and Stuttgart, p. 718.
Reed, L. S. and Archer, M. C. (1976). *J. Chromat.* **121**, 100.
Reeder, S. K. and Park, G. L. (1975). *J. Ass. off. analyt. Chem.* **58**, 595–598.
Reingold, R. N., Picciano, M. F. and Perkins, E. G. (1980). *J. Chromat.* **190**, 237.
Richardson, P. J., Favell, D. J., Gidley, G. C. and Jones, A. D. (1978). *Proc. analyt. Div. chem. Soc.* **15**, 53.
Roe, J. H. (1967). In *The Vitamins*, 2nd edn, Vol. VII (P. György and W. N. Pearson, eds), Academic Press, New York and London, p. 27.
Roels, O. A. and Mahadevan, S. (1967). In The Vitamins, 2nd edn, Vol. VI (P. György and W. N. Pearson, eds), Academic Press, New York and London, p. 139.
Rouseff, R. (1979). In *Liquid Chromatographic Analysis of Food and Beverages*, Vol. 1 (G. Charalambous, ed.), Academic Press, New York and London, p. 161.
Sauberlich, H. E. (1967). In *The Vitamins*, 2nd edn, Vol. VII (P. György and W. N. Pearson, eds), Academic Press, New York and London, p. 169.
Shaikh, B., Huang, H. S. and Zielinski, W. L. (1977). *J. Ass. off. analyt. Chem.* **60**, 137.
Sood, S. P., Sartori, L. E., Wittmer, D. P. and Haney, W. G. (1976) *Analyt. Chem.* **48**, 796.
Stewart, I. (1977). *J. Ass. off. analyt. Chem.* **60**, 132.
Stewart, I. and Wheaton, T. A. (1971). *J. Chromat.* **55**, 325.
Strohecker, R. and Henning, H. M. (1965). *Vitamin Assay – Tested Methods*, Verlag Chemie, Weinheim.
Tanaka, Y., De Luca, H. F. and Ikekawa, N. (1980). *Methods Enzymol.* **67**, 370.
Thompson, J. N. and Hatina, G. (1979). *J. Liquid Chromat.* **2**, 327.
Thompson, J. N. and Maxwell, W. B. (1977). *J. Ass. off. analyt. Chem.* **60**, 766.
Thompson, J. N., Erdody, P. and Maxwell, W. B. (1972). *Analyt. Biochem.* **50**, 267.
Thompson, J. N., Maxwell, W. B. and L'Abbé, M. (1977). *J. Ass. off Analyt. Chem.* **60**, 998.

Thompson, J. N., Hatina, G. and Maxwell, W. B. (1979). In *Proceedings of the 9th Materials Research Symposium*, National Bureau of Standards, Special Publication 519, p. 279.

Thompson, J. N., Hatina, G. and Maxwell, W. B. (1980). *J. Ass. off. analyt. Chem.* **63**, 894.

Toma, R. B. and Tabekhia, M. M. (1979). *J. Fd Sci.* **44**, 263.

Tsukida, K., Kodama, A., Ito, M., Kawamoto, M. and Takahashi, K. (1977). *J. Nutr. Sci. Vitaminol.* **23**, 263.

Tweeten, T. (1979). *Hewlett-Packard Applications Brief on Liquid Chromatography, Foods, Vitamins*, Hewlett-Packard, USA.

Tyler, T. A. and Shrago, R. R. (1980). *J. Liquid Chromat.* **3**, 269.

Vanderslice, J. T. and Maire, C. E. (1980). *J. Chromat.* **196**, 176.

Vanderslice, J. T., Stewart, K. K. and Yarmas, M. M. (1979). *J. Chromat.* **176** 280.

Vanderslice, J. T., Maire, C. E., Doherty, R. F. and Beecher, G. R. (1980). *J. agric. Fd Chem.* **28**, 1145.

Van de Weerdhof, T., Wiersum, M. L. and Reissenweber, H. (1973). *J. Chromat.* **83**, 455.

Van Niekerk, P. J. (1973). *Analyt. Biochem.* **52**, 533.

Van Niekerk, P. J. (1975). MSc thesis, University of South Africa.

Van Niekerk, P. J. and and du Plessis, L. M. (1976). *S. Afr. Fd Rev.* **3**, 167.

Van Niekerk, P. J. and du Plessis, L. M. (1980). *J. Chromat.* **187**, 436,

Van Niekerk, P. J. and Smit, S. C. C. (1980). *J. Am. Oil Chem. Soc.* **57**, 417.

Van Niekerk, P. J., Smit, S. C. C. and Strydom, E. S. P. (1981). To be published.

Weedon, B. C. L. (1971). In *Carotenoids* (O. Isler, Ed.), Birkhäuser Verlag, Basle and Stuttgart, p. 29.

Widicus, W. A. and Kirk, J. R. (1979). *J. Ass. off analyt. Chem.* **62**, 637.

Wiggins, R. A. (1977). *Chemy Ind.* **20**, 841.

Wiggins, R. A. (1979a). In *The Importance of Vitamins to Human Health, Proceedings of the Kellogg Nutrition Symposium*, MTP, Lancaster, p. 73.

Wiggins, R. A. (1979b). In *The Importance of Vitamins to Human Health, Proceedings of the Kellogg Nutrition Symposium*, MTP, Lancaster, p. 9.

Williams, A. K. and Cole, P. D. (1975). *J. agric. Fd Chem.* **23**, 915.

Williams, A. T. R. and Slavin, W. (1977). *Chromat. Newsl.* **5**, 9.

Williams, R. C., Schmit, J. A. and Henry, R. A. (1972). *J. chromatogr. Sci.* **10**, 494.

Williams, R. C., Baker, D. R. and Schmit, J. A. (1973). *J. chromatogr. Sci.* **11**, 618.

Wills, R. H. B., Shaw, C. G. and Day, W. R. (1977) *J. chromatogr. Sci.*, **15**, 262.

Wong, F. F. (1978). *J. agric. Fd Chem.* **26**, 1444.

Yasumoto, K. Tadera, K., Tsuji, H. and Mitsuda, H. (1975). *J. Nutr. Sci. Vitaminol.* **21**, 117.

Zakaria, M., Simpson, K., Brown, P. R. and Krstulovic, A. (1979). *J. Chromat.* **176**, 109.

# 9 Determination of Food Additives

## K. SAAG

Cadbury Schweppes Limited, The Lord Zuckerman Research Centre,
University of Reading, U.K.

## I GENERAL INTRODUCTION

The majority of foods are comprised of carbohydrates, fats, proteins, minerals, vitamins and water. In addition to the natural constituents present in foodstuffs, chemicals may be incorporated, either directly or indirectly, during growing, storage, or processing of the food. These compounds may be described for convenience as "food additives". Food additives can be divided into two major groups, incidental and intentional additives. When they are purposely introduced to aid processing or to preserve or improve the quality of the product, they are called intentional additives. Such materials as colours, flavours, sweeteners, vitamins, mould inhibitors, bactericides, antioxidants and emulsifiers are intentional additives. They are added to the food product in carefully controlled proportions, and the amounts necessary to achieve the desired effect are usually quite small. Incidental additives are those that have been fortuitously incorporated into the food and thus can be considered as food contaminants. Compounds such as mycotoxins, nitrosamines or packaging residues would be included in this group, which will not be considered in this chapter. The quantitative determination of food additives is important both from the point of view of industry and the regulatory bodies.

HPLC has many features which makes it a powerful technique for the determination of food additives. However, as will be seen subsequently, it is not the chromatographic separation of additives that is a problem but rather their extraction and purification from complex food matrices.


## II  ACIDULANTS


Acidulants are employed, both directly and indirectly, for more than 20 separate purposes in food processing. One of their major functions as food additives is to enhance and to modify the flavour of products. In this way, the food is rendered more palatable and hence acceptable to the consumer. Equally important is the ability of food acids to aid the preservation of foods. Shorter times can generally be employed for the sterilization of foods when acidulants are added, since the resistance of most living organisms to heat is decreased at lower pH. Acidulants also act as synergists to the antioxidants added to foods to prevent rancidity and other deleterious reactions. Besides this, acidulants serve other specific functions, such as gelling agents for pectin, as a source of acidity in leavening and as catalysts for inducing inversion of sucrose. In the dairy industry better control can be exercised over the production of cultured dairy products such as cheese, sour cream and yogurt, by monitoring the distinct organic acids produced during fermentation. Analysis of dairy products often requires only precipitation of fats and proteins by the addition of acetonitrile followed by filtration prior to determination. Organic acids of interest to the wine industry – citric, tartaric, malic, succinic, lactic, fumaric and acetic acids – can usually be determined with little or no sample preparation other than filtration. The detection and quantification of organic acids is therefore important for many areas of food processing.

## A Ion-exchange Methods

Ion-exchange chromatography is often the technique of choice for investigating ionic or ionizable compounds usually operated at elevated temperatures with critical control of pH. Heating a resin-packed HPLC column increases its efficiency and decreases column back pressure. Although most separations do not require heated conditions, elevated temperatures can be used to resolve closely eluting compounds and to decrease analysis time. Increased efficiency is also achieved by using lower flow rates of between 0.4 and 1.0 ml min$^{-1}$. The addition of organic solvents as modifiers, up to 30% acetonitrile, to the mobile phase decreases the adsorption of organic compounds on the polystyrene matrix. This is especially useful for the separation of aromatic acids. The addition of 10% acetonitrile causes the aliphatic carboxylic acids to elute in the void volume, and the otherwise highly retained aromatic acids can be eluted and separated.

Ion-exchange methods for organic acids in wine and grape-must have been described by Kaiser (1973) and Rapp and Ziegler (1976), for various mono-, di-, hydroxy- and keto-carboxylic acids. Palmer and List (1973) illustrate the use of a strongly basic anion-exchange Aminex A25 column (75 cm × 4.6 mm, i.d.), operated at 70 °C, with a mobile phase of 1.0 M sodium formate for the separation of 13 food acids in 75 min. Although the resolution for certain acids, e.g. citric and malic, did not achieve base-line separation it was found that adjustments to the eluent concentration could give the desired resolution. Mabrouk (1976) modified and extended the procedure to include the analysis of 31 organic acids. Symmonds (1978) also reports the use of an Aminex A25 column with a mobile phase of 0.9 M sodium formate (pH 7.5) for the separation of galacturonic, lactic, malic, succinic, and tartaric acids in wine with detection by differential refractometry or by UV absorption measurement at 254 nm for shikimic acid. Phenolic acids present were first extracted into diethyl ether from wine saturated with sodium chloride and determined on a LiChrosorb RP-18 column with gradient elution from 10 to 60% methanol in 0.1 M potassium dihydrogen phosphate buffer (pH 2.1) as the eluent. Rapp and Ziegler (1979) have determined sugars, glycerol, ethanol and carboxylic acids in grape-must and wine using a cation-exchange resin Aminex A-8 or Beckman M-72 resin with water–methanol (4 : 1, v/v) as mobile phase and refractometric detection.

Employing cation-exchange resins and eluents of dilute hydrochloric acid (0.001–0.005 M) has enabled Richards (1975) to separate citraconic, fumaric, acrylic and acetic acids as minor components in maleic acid. Using ultraviolet (UV) detection at 210 nm allowed 1 $\mu$g ml$^{-1}$ of each acid except acetic acid (100 $\mu$g ml$^{-1}$) to be determined. Turkelson and Richards (1978) described the separation of various acids from the citric acid cycle using a strong cation-exchange resin Aminex 50W–X4 (30–35 $\mu$m) eluted with 0.001 M hydrochloric acid (Fig. 9.1). Monitoring the column eluate at 210 nm allowed detection limits of 1 p.p.m. or less for *cis*-aconitic, $\alpha$-ketoglutaric and fumaric acids; 2 p.p.m. for oxaloacetic acid; 15 p.p.m. for citric and L-malic acids; 20 p.p.m. for isocitric acid and 30 p.p.m. for succinic acid. While the observed detection limits were established using a 50 $\mu$l injection, as much as 500 $\mu$l has been injected without any significant

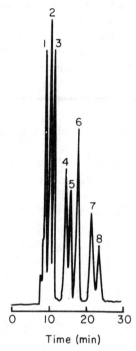

**Fig. 9.1** Separation of citric cycle acids (LC-55 detector). Peaks: 1, *cis*-aconitic, 30 p.p.m.; 2, $\alpha$-ketoglutaric, 190 p.p.m.; 3, oxaloacetic, 260 p.p.m.; 4, citric 1220 p.p.m.; 5, isocitric, 1770 p.p.m.; 6, L-malic, 1650 p.p.m.; 7, fumaric, 9 p.p.m.; 8, succinic, 1340 p.p.m. Flow rate, 1.0 ml min$^{-1}$; UV detection at 210 nm and 0.2 AUFS, 10 mV recorder; injection, 50 $\mu$l; column temperature, ambient; mobile phase, 0.001 M hydrochloric acid. [Redrawn with permission from Turkelson and Richards (1978).]

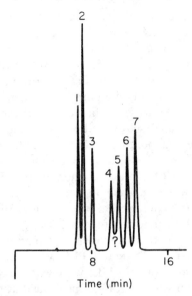

**Fig. 9.2** Analysis of seven organic acids with UV detection at 210 nm. Peaks: 1, citric; 2, tartaric; 3, malic; 4, succinic; 5, lactic; 6, fumaric; 7, acetic. Column, HPLC column for organic acid analysis, 300 mm × 7.8 mm (packed with Aminex HPX-87); eluent, 0.0065 M $H_2SO_4$; temperature, 65 °C; flow rate, 0.8 ml min$^{-1}$. [Redrawn with permission from Bio-Rad Publications (1979).]

peak-broadening effects. Fresh orange and grapefruit juices were also chromatographed after filtering and diluting with deionized water.

Continued use of even dilute hydrochloric acid may, however, eventually pit and corrode the stainless steel components of a pump or column. The problem may be overcome by using dilute sulphuric acid in place of hydrochloric acid (see Fig. 9.2) (Bio-Rad Publications, 1979).

## B Reversed Phase Methods

In recent years reversed phase methods of analysis utilizing buffered eluents or water–methanol mixtures have become common-place for the investigation of food acidulants. The major organic acids in cranberry juice – quinic, malic and citric acids – have been characterized by Coppola, Conrad and Cotter (1978). After dilution and clean up through a disposable column the acids were separated, at ambient temperature, on a $\mu$-Bondapak $C_{18}$ column (300 mm × 4 mm i.d.) and quantified using a differential refractometer. The mobile phase was an aqueous solution of 2.0% potassium dihydrogen phosphate ($KH_2PO_4$) adjusted to pH 2.4 with phosphoric acid. Using a similar phosphate buffer and a LiChrosorb $RP_8$ (10 $\mu$m) column (250 mm × 4.6 mm i.d.), Jeuring, Brands and van Doorninck (1979a) have separated malic and citric acids in apple juice.

Organic acids in fruit juices have also been investigated by Bigliardi, Gherardi and Poli (1979). Employing a $\mu$-Bondapak $C_{18}$ column (300 mm × 4 mm i.d.) and a mobile phase of aqueous phosphoric acid at pH 2.2, these workers effected separation of tartaric, malic and citric acids in 3 min. The results of analyses carried out on orange, lemon, apple and grape juices compared favourably with those obtained by an enzymic method of analysis. Libert (1981) has reported a determination of oxalic acid by reversed phase HPLC using an aqueous mobile phase containing 0.5% potassium dihydrogen phosphate and 5 mM tetrabutyl ammonium hydrogen sulphate (TBA) buffered to pH 2.0 with orthophosphoric acid.

Ong and Nagel (1978) reported the analysis of hydroxycinnamic acid and tartaric acid esters and their glucose esters in grape juice after fractionation on a Polyamide CC-6 column. HPLC was carried out on a Zorbax ODS column (250 mm × 4.6 mm i.d.) with water–acetonitrile (90 : 10, v/v) adjusted to pH 2.6 with orthophosphoric acid as the eluent.

Phenacyl and benzyl derivatives of dicarboxylic acids have been prepared and separated using reversed phase chromatography with UV detection by Grushka, Durst and Kikta (1975). Grushka, Lam and Chassin (1978) have also used pre-column phenacyl derivatization, with fluorescence monitoring, for the analysis of various dicarboxylic and $\alpha$-keto acids. A $C_{18}$ bonded Partisil 10 column (223 mm × 4.2 mm i.d.) operated isocratically with a methanol–water mobile phase was used. Detection limits were found to be of the order of $2 \times 10^{-10}$ mol. It was thought that the use of gradient techniques would greatly facilitate the separation of the $C_2$–$C_{10}$ dicarboxylic acids.

HPLC offers a method for separating and determining aliphatic carboxylic acids directly using ion-exchange and reversed phase columns. Since these acids do not have a significant absorbance in the UV region they have been detected by use of a differential refractometer. However, the availability of variable wavelength UV detectors capable of detection down to 190 nm makes it possible to determine these acids by UV absorption between 190–210 nm.

## III  ANTIOXIDANTS

Antioxidants may be naturally present, such as tocopherols and ascorbic acid, usually in the form of ascorbyl palmitate, or may be formed by processes such as roasting or smoking. A second category are the wholly synthetic antioxidants which are added to foodstuffs in order to retard the onset of rancidity by preventing oxidative degradation of lipids. Most of the important synthetic antioxidants are listed below:

| | |
|---|---|
| BHT | butylated hydroxytoluene |
| BHA | butylated hydroxyanisole |
| TBHQ | mono-tert-butylhydroquinone |
| THBP | 2,4,5-trihydroxybutyrophenone |
| PG | propyl gallate |
| OG | octyl gallate |
| DG | dodecyl gallate |
| Ionox-100 | 4-hydroxymethyl-2,6-di(tert-butyl)phenol |
| NDGA | nordihydroguaiaretic acid |
| TDPA | 3,3'-thiodipropionic acid |

Although many compounds are available the most commonly used are the synthetic phenolic antioxidants BHA, BHT and PG. For regulatory purposes, because of possible toxicity (Branen, 1975) it is often necessary to quantify these antioxidants in various foodstuffs where several antioxidants may be permitted and used either singly or in combination at levels up to 200 p.p.m.

## A Extraction Techniques

Satisfactory extraction of antioxidants from complex foodstuffs is not easy, particularly when antioxidants are present at low levels. Problems are usually associated with incomplete extraction of the antioxidants or with the co-extraction of potentially interfering substances. For quantitative determination, care must be exercised during concentration procedures under vacuum to prevent losses of certain antioxidants, BHA, BHT, TBHQ and Ionox-100, by evaporation. Fortunately, the antioxidants NDGA, THBP, TDPA, PG, OG and DG are relatively polar non-volatile compounds. The use of pure solvents, for extraction purposes, is also crucial as trace impurities may cause antioxidant losses.

Generally techniques of isolating antioxidants from the food matrix have relied on solvent extraction, (Buttery and Stuckey, 1961; Sahasrabudhe, 1964; Schwien, Miller and Conroy 1966; Stoddard, 1972; Pokorny, Čoupek and Pokorny, 1972; Phipps, 1973; Hammond, 1978; King, Joseph and Kissinger, 1980) and to a lesser extent on steam distillation (Anglin *et al.* 1956) and solvent distillation (Keen and Green, 1975). Stuckey and Osborne (1965) recommend that for high fat foods the antioxidant and fat are isolated together, by Soxhlet extraction into petroleum ether, followed by extraction of the antioxidants from the fat. For low fat foods direct steam distillation or solvent extraction of the antioxidants may be employed. However, more interfering substances are obtained when antioxidants are extracted directly from foods and it is usually necessary to include a clean-up step.

The most useful solvents for extracting food antioxidants from fat are acetonitrile and 70–95% water–alcohol mixtures. The fat is usually dissolved in hexane or petroleum ether and the antioxidant is extracted into the polar solvent. The disadvantages of acetonitrile extraction are that BHT recovery is low, and moderately high levels of interfering compounds are co-extracted. The advantage of methanolic or aqueous methanolic extraction of antioxidants, from hexane or a similar hydrocarbon solvent, is that the fat is mostly excluded. Hammond (1978) described a methanolic extraction of a melted fat sample, heated to 40–50 °C, followed by transfer to a deep freeze for a few hours to aid the solidification of any excess fat from the methanol. The methanol layer was then decanted and filtered prior to addition of an internal standard and direct injection.

A procedure for extracting 50–100 p.p.m. of BHA and BHT from vegetable oils has been reported by Phipps (1973). The oil was dissolved in n-heptane and extracted with four portions of dimethyl sulphoxide (DMSO). The combined DMSO extracts were mixed with aqueous 2 M sodium chloride solution and the antioxidants back extracted into petroleum ether for concentration and analysis by thin layer chromatography and HPLC.

## B  Reversed Phase Methods for Synthetic Phenolic Antioxidants

A variety of reversed phase chromatographic procedures have been developed for the analysis of antioxidants, mostly utilizing UV detection at 280 nm. Hammond (1978) reported the separation of five antioxidants: BHT, BHA, and three gallate esters on a $C_{18}$ $\mu$-Bondapak column (300 mm × 4 mm i.d.) using gradient elution with aqueous acetic acid and methanol solvents. A method for the simultaneous determination of nine phenolic antioxidants in oils, lards and shortenings was described by Page (1979). The antioxidants were partitioned from hexane–oil into acetonitrile, concentrated in vacuo, and separated on a LiChrosorb RP-18 (10 $\mu$m) column (250 mm × 3 mm i.d.). Gradient elution from water–acetic acid (95 : 5, v/v) to acetonitrile–acetic acid (95 : 5, v/v) was used at a flow rate of 1.0 ml min$^{-1}$ (Fig. 9.3).

King et al. (1980) reported the extraction of antioxidants from oils and foods prior to analysis on a $\mu$-Bondapak $C_{18}$ (10 $\mu$m) column

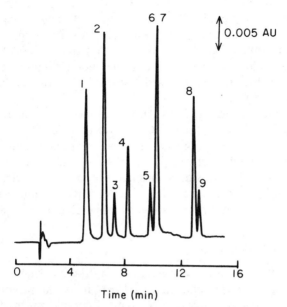

**Fig. 9.3** Chromatographic separation of antioxidant standards, c. 80 ng each antioxidant. Peaks: 1, propyl gallate (PG); 2, 2,4,5-trihydroxybutyrophenone (THBP); 3, tert-butylhydroquinone (TBHQ); 4, nordihydroguaiaretic acid (NDGA); 5, 2- and 3-tert-butyl-4-hydroxyanisole (BHA); 6, Ionox-100; 7, octyl gallate (OG); 8, dodecyl gallate (DG); 9, 3,5-di(tert-butyl)-4-hydroxytoluene (BHT). [Redrawn with permission from Page (1979).]

(150 mm × 4.2 mm i.d.) with methanol–0.1 M ammonium acetate (or 0.01 M phosphate) buffer (1 : 1, v/v) as the mobile phase and amperometric detection. Recoveries of BHA, TBHQ and PG added to samples were greater than 85% in all cases with most in the range 92–102%.

## C Tocopherols and Other Natural Antioxidants

The tocopherols are an important and commonly occurring class of natural antioxidants found in vegetable oils. A number of publications have been reported detailing methods for the determination of tocopherols using mainly microparticulate silica columns. Ultraviolet absorption and fluorescence measurement were the usual means of detection but electrochemical methods may also be used (Loliger and Saucy, 1980).

Thompson and Hatina (1979) have examined tocopherols and tocotrienols from barley, wheat flour, milk and soya bean oil; Carpenter (1979) has investigated tocopherols in edible vegetable oils; van Niekerk and du

Plessis (1980) have examined tocopherols and TBHQ in vegetable oils; and Tiebach and Schramm (1980) have described a preparative HPLC method of $\beta$-tocopherol and cholecalciferol from instant chocolate powder. Pickston (1978) has determined the $\alpha$-tocopherol content of condensed milk and milk substitutes by reversed phase HPLC with a mobile phase of methanol–water (9 : 1, v/v).

Other naturally occurring antioxidants have received comparatively little attention by HPLC with Chang, Ostric-Matijasevic, Hsieh and Huang (1977) investigating the antioxidants present in rosemary and sage. HPLC of 1,3-benzodioxole derivatives, part of the structure of sesamol which is an important natural antioxidant, has been reported by Cole, Crank and Hai Minh (1980). A reversed phase $C_{18}$ $\mu$-Bondapak column was used with methanol–water and acetonitrile–water mobile phases.

## D  Size Exclusion Methods

The use of exclusion chromatography and multi-dimensional HPLC techniques (Apffel, Alfredson and Majors, 1981) holds the promise of direct injection of complex samples with minimal sample pretreatment. Size exclusion columns are currently available which are ideally suited to the separation of small molecular weight (MW) molecules (MW < 2000 daltons) making it possible to analyse directly antioxidants in a fat or oil without prior extraction. For example, Doeden, Bowers and Ingala (1979) have investigated BHA, BHT and TBHQ present in edible fats and oils using gel permeation chromatography (GPC). No isolation stage was necessary as the oils were simply diluted in chloroform and a 0.25 ml aliquot of the filtered sample injected. Pokorny *et al.* (1972) have also described the use of GPC for the determination of the commonly used antioxidants ascorbyl palmitate, BHA, BHT, PG, DG and NDGA in edible oils and fats. A preliminary extraction of the oil in heptane using 80% aqueous ethanol was first performed to obtain an extract for GPC analysis.

Antioxidants have also been incorporated into food packaging materials. In the polymer matrix, however, antioxidants themselves are subject to degradation, especially during processing or on exposure to UV radiation. Phenolic antioxidants and their transformation products formed in polythene when subject to heat processing and UV radiation have been investigated by Lichtenthaler and Ranfelt (1978). Separation of antioxidants from transformation products was achieved on a Partisil (5 $\mu$m) column (250 mm × 4 mm i.d.) using *n*-hexane–dichloromethane gradient elution.

## IV ARTIFICIAL SWEETENERS

Investigations into the toxicity of artificial sweeteners have raised doubts about their safety. The use of cyclamate has been prohibited in many countries since 1970, and more recently there have been reports expressing doubts on the safety of saccharin. Impurities present in commercial saccharin preparations have also been implicated as being toxic and their significance has been studied (British Industrial Biological Research Association, 1973; Renwick and Ball, 1977; Lederer, 1977).

### A Saccharin and Saccharin Impurities

Saccharin has been quantitatively determined by a variety of HPLC methods. Nelson (1973) reported an HPLC method for the quantification of saccharin and other food additives using an anion–exchange resin with an aqueous borate buffer (pH 9.2) containing sodium nitrate. Reversed phase techniques, however, are the most usual means of determining saccharin in foodstores and beverages, with elution effected by aqueous phosphate buffers or aqueous acetic acid mobile phases. Smyly, Woodward and Conrad (1976) described a method for the determination of saccharin, sodium benzoate and caffeine in beverages using a reversed phase column and a mobile phase of water–glacial acetic acid (95 : 5, v/v). Tenenbaum and Martin (1977) reported the determination of saccharin in wines and proprietary drugs by using a reversed phase $C_{18}$ column (250 mm × 4 mm i.d.) with a phosphate buffer mobile phase at a flow rate of 1.2 ml min$^{-1}$. Saccharin was isolated from interfering substances by conversion to its acid form by the addition of sulphuric acid, with subsequent extraction into diethyl ether. Woodward, Heffelinger and Ruggles (1979) reported the results of a collaborative study for the determination of sodium saccharin, sodium benzoate and caffeine in beverages using a $\mu$-Bondapak $C_{18}$ column (300 mm × 4 mm i.d.) and a mobile phase of water–glacial acetic acid (80 : 20 v/v) buffered to pH 3.0 with saturated sodium acetate solution. Slight modification of the acetic acid concentration or the addition of 0–2% propan-2-ol to the mobile phase resolved any interference from sorbate or artificial colours. The HPLC methods described have determined saccharin in solution or in easily solubilizable forms; Eng, Calayan and Talmage (1977) described the determination of saccharin in chewing gum. The gum was dispersed in toluene and the saccharin quantiatively isolated by extraction with water. Analysis was carried out on the filtered aqueous phase using a $\mu$-Bondapak $C_{18}$ column with water–glacial acetic acid (95 : 5, v/v) mobile phase at a flow rate of 2.0 ml min$^{-1}$.

The detection and quantification of o- and p-sulphamoylbenzoic acids in commercial saccharin has been reported by Nelson (1976). These acid impurities were isolated by methanolic extraction of commercial saccharin and were separated on a strong anion–exchange (SAX) column (1 m × 2.1 mm i.d.) using an aqueous borate buffer (pH 9.2) as mobile phase. Levels of detection were estimated as 8 p.p.m. for the *ortho* isomer and 2.5 p.p.m. of the *para* isomer. Contaminants in saccharin preparations have also been studied by Szokolay (1980) on a Zorbax CN bonded phase with a mobile phase of water–acetic acid (95 : 5, v/v) (Fig. 9.4). The

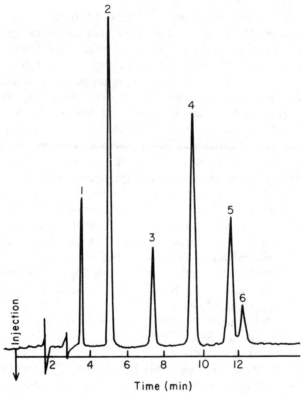

Time (min)

**Fig. 9.4** HPLC chromatogram of saccharin and its impurities. Peaks: 1, saccharin; 2, 2-sulphamoylbenzoic acid (*o*-SBA); 3, 4-sulphamoylbenzoic acid (*p*-SBA); 4, 4-hydroxybenzoic acid (internal standard); 5, toluene 2-sulphomamide (*o*-TS); 6, toluene 4-sulphonamide (*p*-TS). Column, Zorbax CN (250 mm × 4.6 mm i.d.); mobile phase, water–glacial acetic acid (95 : 5, v/v); flow rate, 1.7 ml min$^{-1}$; Detector, UV at 268 nm, 0.05 AUFS. [Redrawn with permission from Szokolay (1980).]

sample of saccharin was dissolved in 1% disodium hydrogen phosphate solution and amide impurities were extracted into dichloromethane. Acid impurities were isolated by direct methanolic extraction of saccharin.

## B Polyhydric Alcohols

Although classification of polyhydric alcohols as "artificial" is probably inappropriate, their inclusion here is justifiable on the grounds of their addition to diabetic and dietetic foods. However, the analysis of foodstuffs like fruits and dietetic foods, where sorbitol occurs in the presence of its parent sugar glucose and other monosaccharides, has not been without difficulties. Woidich, Pfannhauser and Blaicher (1978) reported a procedure for the determination of sorbitol in the presence of glucose; however, two different systems were required to determine fructose, glucose, sorbitol, sucrose, maltose and lactose. The separation and quantification of sorbitol from other polyhydric alcohols and carbohydrates using a single solvent system has been reported by Dokladalova, Barton and Mackenzie (1980). An Aminex carbohydrate HPX-87 column (250 mm × 4 mm i.d.) thermostated at 80 °C, was employed with water as the mobile phase and detection by differential refractometry. Xylitol was reported as the only interference in sorbitol quantification. Brandao, Richmond, Gray, Morton and Stine (1980) and Richmond, Brandao, Gray, Markakis and Stine (1981) described the separation and quantification of sorbitol from mono- and disaccharides using a Waters μ-Bondapak/carbohydrate column (300 mm × 4.2 mm i.d.) and a mobile phase of acetonitrile–water (85 : 15, v/v). Improved resolution of glucose and sorbitol was achieved by placing a Partisil PXS 10/25 PAC column (250 mm × 4.6 mm i.d.) in tandem ahead of the Waters column. A ternary mobile phase of acetonitrile–water–ethanol (80 : 15 : 5, v/v/v) was used at a flow rate of 1.8 ml min$^{-1}$.

## C Potential Sweetening Agents

HPLC has also been used for the analysis of other potential sweetening agents. Fox, Anthony and Lau (1976) described the analysis of aspartame in soft drink formulations and solid food products, using a strong cation–exchange (SCX) resin with 0.1 M citric acid and 0.5 M sodium perchlorate, containing sodium hydroxide to pH 4.7, as the eluent. Analysis of the non-nutritive sweetening agent neohesperidin dihydrochalcone (NHDC) has been reported by Schwarzenbach (1976) and Fisher (1977). Reversed phase $C_{18}$ columns were utilized with water–methanol (3 : 2, v/v) or water–acetonitrile (3 : 1, v/v) mobile phases with detection at

280 nm. A sample of yoghurt, containing 75 p.p.m. of NHDC, was extracted by mixing with acetone and then centrifuged to give a clear acetone solution which was injected directly; chewing gum was first dissolved in chloroform and extracted with water prior to injection. Acesulpham-K was determined (Grosspietsch and Hachenberg, 1980) by direct injection of ultrafiltered aqueous extracts of the samples (beverages, foods and cosmetics) on to a LiChrosorb RP-18 (10 $\mu$m) column. Elution was effected with a mobile phase of water–methanol (9 : 1, v/v) containing 0.01 M tetrabutylammonium hydrogen sulphate and detection by UV absorption at 227 nm.

## V  FLAVOUR COMPOUNDS

It is generally recognized that the resolution afforded by gas–liquid chromatography (GLC) for the separation of volatile flavour compounds remains unsurpassed. High performance liquid chromatography (HPLC) is, however, ideally suited to the analysis of many flavour components in food and beverages, and sometimes shows significant advantages over GLC. Both non-volatile and volatile components may be examined and since the analyses are generally performed at ambient temperature, the destructive temperatures needed for GLC are avoided and degradation products are not encountered.

The isolation of trace components of flavours has been pursued with increasing sophistication over a period of many years. The development of GLC coupled to mass spectrometry has led to the identification of the important constituents of even the complex flavours formed during fermentation and/or heating. However, as the techniques have become more sophisticated the problems have correspondingly appeared to be more complex, with the desired constituents found to be obscured in chromatograms under far larger quantities of compounds of little interest. The problem of isolating the components of a very complex flavour can be approached by the sequential use of separation techniques in which each step uses a different set of physical properties to effect separation. HPLC, using normal phase and/or bonded-phase columns, can be utilized to good effect as a technique for pre-fractionation of a complex flavour mixture, without undue exposure to light and air, prior to optimal separation by GLC. In this case the fraction(s) of interest must be eluted from the HPLC column with a solvent compatible to gas chromatographic analysis.

## A  Bitter Compounds

### 1  *Hop Flavour Components*

Characterization of the principal bittering and flavouring components of
hops and beer has received considerable attention by the brewing industry.
The proportions of α-acids (humulones) and β-acids (lupulones) in hops
and the amount of iso-α-acids formed during processing are of importance
when assessng the final quality of a beer. The chromatography of hop resin
components has been complicated by the wide polarity range of products
formed during the brewing process. Gas chromatography of the trimethylsilyl ethers of many of the compounds has been attempted, but the inherent
advantages of HPLC make it well suited to the analysis. A preliminary
separation of the complex components of beer may be made by solvent
extraction. A 2,2,4-trimethylpentane extract of acidified beer (adjusted to
pH 2.0 with hydrochloric acid) contains the isohumulones and non-
oxidized hop resin components, while a chloroform extract contains the
oxidized hop resins and non-hop-derived compounds.

Initial attempts at the separation of these hop resin compounds by
HPLC were primarily concerned with the used of silica as column packing
material, e.g. Molyneux and Wong (1973), Palamand and Aldenhoff
(1973), Siebert (1976) and Otter and Taylor (1977). Mobile phases have
included 2,2,4-trimethylpentane–ethylacetate or 2,2,4-trimethylpentane–
chloroform mixtures in various proportions with UV detection at 254
and 280 nm. Unfortunately none of these attempts was entirely successful and frequently resulted in poor resolutions and badly tailed peaks.
These problems were overcome by Gill (1979) who separated hop resins
on a 5 μm microparticulate silica column with a mobile phase of light
petroleum (60–80 °C)–chloroform (9 : 1, v/v) containing di-*n*-butyl-
ammonium acetate (0.1 M) at a flow rate of 2.5 ml min⁻¹. The addition
of the disubstituted ammonium salt was thought to improve resolution
and peak shape by suppressing the polar character of the hop acids. Conrad
and Fallick (1974) described the separation of nucleotides in beer using an
amino bonded phase with gradient elution using aqueous phosphate buffer.
Nucleosides were also investigated using a CX/Corasil cation-exchange
resin eluted isocratically with aqueous 0.03 M phosphate buffer as the
eluent. Hop acids were separated on a μ-Bondapak C₁₈ column
(300 mm × 4 mm i.d.) with methanol/water gradient elution at a flow rate
of 2.0 ml min⁻¹. Otter and Taylor (1977, 1978) reported the investigation
of α-acid conversion during the hop boiling process using a weak exchange
bonded phase (phenyl diethanolamine), operated at 60 °C, with
methanol–water (60 : 40, v/v) containing 1 mM citric acid as the mobile

phase at a flow rate of 0.5 ml min$^{-1}$. Verzele and De Potter (1978) have evaluated several packing materials for the analysis of hop bitter compounds. Separations obtained using amino-cyano and ion–exchange bonded phases were less satisfactory than those obtained using 5 and 10 $\mu$m ODS phases. The compounds were eluted with methanol–water (59 : 41, v/v) containing sodium acetate/acetic acid buffer (0.2 M, pH 7.0) at a flow rate of 1.0 ml min$^{-1}$ and the eluate was monitored at 334 nm. Complete resolution of cohumulone, adhumulone, humulone, colupulone and lupulone was effected in c. 80 min. For routine analysis, a faster separation (10 min) was achieved with methanol–water–acetic acid (85 : 15 : 1, v/v/v) although humulone and adhumulone were not separated. Whit and Cuzner (1979) reported the determination of isohumulone and isocohumulone in beer by reversed phase chromatography. The compounds were examined on a $\mu$-Bondapak C$_{18}$ phase with methanol–water gradient elution containing 5 mM tetrabutylammonium phosphate at a flow rate of 2.1 ml min$^{-1}$.

HPLC has also been used to obtain profiles of various types of beer to study the effect of hops on the flavour and the effect of ageing on the components of beer. For example, determination of nucleosides, purines, pyrimidines, nucleotides, polyhydric phenols and pyrazines in wort and beer at various stages during the brewing process and on storage has been described by Qureshi, Burger and Prentice (1979a,b). Reversed phase $\mu$-Bondapak C$_{18}$ phases were employed with an appropriate mobile phase composition of water–methanol–acetic acid–tetrabutylammonium phosphate. Detection was by UV absorption at 280 or 254 nm.

## 2   Xanthine Alkaloids

The naturally occurring alkaloids caffeine, theobromine and quinine represent important flavour constituents of tea, coffee, cocoa and certain soft drinks. Published methods for the determination of these alkaloids in foodstuffs by HPLC have included use of normal phase adsorption and ion-exchange techniques. For example, Wildanger (1975) has investigated caffeine, theophylline and theobromine in chocolate using 5 $\mu$m silica gel; Madison, Kazarek and Damo (1976) have studied caffeine in coffee and Van Duijn and Van der Stegen (1979) have analysed caffeine and trigonelline in tea and coffee using strong cation exchange (SCX) resins. Reversed phase chromatography, however, on C$_8$ and C$_{18}$ bonded phases using water–methanol or water–acetonitrile mobile phases is the most popular approach to alkaloid analyses. For example, Kreisser and Martin (1978) have analysed caffeine in cocoa and chocolate products, Timbie, Sechrist and Keeney (1978) have investigated theobromine and caffeine in cocoa beans while Juergens and Riessner (1980) have determined caffeine, theobromine and theophylline in tea, coffee and cola beverages. Other components of tea have been studied by Hoefler and Coggon (1976). These workers have investigated theoflavins by direct injection of a tea infusion on to a reversed phase $\mu$-Bondapak $C_{18}$ (10 $\mu$m) column (300 mm × 4 mm i.d.) eluted with a mobile phase of water–acetone–acetic acid (139 : 60 : 1, v/v/v). For analysis of aqueous samples filtration through an 0.45 $\mu$m cellulose filter was usually the only sample work-up required. For analysis of chocolate products, however, sample preparation (Kreiser and Martin, 1978) involved removal of the fat component into petroleum ether prior to extraction of the alkaloids into boiling water. Separation was achieved on a $\mu$-Bondapak $C_{18}$ (10 $\mu$m) column (300 mm × 4 mm i.d.) eluted isocratically with a mobile phase of water–methanol–acetic acid (79 : 20 : 1, v/v/v). Apffel et al. (1981) have reported the determination of caffeine and theophylline in biological fluids. The technique of multi-dimensional chromatography was described whereby a microparticulate aqueous-compatible steric exclusion column was used as the primary separation step coupled to either reversed phase, normal phase or ion-exchange columns as the secondary stage. This allowed direct injection and analysis of samples in a complex matrix eliminating the need for prior clean-up.

Quinine analysis in soft drinks has been reported by Frischkorn and Frischkorn (1976) using silica as the stationary phase and eluting with methanol–concentrated ammonium hydroxide (99.5 : 0.5, v/v) as the mobile phase. Quinine, however, had to be extracted from aqueous samples into diethyl ether prior to determination. In a more recent method, Jeuring, van den Hoeven, van Doorninck and ten Broeke (1979b) employed direct injection of the beverage on to a LiChrosorb RP-8 (10 $\mu$m) column (250 mm × 4.6 mm i.d.) with acetonitrile–water (3 : 1, v/v) as the mobile phase containing 5 mM sodium lauryl sulphate and perchloric acid to pH 3, at a flow rate of 1.0 ml min$^{-1}$. Quantification was by UV absorption measurement at 250 nm and fluorescence emission measurement at 435 nm with excitation at 350 nm.

## 3 Citrus Bitter Components

Another naturally occurring bitter substance is the glycoside naringin which occurs in grapefruit and some other citrus fruits. A further bitter component of grapefruit, and several varieties of oranges, is limonin, an intensely bitter triterpenoid dilactone. While some bitterness is desirable in grapefruit juice excessive bitterness is considered objectionable and any bitterness in orange juice is generally undesirable. Several publications for the determination of naringin and limonin have been reported. For example limonin has been determined in grapefruit juice (Fisher, 1975) and citrus juices (Fisher, 1978). Naringin has been determined in grapefruit juice (Fisher and Wheaton, 1976) while limonin and related limonoids in citrus juices have been investigated by Rouseff and Fisher (1980).

Initial attempts of limonin analysis by Fisher (1975) employed a $\mu$-Porasil column (300 mm × 4 mm i.d.) eluted by chloroform–acetonitrile (95 : 5, v/v) and monitored by refractive index measurement. However, the build up of highly polar grapefruit constituents on the column and the low sensitivity and poor thermal stability of the refractive index detector necessitated the investigation of an alternative method. Fisher (1978) utilized a $\mu$-Bondapak CN (10 $\mu$m) column (300 mm × 4 mm i.d.) operated in the reversed phase mode with methanol–water (33 : 65 or 40 : 60, v/v) as the eluent and UV detection at 210 nm. Utilizing these mobile phases, the cyano-bonded phase separated limonin from other citrus juice components better than a reversed phase $C_{18}$ column. However, the build up of irreversibly retained components reduced the column's usefulness for routine analysis after approximately 200 injections. A chromatographic method for separating and quantifying limonin and related liminoids in citrus juice has been reported by Rouseff and Fisher (1980). A Zorbax CN (5 $\mu$m) column (250 mm × 4.6 mm i.d.) was used, operated in the normal phase mode at 40 °C with propan-2-ol–hexane–methanol (12 : 11 : 2, v/v/v) as the mobile phase at a flow rate of 1.0 ml min$^{-1}$. Detection was by UV absorption measurement at 207 nm. When both naringin and limonin determinations were required the limonin was first extracted into chloroform and the naringin analysed on a $\mu$-Bondapak $C_{18}$ column (300 mm × 4 mm i.d.) with water–acetonitrile (8 : 2, v/v) as the mobile phase (Fisher and Wheaton, 1976).

## 4 Root Bitter Components

Characterization of the bitter components of the Gentiana root has been reported by Quercia, Battaglino, Pierini and Turchetto (1980), using a reversed phase ODS column (250 mm × 2.6 mm i.d.) operated at 60 °C with gradient elution using 0.01 M phosphate buffer (pH 5) as the primary eluent and methanol as the secondary eluent.

## B Essential Oils

While HPLC cannot normally be considered for the total analysis of volatile oils, the technique can be used to good effect for purposes of routine quality control. Many essential oils possess major components with aromatic structures and a criterion of acceptability may be based on these constituents with suitable chromatographic properties that can be detected by UV or fluorescence measurement.

Characterization of citrus fruit essential oils has been reported by Schmit, Williams and Henry (1973). A combination of gel permeation chromatography (GPC), reversed phase partition chromatography on a Permaphase ODS column (1 m × 2.1 mm i.d.) operated at 60 °C, and normal partition separation on Permaphase ETH, operated at ambient temperature, were described. Latz and Ernes (1978) have used selective fluorescence detection of citrus oil components after separation by adsorption chromatography. After dilution with chloroform, lime, lemon, grapefruit and bergamot oils were investigated on a $\mu$-Porasil column (300 mm × 4 mm i.d.) with a linear gradient of hexane–chloroform (8 : 2, v/v) to 100% chloroform at a flow rate of 1.0 ml min$^{-1}$. Ross (1978) has examined a number of volatile oils and flavour compounds on reversed phase (Hypersil-SAS) and normal phase (Partisil-5) packing materials. Constituents of volatile oils, e.g. vanillin, methyl salicylate, eugenol, thymol and ylang-ylang oil constituents: benzaldehyde, methyl anthranilate, benzyl acetate and p-methylanisole were separated on a Hypersil-SAS column eluted with methanol–water (1 : 1, v/v) at a flow rate of 2.0 ml min$^{-1}$. Measurement was by UV monitoring at 260 nm coupled with fluorescence detection for methyl anthranilate. Carvone was well separated from neral and geranial, which tended to co-elute from a partisil-5 column using n-heptane–acetonitrile (99 : 1, v/v) as the mobile phase. Compounds were readily detected by UV monitoring at 242 nm. Menthone, carbone, neral and geranial were separated and detected as their 2,4-dinitrophenylhydrazine derivatives from a Partisil-5 column eluted by n-heptane–ethyl acetate (19 : 1, v/v) with UV monitoring at 370 nm. Methyl anthranilate has also been determined in grape beverages by Rhys-Williams and Slavin (1977) using a reversed phase ODS column (250 mm × 2.6 mm i.d.) operated at 60 °C, eluted by water–acetonitrile (97 : 3, v/v) containing 35 mM phosphate buffer (pH 6). Detection was by fluorescence emission monitoring at 430 nm, excitation at 330 nm, and by UV absorption measurement at 217 nm. Jones, Clark and Iacobucci (1979) have separated a model mixture containing limonene, carvone, and six terpene alcohols on a Partisil 10-PXS (10 $\mu$m) column (250 mm × 4.6 mm i.d.). Eluent compositions of hexane–ethyl acetate (9 : 1, v/v) or dichloromethane–ethyl acetate (39 : 1, v/v) were found to be most suitable, with flow rates of 2.0 ml min$^{-1}$. Solutes were detected with use of a differential refractometer (Fig. 9.5). Shu, Walradt and Taylor (1975) have investigated several commercially available chromatographic packing materials for the determination of bergapten, a photoxic furocoumarin, in citrus oils. The most satisfactory results were obtained on

246 K. SAAG

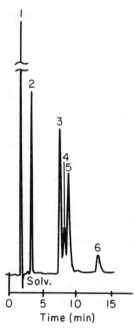

**Fig. 9.5** HPLC separation of a 2 μl mixture with ethyl acetate–hexane (10 : 90, v/v). Peaks 1, limonene; 2, carvone; 3, *trans*- and *cis*-*p*-mentha-1(7),8-dien-2-ol; 4, *trans*- and *cis*-carveol; 5, *trans*-*p*-mentha-2, 8-dien-1-ol; 6, *cis*-*p*-mentha-2,8-dien-1-ol. [Redrawn with permission from Jones *et al.* (1979).]

a Zorbax-Sil column (250 mm × 2.1 mm i.d.) with a mobile phase of 2,2,4-trimethylpentane–ethylacetate–propan-2-ol (80 : 1 : 1, v/v/v) at a flow rate of 0.4 ml min$^{-1}$, with UV detection at 254 nm. A natural constituent of calmus oil, β-asarone, has been studied by Micali, Curro and Calabro (1980) using a reversed phase ODS column (250 mm × 2.6 mm i.d.), thermostated at 65 °C, with methanol–water (31 : 19, v/v) mobile phase at a flow rate of 0.75 ml min$^{-1}$.

## C  Pungency and Aroma Compounds

Hotness is a property associated with spices and is also referred to as pungency. The main compound responsible for the pungent flavour of black pepper is the non-volatile compound piperine.

Verzele, Mussche and Qureschi (1979) have reported the analysis of the pungent components of pepper. Piperine was well resolved from other compounds on a nitro silica gel packing material eluted with a mobile phase of dichloromethane–methanol (200 : 9, v/v), with detection at 280 nm.

Determination of capsaicinoids, the important flavouring and pungency compounds in paprikas, chilli peppers and red peppers, has been described by Woodbury (1980). Capsaicinoids were isolated by extraction at 60 °C for 5 h using 95% ethanol saturated with sodium acetate. Separation was effected by injection of the sample on to a LiChrosorb RP-18 (10 $\mu$m) column (250 mm × 4.6 mm i.d.) eluted with water–acetonitrile–dioxane–2 M perchloric acid (500 : 30 : 20 : 4, v/v/v/v) as the primary|eluent and methanol–dioxane–acetonitrile (8 : 2 : 1, v/v/v) as the secondary eluent used either isocratically or with gradient elution. Detection was by monitoring fluorescence emission at 320 nm, with excitation at 288 nm.

It is generally recognized that pyrazine compounds contribute significantly to the roasted or cooked flavour of foods. Numerous foodstuffs, including cocoa products, coffee, peanuts and potato products have been found to contain pyrazines (Maga and Sizer, 1973). The use of synthesized methoxypyrazines to enhance the flavour of various potato products has already been reported (Guadagni, Buttery, Seifert and Venstrom, 1971).

Qureshi, Prentice and Burger (1979c) described the application of HPLC to the separation of pyrazines, purines, pyrimidines, nucleosides, nucleotides and polyphenols on a reversed phase $\mu$-Bondapak $C_{18}$ column (300 mm × 4 mm i.d.). Elution was effected using various proportions of water, methanol, acetic acid and tetrabutylammonium phosphate, with UV detection at 254 or 280 nm. Products obtained from a model browning reaction system, cysteamine/D-glucose/water, have been investigated by Mihara and Shibamoto (1980). A dichloromethane extract of the reaction mixture was fractionated into seven portions using HPLC on a silica gel LiChrosorb SI 100 (10 $\mu$m) column (250 mm × 4 mm i.d.) with gradient elution from hexane–dichloromethane (98 : 2, v/v) to ethanol–hexane–dichloromethane (78 : 20 : 2, v/v/v) at a flow rate of 1.0 ml min$^{-1}$.

The formation of 5-hydroxymethyl-2-furaldehyde (5-HMF) has been determined during heat processing of tomato paste (Allen and Chin, 1980) and during caramelization (Alfonso, Martin and Dyer, 1980). Reversed phase chromatography was employed with isocratic elution using water or by gradient elution using methanol–water. Alfonso et al. (1980) also reported the separation of other flavour compounds; theobromine, caffeine, vanillin, benzaldehyde, ethyl vanillin, coumarin, anethole, cinnaldehyde and methyl salicylate, in less than 20 min, using the same methanol–water gradient system (Fig. 9.6).

Nucleotides are important flavour compounds that are known to enhance and modify the organoleptic properties of foods. They have been investigated using primarily weak and strong anion-exchangers with phosphate buffer as the mobile phase. For example, Bennett (1977) used a Partisil 10-SAX anion-exchange column, operated at ambient tempera-

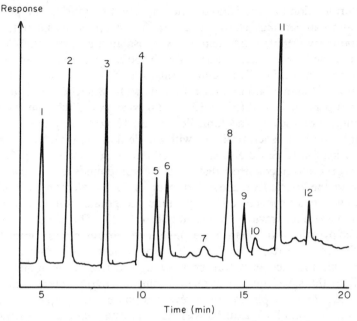

**Fig. 9.6** Chromatogram of the following compounds: 1, 5-(hydroxymethyl-2-furaldehyde; 2, 2-furaldehyde; 3, theobromine; 4, 5-methyl-2-furalde-hyde; 5, caffeine; 6, vanillin; 7, benzaldehyde, 8, ethyl vanillin; 9, coumarin; 10, anethole; 11, cinnamaldehyde; 12, methyl salicylate. Column, 10 μm Partisil PXS 10/25 (250 mm × 4.6 mm i.d.); methanol–water gradient (10% methanol in water for 5 min, programmed to 30% methanol in water for 4 min, and programmed from 30% to 60% methanol in water for 4 min); flow rate, 2 ml min⁻¹.
[Redrawn with permission from Alfonso *et al.* (1980).]

ture, to examine nucleotides in citrus juices. Twelve nucleotides were eluted in 45 min with gradient elution from aqueous 22.5 mM potassium dihydrogen phosphate (pH 3.74) to aqueous 675 mM potassium dihydrogen phosphate (pH 4.77) at a flow rate of 0.9 ml min⁻¹. Detection was by UV absorbance measurement at 254 nm. Heyland and Moll (1977) also report the use of ion-exchange, operated at 60 °C, for the separation of nucleotides using a buffered ammonium hydrogen phosphate mobile phase.

## VI  PRESERVATIVES

Preservatives are substances added to foods and beverages to prevent or inhibit microbial growth. A number of different compounds are used depending on the food product and the spoilage organism involved. Benzoic acid, sorbic acid, propionic acid, and the methyl, ethyl and propyl esters of p-hydroxybenzoic acid (parabens) represent the most commonly used preservatives in the food industry. The most widely used preservatives in fruit juices and soft drinks, apart from sulphur dioxide, are sorbic acid and benzoic acid either individually or as mixtures. A variety of methods are available for the analysis of benzoic and sorbic acids, usually involving extraction and the preparation of volatile derivatives. They are, however, relatively few methods available that permit simultaneous direct determination of both benzoic and sorbic acids. For example, direct quantitative spectrophotometric analysis of fruit juices is generally impracticable due to the presence of other UV absorbing compounds and if both acids are present in the same sample, overlap of the broad and intense sorbate absorbance (256 nm) interferes with the benzoate determination (225 nm) (Gutfinger, Ashkenazy and Letan, 1976).

## A  Ion-exchange Methods

Preservatives present in beverages such as fruit juices, soft drinks or wine can usually be determined directly using HPLC with little or no sample pretreatment other than filtration. Several methods of HPLC analysis using ion-exchange columns with aqueous borate or phosphate buffers have been published; for example, Nelson (1973) has determined sodium benzoate, saccharin and other additives in soft drink beverages, McCalla, Mark and Kipp (1977) have investigated sorbic acid in wine, and Bennett and Petrus (1977) have determined benzoic and sorbic acids in citrus juices. Benzoic and sorbic acids have also been separated on a silica column by Wildanger (1973), but the composition of the mobile phase, largely 2,2,4-trimethylpentane and diethyl ether, would preclude the direct injection of aqueous samples.

## B  Reversed Phase Methods

Reversed phase systems have advantages over the use of silica gel in the ease of equilibration and in that the column may be cleaned from accumulated contaminants by simply washing with methanol.

Smyly et al. (1976) report the separation of sodium benzoate, saccharin and caffeine in soft drinks and other beverages using a reversed phase $\mu$-Bondapak $C_{18}$ column (300 mm $\times$ 4 mm i.d.) and a mobile phase of water–glacial acetic acid (99 : 5, v/v) at a flow rate of 2.0 ml min$^{-1}$. Unfortunately sorbic acid was reported to interfere in the determination and was not totally resolved from benzoic acid. A similar interference, in the determination of sorbic acid in wine, was avoided by separating the two acids on a LiChrosorb RP-8 column with water–propan-2-ol (7 : 3, v/v) as the eluant (Eisenbeiss, Weber and Ehlerding, 1977).

Leuenberger, Gauch and Baumgartner (1979) report the analyses of benzoic and sorbic acids, methyl, ethyl and propyl hydroxybenzoates, and saccharin in beverages and foods. Fatty samples such as margarine and butter were dissolved in diethyl ether and extracted with 0.1 M sodium hydroxide solution. The basic aqueous extract was acidified with sulphuric acid, diluted to volume with methanol and filtered prior to analysis. Foodstuffs such as cheese, cakes and yoghurts were first homogenized in 0.25 M sulphuric acid and the preservatives and saccharin isolated by retention on an Extrelut column, containing kieselguhr, followed by elution with chloroform–propan-2-ol (9 : 1 v/v) and concentrated by evaporation. Analyses were carried out on a $\mu$-Bondapak $C_{18}$ column using isocratic and gradient elution with an aqueous phosphate buffer as the primary eluent and methanol as the secondary eluent.

Various preservatives have been studied by Hild and Gertz (1980) on a LiChrosorb RP-18 (7 $\mu$m) column (250 mm × 4.6 mm i.d.). The mobile phase of aqueous 5 mM ammonium acetate/acetic acid buffer (pH 4.4)–acetonitrile (4 : 1, v/v), was monitored at 232 nm for sorbic, benzoic and hydroxybenzoic acids and at 225 nm for formic and propionic acids. Recoveries of these acids from several foodstuffs ranged from 76 to 101%. Archer (1980) reported the analysis of benzoic and sorbic acids in orange juice using a $\mu$-Bondapak phenyl column (300 mm × 3.9 mm i.d.) and a mobile phase of water–tetrahydrofuran–methanol–acetic acid (68 : 20 : 10 : 2, v/v/v/v) monitored at 270 nm.

Carnevale (1980) described the simultaneous determination of ascorbic, sorbic and benzoic acids in citrus juices using a $\mu$-Bondapak-CN column (300 mm × 4 mm i.d.) with 2% acetic acid–methanol (19 : 1, v/v) as the mobile phase at a flow rate of 1.5 ml min$^{-1}$. The analyses were also carried out on a $\mu$-Bondapak-NH$_2$ column with a range of dilute acetic acid–methanol mixtures. Although sorbic and benzoic acids were separated, ascorbic acid was irreversibly absorbed on to the column and could not be removed by clean-up procedures which were not destructive to the column packing.

Deming and Turoff (1978) have investigated benzoic acid, 2-aminobenzoic acid, 4-aminobenzoic acid, 4-hydroxybenzoic acid, and 1,4-benzenedicarboxylic acid on a reversed phase $\mu$-Bondapak $C_{18}$ column (300 mm × 4 mm i.d.) and have arrived at a semi-empirical scheme for the optimization of chromatographic resolution. Optimum resolution of the five weak organic acids was achieved in 30 min using a mobile phase of water–1 M acetic acid–1 M sodium hydroxide–1 M sodium chloride (880 : 60 : 21 : 39, v/v/v/v) at a flow rate of 2.0 ml min$^{-1}$.

Numerous methods for the analysis of fungicides, pesticides and antibacterial agents have been published; for example, Norman and Fousse (1978) have investigated imazilil residues in citrus fruit, Beljaars and Rondags (1979) have studied chloramine-T in ice-cream, minced meat and shrimp, and Isshiki, Tsumura and Watanabe (1980) have determined thiabendozole in banana and citrus fruits. Reversed phase techniques were reported with LiChrosorb RP-8 or RP-18 materials and mobile phases of acetonitrile–2 mM sodium chloride/phosphate buffer (47 : 53, v/v); acetonitrile–water (1 : 9, v/v) or methanol–0.28% ammonium hydroxide (60 : 40, v/v), with UV and fluorescence detection.

In general the problems encountered in determining trace levels of fungicide and antibacterial agents in foods are similar to those encountered in pesticide residue analysis. The compounds are only present at extremely low levels and considerable concentration after extraction is required, which accentuates the problems of co-extracted materials.

## REFERENCES

Alfonso, F. C., Martin, G. E. and Dyer, R. H. (1980). *J. Ass. off. analyt. Chem.* **63**, 1310.
Allen, B. H. and Chin, H. B. (1980) *J. Ass. off. analyt. Chem.* **63**, 1074.
Anglin, C., Mahon, J. H. and Chapman, R. A. (1956). *J. agric. Fd Chem.* **4**, 1018.
Apffel, J. A., Alfredson, J. V. and Majors, R. E. (1981). *J. Chromat.* **206**, 43.
Archer, A. W. (1980). *Analyst, Lond.* **105**, 407.
Beljaars, P. J. and Rondags, T. M. M. (1979). *J. Ass. off. analyt. Chem.* **62**, 1087.
Bennett, M. C. (1977). *J. agric. Fd Chem.* **25**, 219.
Bennett, M. C. and Petrus, D. R. (1977). *J. Fd Sci.* **42**, 1220.
Bigliardi, D., Gherardi, S. and Poli, M. (1979). *Indust. Conserve* **54**, 209.
Bio-Rad Publications (1979). *The Liquid Chromatographer*, No. 2EG.
Brandao, S. C. C., Richmond, M. L., Gray, J. I., Morton, I. D. and Stine, C. M. (1980). *J. Fd Sci.* **45**, 1492.
Branen, A. L. (1975). *J. Am Oil Chem. Soc.* **52**, 59.
British Industrial Biological Research Association (1973). *BIBRA Information Bulletin* **12**, 323
Buttery, R. G. and Stuckey, B. N. (1961). *J. agric. Fd Chem.* **9**, 283.
Carnevale, J. (1980). *Fd Technol. Aust.* **32**, 302.
Carpenter, A. P. (1979). *J. Am. Oil Chem. Soc.* **56**, 668.
Chang, S. S., Ostric-Matijasevic, B., Hsieh, O. A. L. and Huang, C. L. (1977). *J. Fd Sci.* **42**, 1102.
Cole, E. R., Crank, G. and Hai Minh, H. T. (1980). *J. Chromat.* **193**, 19.
Conrad, E. C. and Fallick, G. J. (1974). *Brewers Digest* **49**, 72.
Coppola, E. D., Conrad, E. C. and Cotter, R. (1978). *J. Ass. off. analyt. Chem.* **61**, 1490.
Deming, S. N. and Turoff, M. L. H. (1978). *Analyt. Chem.* **50**, 546.

Doeden, W. G., Bowers, R. H. and Ingala, A. C. (1979). *J. Am. Oil Chem. Soc.* **56**, 12.

Dokladolova, J., Barton, A. Y. and Mackenzie, E. A. (1980). *J. Ass. off. analyt. Chem.* **63**, 664.

Eisenbeiss, F., Weber, M. and Ehlerding, S. (1977). *Chromatographia* **10**, 262.

Eng, M.-Y., Calayan, C. and Talmage, J. M. (1977). *J. Fd Sci.* **42**, 1060.

Fisher, J. F. (1975). *J. agric. Fd Chem.* **23**, 1199.

Fisher, J. F. (1977). *J. agric. Fd Chem.* **25**, 682.

Fisher, J. F. (1978). *J. agric. Fd Chem.* **26**, 497.

Fisher, J. F. and Wheaton, T. A. (1976). *J. agric. Fd Chem.* **24**, 898.

Fox, L., Anthony, G. D. and Lau, E. P. K. (1976). *J. Ass. off. analyt. Chem.* **59**. 1048.

Frischkorn, C. G. B. and Frischkorn, H. E. (1976). *Z. Lebensmittelunters. u-Forsch.* **162**, 273.

Gill, R. (1979). *J. Inst. Brew.* **85**, 15.

Grosspietsch, H. and Hachenberg, H. (1980). *Z. Lebensmittelunters. u-Forsch.* **171**, 41.

Grushka, E., Durst, D. H. and Kikta, E. J. (1975). *J. Chromat.* **112**, 673.

Grushka, E., Lam, S. and Chassin, J. (1978). *Analyt. Chem.* **50**, 1398.

Guadagni, D. G., Buttery, R. G., Seifert, R. M. and Venstrom, D. W. (1971). *J. Fd Sci.* **36**, 363.

Gutfinger, T., Ashkenazy, R. and Letan, A. (1976). *Analyst, Lond.* **101**, 49.

Hammond, K. J. (1978). *J. Ass. off. publ. Analysts* **16**, 17.

Heyland, S. and Moll, H. (1977). *Mitt. Geb. Lebensmittelunters. Hyg.* **68**, 72.

Hild, J. and Gertz, C. (1980). *Z. Lebensmittelunters. u-Forsch.* **170**, 110.

Hoefler, A. C. and Coggon, P. (1976). *J. Chromat.* **129**, 460.

Isshiki, K., Tsumura, S. and Watanabe, T. (1980). *J. Ass. off. analyt. Chem.* **63**, 747.

Jeuring, H. J., Brands, A. and van Doorninck, P. (1979a). *Z. Lebensmittelunters. u-Forsch.* **168**, 185.

Jeuring, H. J., van den Hoeven, W., van Doorninck, P. and ten Broeke, R. (1979b). *Z. Lebensmittelunters. u-Forsch.* **169**, 281.

Jones, B. B., Clark, B. C. and Iacobucci, G. A. (1979). *J. Chromat.* **178**, 575.

Juergens, U. and Riessner, R. (1980). *Dt. LebensmittRdsch.* **76**, 39.

Kaiser, U. J. (1973) *Chromatographia* **6**, 387.

Keen, G. and Green, M. S. (1975). *J. Ass. off. publ. Analysts* **13**, 99.

King, W. P., Joseph, K. T. and Kissinger, P. T. (1980). *J. Ass. off. analyt. Chem.* **63**, 137.

Kreiser, W. R. and Martin, R. A. (1978). *J. Ass. off. analyt. Chem.* **61**, 1424.

Latz, H. W. and Ernes, D. A. (1978). *J. Chromat.* **166**, 189.

Lederer, J. (1977). *Med. Nutr.* **13**, 23.

Leuenberger, U., Gauch, R. and Baumgartner, E. (1979). *J. Chromat.* **173**, 343.

Libert, B. (1981). *J. Chromat.* **210**, 540.

Lichenthaler, R. G. and Ranfelt, F. (1978). *J. Chromat.* **149**, 553.

Loliger, J. and Saucy, F. (1980). *Z. Lebensmittelunters. u-Forsch.* **170**, 413.

McCalla, M. A., Mark, F. G. and Kipp, W. H. (1977). *J. Ass. off. analyt. Chem.* **60**, 71.

Mabrouk, A. F. (1976). *Abstr. Pap. Am. chem. Soc.* **172**, AGFD 111.

Madison, B. L. Kozarek, W. J. and Damo, C. P. (1976). *J. Ass. off. analyt. Chem.* **59**, 1258.

Maga, J. A. and Sizer, C. E. (1973). *J. agric. Fd Chem.* **21**, 22.
Micali, G., Curro, P. and Calabro, G. (1980). *J. Chromat.* **194**, 245.
Mihara, S. and Shibamoto, T. (1980). *J. agric. Fd Chem.* **28**, 62.
Molyneux, R. J. and Wong, Y. I. (1973). *J. agric. Fd Chem.* **21**, 531.
Nelson, J. J. (1973). *J chromatogr. Sci.* **11**, 28.
Nelson, J. J. (1976). *J. Ass. off analyt. Chem.* **59**, 243.
Norman, S. M. and Fouse, D. C. (1978). *J. Ass off. analyt. Chem.* **61**, 1469.
Ong, B. Y. and Nagel, C. W. (1978). *J. Chromat.* **157**, 345.
Otter, G. E. and Taylor, L. (1977). *J. Chromat.* **133**, 97.
Otter, G. E. and Taylor, L. (1978). *J. |Inst. Brew.* **84**, 160.
Page, D. B. (1979). *J. Ass. off. |analyt. Chem.* **62**, 1239.
Palamand, S. R. and Aldenhoff, J. M. (1973). *J. agric. Fd Chem.* **21**, 535.
Palmer, J. K. and List, D. M. (1973). *J. agric. Fd Chem.* **21**, 903.
Phipps, A. M. (1973). *J. Am. Oil Chem. Soc.* **50**, 21.
Pickston, L. (1978). *N.Z. J. Sci.* **21**, 383.
Pokorny, S., Čoupek, J. and Pokorny, J. (1972). *J. Chromat.* **71**, 576.
Qureshi, A. A., Burger, W. C. and Prentice, N. (1979a). *J. Am. Soc. Brew. Chem.* **37**, 153.
Qureshi, A. A., Burger, W. C. and Prentice, N. (1979b). *J. Am. Soc. Brew. Chem.* **37**, 161.
Qureshi, A. A., Prentice, N. and Burger, W. C. (1979c). *J. Chromat.* **170**, 343.
Quercia, V., Battaglino, G., Pierini, N. and Turchetto, L. (1980). *J. Chromat.* **193**, 163.
Rapp, A. and Ziegler, A. (1976). *Chromatographia* **9**, 148.
Rapp, A. and Ziegler, A. (1979).| *Dt. Lebensmitt Rdsch.* **76**, 396.
Renwick, A. G. and Ball, L. (1977). *Trans. biochem. Soc.* **5**, 1363.
Rhys-Williams, A. T. and Slavin, W. (1977). *J. agric. Fd Chem.* **25**, 756.
Richards, M. (1975). *J. Chromat.* **115**, 259.
Richmond, M. L., Brandao, S. C. C., Gray, I. J., Markakis, P. and Stine, C. M. (1981). *J. agric. Fd Chem.* **29**, 4.
Ross, M. S. F. (1978). *J. Chromat.* **160**, 199.
Rouseff, R. L. and Fisher, J. F. (1980). *Analyt. Chem.* **52**, 1228.
Sahasrabudhe, M. R. (1964). *J. Ass. off. agric. Chem.* **47**, 888.
Schmit, J. A., Williams, R. C. and Henry, R. A. (1973). *J. agric. Fd Chem.* **21**, 551.
Schwarzenbach, R. (1976). *J. Chromat.* **129**, 31.
Schwien, W. G., Miller, B. J. and Conroy, H. W. (1966). *J. Ass. off. agric. Chem.* **49**, 809.
Shu, C. K., Walradt, J. P. and Taylor, W. I. (1975). *J. Chromat.* **106**, 271.
Siebert, K. J. (1976). *Proc. Am. Soc. Brew. Chem.* **34**, 79.
Smyly, D. S., Woodward, B. B. and Conrad, E. C. (1976). *J. Ass. off. analyt. Chem.* **59**, 14.
Stoddard, E. E. (1972). *J. Ass. off. analyt. Chem.* **55**, 1081.
Stuckey, B. N. and Osborne, C. E. (1965). *J. Am. Oil Chem. Soc.* **42**, 228.
Symmonds, P. (1978). *Annls Nutr. Aliment.* **32**, 957.
Szokolay, A. M. (1980). *J. Chromat.* **187**, 249.
Tenenbaum, M. and Martin, G. E. (1977). *J. Ass. off. analyt. Chem.* **60**, 1321.
Thompson, J. N. and Hatina, G. (1979). *L. Liquid Chromat.* **2**, 327.
Tiebach, R. K. D. and Schramm, M. (1980). *Chromatographia* **13**, 403.
Timbie, D. J., Sechrist, L. and Keeney, P. G. (1978). *J. Fd Sci.* **43**, 560.
Turkelson, V. T. and Richards, M. (1978). *Analyt. Chem.* **50**, 1420.

Van Duijn, J. and Van der Stegen, G. H. D. (1979). *J. Chromat.* **179**, 199.
Van Niekerk, P. J. and du Plessis, L. M. (1980). *J. Chromat.* **187**, 436.
Verzele, M. and De Potter, M. (1978). *J. Chromat.* **166**, 320.
Verzele, M., Mussche, P. and Qureshi, S. A. (1979). *J. Chromat.* **172**, 493.
Whitt, J. T. and Cuzner, J. (1979). *J. Am. Soc. Brew. Chem.* **37**, 41.
Wildanger, W. A. (1973). *Chromatographia* **6**, 381.
Wildanger, W. A. (1975). *J. Chromat.* **114**, 480.
Woidich, H., Pfannhauser, W. and Blaicher, G. (1978). *Lebensmittelchem Gerichtl. Chem.* **32**, 74.
Woodbury, J. E. (1980). *J. Ass. off. analyt. Chem.* **63**, 556.
Woodward, B. B., Heffelinger, G. P. and Ruggles, D. I. (1979). *J. Ass. off. analyt. Chem.* **62**, 1011.

# 10 Determination of Food Colourants

## K. SAAG
Cadbury Schweppes Limited, The Lord Zuckerman Research Centre, University of Reading, U.K.

## I GENERAL INTRODUCTION

Food colours of both natural and synthetic origin are used extensively in processed foods and beverages. They serve to supplement and enhance natural colour destroyed during processing or storage, and substantially increase the acceptability and appeal of foodstuffs where no natural colour exists, e.g. soft drinks and ice-cream. The indiscriminate use of colouring matter, however, can mask damage or disguise poor products.

The colourants which are permitted for food use may be divided into three categories: (1) synthetic organic dyes, (2) natural and nature-identical colours and (3) inorganic pigments (not covered in this review).

Over the years all of the originally permitted fat-soluble synthetic dyes have been deleted from the approved list and only water-soluble colours are now permitted. The synthetic or artificial colours used in food products are predominantly azo and triarylmethane dyes. The range of shades covered by the azo group of food colours is diverse and includes red, orange yellow, brown and blue-black. These azo dyes are usually prepared by coupling of a diazotized sulphanilic acid to a phenolic sulphonic acid moiety. Both intermediates, unfortunately, often contain impurities and coupling reactions may result in the formation of unwanted products.

Triarylmethane colours are distinguished by their brilliance of colour and high tinctorial strength, but have poor light fastness. Also of importance in food colouring are three individually classified synthetic dyes: erythrosine, quinoline yellow and indigo carmine.

In recent years concern has been expressed about the safety of certain synthetic dyes, and this has prompted an increased consideration of the use of natural pigments as food colourants. Natural and nature-identical colours are in some cases single compounds, but more usually mixtures of several coloured as well as non-coloured compounds. Their composition may vary depending on their source and method of preparation. Natural and nature-identical colours may be considered as being derived from three main sources: (1) those natural colours extracted from natural foodstuffs by physical processes, e.g. beetroot, (2) those natural colours extracted after chemical modification of natural foodstuffs, e.g. caramel, and (3) chemically synthesized nature-identical colours, e.g. canthaxanthin (red), apo-carotenal (orange red) and beta-carotene (yellow-orange). Although the range of nature-identical shades is restricted, skilful blending can yield reasonable results for a moderate range of shades. There is still however, a gap to be filled in the blue region of the spectrum.

To ensure compliance with regulatory requirements, analytical methods are required to determine the nature and the concentration of a colourant in a food product, to ascertain the impurities present and to study the interaction between food components and colourants during processing. High performance liquid chromatography (HPLC) is proving to be an invaluable technique which allows qualitative and quantitative analysis of dyes after extraction from a food matrix.

## II  SYNTHETIC ORGANIC DYES

### A  Extraction Techniques

Numerous methods have been reported for the isolation and purification of water-soluble acid dyes from foods. These include adsorption column chromatography (Yanuka, Shalon, Weissenberg and Nir-Grosfeld, 1963), ion-exchange (Dolinsky and Stein, 1962; BFMIRA, 1963; Takeshita, Yamashita and Itoh, 1972), wool dyeing (Stanley and Kirk, 1963), and solvent extraction (BFMIRA, 1963). Column chromatography using polyamide powder (Lehmann, Collet, Hahn and Ashworth, 1970; Gilhooley, Hoodless, Pitman and Thomson, 1972) gives results superior to those of solvent extraction and wool dyeing.

The wool dyeing technique has several disadvantages. A number of dyes are taken up slowly from the hot acidic solution and the adsorbed dyes often undergo changes during the stripping from the wool with hot ammonium hydroxide solution. Extraction of dyes by adsorption with polyamide retains all acid dyes and, unlike the wool dyeing technique, this procedure does not alter the chemical composition of the dye. Foodstuffs which are completely soluble in water, e.g. jams, marmalades, jellies and boiled sweets, are dissolved in water and acidified with acetic acid. The solution is stirred with polyamide powder and transferred to a micro-column (150 mm × 15 mm) which is washed with hot water and acetone. The water removes sugars, acids and flavouring materials and acetone removes any basic dyes, water-soluble carotenoids and some anthocyanins. The adsorbed acid dyes are then washed from the column with methanolic sodium hydroxide. The eluate is acidified to pH 5–6 with methanol–acetic acid (1 : 1 v/v) and diluted with water. The dyes are further purified by another treatment on a polyamide column. The final solution is acidified with acetic acid and concentrated *in vacuo* to about 1 ml, ready for analysis. Protein-rich dairy products, e.g. ice cream, cheese and yoghurt, are treated with acetone to coagulate the proteins, which adsorb the acid dyes. The protein is ground with sand and celite and transferred to a tube and the dyes are eluted with methanol–ammonium hydroxide followed by treatment on a polyamide column.

Graichen (1975) reported the use of Amberlite LA-2, a liquid anion-exchange resin, to extract colours from food. Uematsu, Kurita and Hamada (1979) described the use of Amberlite XAD-2 resin as an adsor-bent for the isolation of 11 water-soluble food dyes and quoted recoveries of between 90.0 and 100.1%. Puttemans, Dryon and Massart (1980) have quantitatively extracted synthetic colouring materials as ion-pairs with tri-*n*-octylamine (TOA) in chloroform or *n*-heptane. The best conditions for extraction were 0.1 M TOA in chloroform as the organic phase and a buffer solution of pH 5 as the aqueous phase. Van Peteghem and Bijl (1981) have also described an ion-pairing technique for the extraction of water-soluble food dyes. These workers have used cetyltrimethylam-monium bromide at concentrations ranging from 10 to 50 mM to effect recovery of dyes from acid buffer solution (pH 2.5) into dichloromethane. For purposes of quantification, when incomplete extraction of food colour has been observed using conventional techniques, Boley, Crosby and Roper (1979) and Boley *et al.* (1980) recommend the use of a preliminary enzymic digestion to release the dye, prior to extraction and separation by HPLC.

## B Ion-exchange Methods

Intermediates and impurities present in synthetic food colourants have been characterized by many workers using ion-exchange chromatography, e.g. FD&C Red No. 40 (Singh, 1974a; Bailey and Cox, 1976; Singh and Adams, 1979; Cox and Reed, 1981) (Fig. 10.1), FD&C Yellow No. 6 (Singh, 1974b; Bailey and Cox, 1975; Marmion, 1977; Singh and Adams, 1979; Cox, 1980), FD&C Blue No. 2 (Singh, 1975; Bailey, 1980), FD&C Red No. 2 (Singh, 1977a) and Orange B (Singh, 1977b). The references

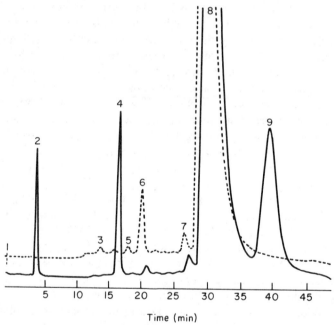

Time (min)

**Fig. 10.1** Chromatogram of FD&C Red No. 40. Peaks; 1, injection; 2, Cresidine sulphonic acid (CSA); 3, unknown; 4, Schaeffer's salt (SS); 5, unknown; 6, 4,4'-diazoamino-bis(5-methoxy)-2-methylbenzenesulphonic acid (DMMA); 7, unknown; 8, FD&C Red No. 40; 9, 6,6'-oxybis(2-naphthalenesulphonic acid) (DONS). Top tracing from 382 nm detector; bottom tracing from 254 nm detector. Column, DuPont Zipax SAX (100 cm × 2.1 mm i.d.); primary eluent, 0.01 M disodium tetraborate solution; secondary eluent, 0.01 M disodium tetraborate solution containing 0.2 M sodium perchlorate. [Redrawn with permission from Cox and Reed (1981).]

cited above make use of strong anion-exchange (SAX) columns (1 m × 2.1 mm i.d.), operated at ambient temperature. In all cases, the technique of gradient elution has been used where the primary eluent was an aqueous solution of 0.01 M disodium tetraborate and the secondary eluent was an aqueous solution of 0.01 M disodium tetraborate containing sodium percholate at a concentration of 0.20–0.5 M. Flow rates of between 0.5 and 1.0 ml min$^{-1}$ were most commonly used and detection was by UV absorbance measurement at 254 nm, usually connected in series with a colorimeter.

A major problem in ion-exchange chromatography has been the heat conditioning of new SAX columns. Normally this involves heating the column for 50 h at 50 °C whilst pumping the primary eluent. As this conditioning time may vary from column to column, no two heat-conditioned columns will be identical. Several time-consuming blank gradient runs are then required to remove UV-absorbing material from the column, followed by a 10–15 min equilibration using the primary eluent. Bailey and Cox (1976) mention the use of an improved SAX column that does not require the heat-conditioning treatment. The columns also suffer from a lack of reproducibility attributable to degradation of the resin over 6 months of continuous use (Singh, 1974b). Attina and Ciranni (1977) reported the use of a strong anion-exchange (SAX) column (1 m × 2.1 mm i.d.) with a mobile phase of dilute hydrochloric acid and sodium citrate (pH 4) for the investigation of a number of synthetic dyes (E102, E104, E110, E123, E126 and E131) used in food and in pharmaceutical formulations. Azo dyes and sulphonated intermediates have also been analysed using weak anion-exchange (WAX) columns with 0.01 M aqueous citric acid (pH 2.8) as the mobile phase (Passarelli and Jacobs, 1975). These workers also reported the separation of five anthraquinone dyes in 90 min using a "Zipax" prepacked hydrocarbon polymer (HCP) column with a water–ethanol (50 : 50, v/v) mobile phase at 50 °C.

## C  Reversed Phase Methods

Non-ionic substances can usually be chromatographed on reversed phase media in the absence of buffers. However, because of their hydrophilic nature most dyestuffs, and other ionic substances, frequently require buffered eluents to achieve the desired selectivity, retention and peak symmetry. Chromatography of ionic substances with unbuffered eluents generally leads to poor results because of band-broadening and low retention. The mobile phase may alter the affinity of ionic compounds for the stationary phase by three possible mechanisms: (1) buffers may be used to suppress ionization; (2) the presence of ion-pairing reagents may lead to the formation of hydrophobic ion-pairs and hence increase solute retention (Knox and Laird, 1976); (3) buffers may simply reduce the solubility of the solute in the mobile phase by a "salting-out effect", also known as solubility suppression (Bailey, 1980).

Reversed phase gradient elution chromatography with buffered eluents has produced resolution of dye components in a relatively short time. FD&C Blue No. 2, its intermediates and decomposition products have been analysed (Bailey, 1980) using gradient elution (Fig. 10.2). Martin, Tenenbaum, Alfonso and Dyer (1978) have investigated various synthetic acid-fast dyes: FD&C Blue No. 1, FD&C Blue No. 2, FD&C Red No. 2, FD&C Red No. 3, FD&C Red No. 4, FD&C Red No. 40, FD&C Yellow

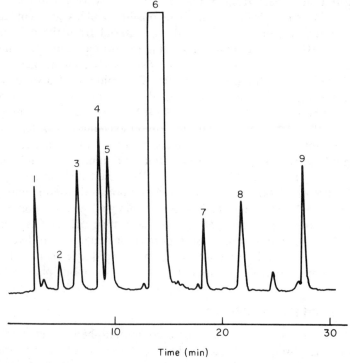

**Fig. 10.2** Liquid chromatogram of FD&C Blue No. 2 spiked with possible contaminants. Peaks: 1, 5-sulphoanthranilic acid; 2, isatin-4-sulphonic acid; 3, isatin-5-sulphonic acid; 4, isatin-7-sulphonic acid; 5, anthranilic acid; 6, FD&C Blue No. 2; 7,5,7'-disulphoindigo; 8, isatin; 9, 5-sulphoindigo. Altex Ultrasphere (5 $\mu$m) ODS column (250 mm × 4.6 mm i.d.) detection at 254 nm; primary eluent water–acetonitrile–ammonium acetate (99.5 : 0.5 : 1.5, v/v/m); secondary eluent, water–acetonitrile–ammonium acetate (50 : 50 : 1.5, v/v/m); flow rate, 1.0 ml mn$^{-1}$.
[Redrawn with permission from Bailey (1980).]

No. 5, FD&C Yellow No. 6 and FD&C Green No. 3, as found in alcoholic beverages. Gradient elution was again used on a RP-8, 10 $\mu$m particle size column (250 mm $\times$ 4.6 mm i.d.) in which the primary eluent was aqueous 0.01 M potassium dihydrogen orthophosphate and the secondary eluent methanol. Wolfgang (1978) has separated the azo dyes E110 and E111 on LiChrosorb RP-2, RP-8 or RP-18 with 0.021 M phosphate buffer (pH 5.63, 6.85 or 8.08)–methanol (10 : 3, v/v) as mobile phase with detection at 480 nm.

Efficient separations of some aromatic sulphonic acids and dyestuffs have been achieved using reversed phase ion-pair chromatography. Retention of solutes and the selectivity, to a certain extent, can be controlled by adjusting the type and concentration of ion-pairing reagent added (usually a tetralkylammonium salt such as cetyltrimethylammonium bromide or tetrabutyl-, tetraethyl- and tetramethylammonium phosphate or sulphate) and by selection of the type and concentration of the organic solvent in the mobile phase. Addition of inorganic electrolytes in minor concentrations to the mobile phase or adjustment of its pH can also be used to improve the separation.

Knox and Laird (1976) demonstrated the separation of various sulphonic acids and derived azo dyes (Sunset Yellow, Triazine, Carmosine, Tartrazine, Ponceau MX, Amaranth and Ponceau 4R) using a 7 $\mu$m SAS silica column packing (Wolfson Unit reversed phase material) and a water–propanol (5 : 2, v/v) |eluent containing about 1% (m/v) cetyltrimethylammonium bromide (cetramide). Resolution of the dyestuffs and intermediates was also achieved on a silica gel Partisil column (100 or 120 mm $\times$ 5 mm i.d.) using a much less hydrophilic mixture of water–propanol (1 : 3, v/v) again containing 0.5–2.0% (m/v) cetrimide.

Thirty-three dyestuffs including red, orange, yellow, green, blue, brown and black colours have been investigated by Chudy, Crosby and Patel (1978). By using a reversed phase column (120 mm $\times$ 4.6 mm i.d.) packed with 4.6 $\mu$m SAS-Hypersil and an eluent composition of propan-2-ol–water–cetrimide–glacial acetic acid (41 : 59 : 0.25 : 0.25, v/v/m/v), nine red dyes were resolved in 20 min, with a tenth, Ponceau 4R, eluted as a broad band after 37 min (Fig. 10.3).

Tartrazine has been studied using the reversed phase mode by Wittmer, Nuessle and Haney (1975). These workers used a 10 $\mu$m particle size $\mu$-Bondapak $C_{18}$ column (300 mm $\times$ 4 mm i.d.) and a mobile phase of methanol–water–formic acid (400 : 400 : 1, v/v/v) with ion-pairing reagents such as tetrabutylammonium hydroxide (TBAH), tetraethylammonium hydroxide (TEAH) and tridecylamine (TDA) and various combinations of these reagents at concentrations of between $3 \times 10^{-4}$ M and $1 \times 10^{-3}$ M, to effect separation of tartrazine from its synthetic intermedi-

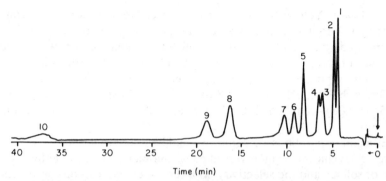

**Fig. 10.3** Separation of 10 red dyes. 5 μl of a mixture of 0.1% (m/v) solutions of each dye were injected. Peaks: 1, FD&C Red No. 40 (Allura Red); 2, Red 6B; 3, FD&C Red No. 3 (erythrosine); 4, Fast Red E; 5, Red 2G; 6, Carmoisine (E122); 7, FD&C Red No. 2 (Amaranth); 8, FD&C Red 4 (Ponceau SX); 9, Ponceau 6R; 10, Ponceau 4R. [Redrawn with permission from Chudy *et al.* (1978).]

ates sulphanilic acid and pyrazolone T in 8 min. The food dyes FD&C Yellow No. 6 (synonyms E110, Sunset Yellow), E111, and Ponceau 4R (E124), extracted from canned fish, have been separated on a LiChrosorb RP-8 column with a mobile phase of water–acetone (4 : 1, v/v) containing tetrabutylammonium chloride (0.2 g l$^{-1}$) (Aitzetmüller and Arzberger, 1979).

The work-up procedure involved removing residual oil by washing with chloroform followed by deproteination by boiling the homogenized sample, cooling and filtering. The residue was then washed with ammonium hydroxide–water (1 : 49, v/v) and the combined filtrate washings concentrated *in vacuo*. This concentrate was applied to a column of Sephadex LH-20 from which the dyes were eluted with water, and re-concentrated before injection on to the HPLC column.

A reversed phase system employing ion-pairing techniques has been reported by Masiala-Tsobo (1979) for the separation of 11 synthetic dyes including tartrazine, Orange GGN and Ponceau 4R. Dyes were eluted isocratically from a Nucleosil 10 μm C$_8$ column (250 mm × 4 mm i.d.) with water–methanol (11 : 9, v/v) containing 5 mM tetrabutylammonium phosphate (TBAP), at a flow rate of 1.0 ml min$^{-1}$. Lawrence, Lancaster and Conacher (1981) have also reported a reversed phase method, using various methanol–water ratios (45 : 55 or 60 : 40 v/v) each containing 5 mM TBAP, for the determination of 12 synthetic food colours. Solutions containing a series of food dyes were analysed at a compromise wavelength of 480 nm for the red, orange, yellow colours and 610 nm for the blue and green colours. An increase in flow rate during the run was employed to accelerate the elution of Fast Green FCF and Brilliant Blue FCF (Fig. 10.4). Puttemans, Dryon and Massart (1981) have examined the effect of methanol concentration and tetrabutylammonium ion-pair reagent concentration on the capacity ratios of various food dyes: E102, E103, E104, E105, E110, E111, E1232, E123, E124, E127, E131, E132, E142 and E151.

Boley et al. (1980) gave examples of the extraction of synthetic dyes from water-soluble and water-insoluble foods and their separation on reversed phase columns employing methanol–water–cetrimide mobile phases. Dyes found to be irreversibly bound to foodstuffs were released by a preliminary enzyme digestion stage prior to extraction.

The incorporation of a strong inorganic electrolyte into the mobile phase can lead to efficient and highly selective separations of aromatic sulphonic acids on reversed phase columns (Jandera and Churacek, 1980). With the addition of a strong electrolyte in a relatively high concentration to the eluent, usually 0.1 M or more, strongly polar and ionic compounds are retained and can be eluted as narrow symmetrical peaks. An increase in

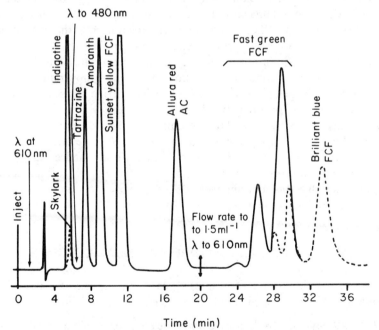

**Fig. 10.4** Chromatogram of eight colours. Mobile phase, methanol–water (45 : 55, v/v) containing 0.005 M TBAP; flow rate, 1.0 ml min$^{-1}$ up to 20.0 min, then changed to 1.5 ml min$^{-1}$. Wavelength changes were made as indicated. Skylark and Brilliant Blue FCF peaks are in dashed lines. [Redrawn with permission from Lawrence et al. (1981).]

the concentration of the electrolyte or a decrease in the concentration of the organic solvent promotes retention on reversed phase columns.

Utilizing this technique with gradient elution, where the primary eluent was aqueous 0.4 M sodium sulphate and the secondary eluent methanol–water (40 : 60, v/v) on a $C_{18}$ (10 $\mu$m) column (300 mm × 4.2 mm i.d.) has enabled a series of 10 naphthalenesulphonic acids to be analysed in less than 30 min.

## D   Amino Bonded-Phase Methods

An amino bonded phase has been used by Steuerle (1979) for the determination of dyes in confectionery, liqueurs, red wine and mock caviar. A sample enrichment technique was used to concentrate a solution of the dyes (e.g. 1 ml) on the top of a LiChrosorb $NH_2$ (10 $\mu$m) column (250 mm × 4 mm i.d.). A mobile phase of acetonitrile–water–glacial acetic acid–(ammonium hydroxide, 26.3%) (500 : 380 : 15 : 14, v/v/v/v) was applied to elute the dyes which were detected at their wavelength of maximum absorption. Jones, Hoar and Sellings (1978) have investigated the six major dye components and three non-dye components of Brown FK on amino bonded and ODS columns. Improved resolution of the disubstituted dye components and of the diamino starting materials was achieved on the reversed phase column using linear gradient elution from 5 to 40% acetonitrile in water containing sodium dihydrogen phosphate (1.2 g l$^{-1}$) and sodium hydrogen phosphate (2.4 g l$^{-1}$). Tartrazine has been quantified in a variety of food products by Hurst, McKim and Martin (1981). These workers have used a Zorbax $NH_2$ column (250 mm × 4.6 mm i.d.) with a mobile phase of 0.7 M sodium acetate at pH 5.0 to elute tartrazine in less than 10 min. Recoveries of 93–97% from three food matrices suggest that the method is accurate. Samples containing greater than 5% fat were ground in a mill or blender and were defatted by extraction with petroleum ether prior to dye extraction.

## III   NATURAL AND NATURE-IDENTICAL COLOURS

Plants are a major source of naturally occurring pigments for use as food colourants. These plant pigments may be classified as follows:

(1) Anthocyanins (red, orange and blue) are responsible for the colours of many vegetables and fruits, e.g. grapes, strawberries, raspberries, blueberries, cranberries, apples and roses.

(2) Betacyanins (red) are found in red beets, cactus plants, red chard and bougainvillea flowers.

(3) Carotenoids (red, orange and yellow) are a large group of compounds which are widely distributed in animal and vegetable products. They are responsible for the colouration in carrots, red tomatoes, apricots, orange juice, paprika, marigold petals and lobsters.

(4) Chlorophylls (green and olive green) are the pigments responsible for the colour of green leaves, vegetables and some fruits.

The use of natural pigments to colour foodstuffs has been limited due mainly to the fact that they are not as stable as artificial colours. Anthocyanins are pH sensitive, becoming bluish and unstable at a pH greater than 4.5. Usage is restricted to mainly medium-acid products such as soft drinks, wines, jams and jellies. Carotenoids are sensitive to oxidation and are readily bleached when exposed to oxygen and sunlight. However, the use of antioxidants such as ascorbic acid and tocopherols can reduce this oxidative degradation substantially. Carotenoids are used in both fat- and water-based foods. Fat-based products include butter, margarine and processed cheese, while water-based applications include ice-cream and soft drinks. The various chlorophylls used as food colourants include chlorophyll itself and copper-complexed derivatives. These compounds exhibit fairly good heat and light stabilities but their stabilities towards acids and alkalis are poor. Chlorophylls, for example, are easily degraded in acid solution to olive-brown compounds known as pheophytins. Their main uses are in canned products, soups and confectionery.

## A  Extraction Techniques

The quantitative extraction and determination of natural pigments is not easy due mainly to their innate physicochemical properties. Carotenoids, for example, are readily oxidized, sensitive to light and thermally unstable. The extraction of carotenoids must, therefore, be performed under low light intensity and, where practical, under nitrogen. Stewart (1977) has extracted carotenoids from fruit juice by adsorption on to magnesia (Fisher S-120 infusorial earth). The fruit juice was drawn through a layer of magnesia which was then extracted with dichloromethane–methanol (1 : 1, v/v) to remove adsorbed carotenoids.

Chlorophyll pigments have been isolated (Schwartz, Woo and von Elbe, 1981) by extraction with acetone. Diethyl ether was then added and the mixture washed with 5% sodium sulphate solution. The aqueous layer was then removed and washed with further aliquots of ether and the combined ether extracts dried over anhydrous sodium sulphate. Concentration under a stream of nitrogen afforded a residue suitable for analysis by HPLC, after being redissolved in a small volume of solvent. Red beet pigments have been isolated by adsorption from acid medium (pH 1) on to Dowex

50W-X4 resin (H⁺ form) from which they were eluted with water (Andrey, 1979).

## B  Chlorophyll and Carotenoid Pigments

HPLC allows separation of complex pigment mixtures under reproducible conditions without due exposure to air or light. Sources of naturally occurring pigments that have been studied by HPLC include spinach and the brown sea diatom *Nitzschia closterium* which represent systems containing chlorophyll and carotenoid pigments (Eskins, Scholfield and Dutton, 1977). HPLC was carried out using reversed phase partition chromatography on two (61 cm × 7 mm i.d.) stainless steel columns in series packed with 37–75 $\mu$m Bondapak $C_{18}$–Porasil B. A programmed gradient elution of aqueous methanol to ether in methanol was used to elute seven pigments: chlorophyll *c*, fucoxanthin, neofucoxanthin, diadinoxanthin, diatoxanthin, chlorophyll *a*, carotene and pheophytin in a total of 274 min. A gradient elution of 98% methanol and then 50% ethyl acetate in methanol enabled seven spinach pigments – neoxanthin, violaxanthin, chlorophyll *b*, lutein, chlorophyll *a*, pheophytin and carotene – to be eluted in 96 min. Semi-preparative columns were used in order to obtain large quantities of plant pigments for subsequent work. Analysis on analytical columns reduced the total elution time to less than 1 h. Iriyama, Yoshiura and Shiraki (1978) reported a microscale method for the separation of chlorophylls using a silica gel (0.5 $\mu$m) column (65 mm × 0.5 mm i.d.), and an eluent composition of hexane–propan-2-ol (99 : 1 or 98 : 2, v/v) at a flow rate of 16 $\mu$l min⁻¹. These techniques were time-consuming for routine analyses, requiring 75 min for the microscale separation and over 4 h for the preparative procedure. Braumann and Grimme (1979) shortened the analysis time to about 40 min and resolved 15 different components from the green algae *Chlorella fusca* by using a $C_{18}$ reversed phase column (300 mm × 3 mm i.d.) and a step-gradient technique.

Shoaf (1978) has reported the separation of chlorophyll *a* and chlorophyll *b* in less than 25 min, using a reversed phase column with methanol–water (95 : 5, v/v) as a solvent. A procedure for monitoring changes in chlorophyll during processing of spinach has been described by Schwartz *et al.* (1981). Chlorophylls *a* and *b* were the only pigments detected in fresh spinach. In canned spinach, almost all the chlorophyll had been converted to pheophytins and pyropheophytins. The pigments were eluted from a $\mu$-Bondapak $C_{18}$ column with gradient elution where the primary eluent was methanol–water (3 : 1, v/v) and the secondary eluent was ethylacetate, at a flow rate of 2.0 ml min⁻¹.

Hajibrahim, Tibbetts, Watts, Maxwell and Eglinton (1978) have applied HPLC to the analysis of carotenoid mixtures. The separations were obtained on silica columns and pigment components were eluted with a gradient of acetone in hexane.

## C  Anthocyanin and Betacyanin Pigments

A publication by Wulf and Nagel (1976) on the separation of phenolic acids and flavonoids using reversed phase material was soon followed by methods for separating anthocyanidins (Adamovics and Stermitz, 1976; Wilkinson, Sweeny and Iacobucci, 1977). Plant extract hydrolysates were eluted from a $\mu$-Bondapak $C_{18}$ column with an appropriate mixture of water–methanol–acetic acid (71 : 19 : 10, or 75 : 20 : 5, v/v/v) as the mobile phase. A modification of the original procedure for the separation of flavonoids and phenolic acids enabled Wulf and Nagel (1978) to separate 21 Cabernet Sauvignon grape pigments. The method used a LiChrosorb Si-60 column treated with octadecyltrichlorosilane with elution of pigments effected by a linear gradient between aqueous formic acid and aqueous methanolic formic acid solutions (Fig. 10.5).

A reversed phase preparatory HPLC method has been developed to obtain crystalline betacyanin pigments, betanin and betanidin which are the principal red pigments in beet colour (Schwartz and Elbe, 1980). Analysis of the betacyanin residue was carried out on a $\mu$-Bondapak $C_{18}$ column, which was developed isocratically with methanol–0.05 M potassium dihydrogen orthophosphate (18 : 22, v/v) adjusted to pH 2.75 with orthophosphoric acid. The eluate was monitored at 535 nm. The more polar glucosides betanin and isobetanin were eluted first, followed by the less polar aglycons betanidin and isobetanidin, in a total analysis time of 20 min.

Red beet betacyanins and betaxanthins have also been investigated by Vincent and Scholz (1978). Isocratic and gradient elution techniques were employed on $C_{18}$ reversed phase microparticulate columns employing water–methanol mixtures containing ion-pairing reagents such as tetrabutylammonium phosphate or aliphatic sulphonic acids (Waters Associates PIC A and PIC B reagents).

In conclusion it is only reasonable that natural and nature-identical colours added to our food should eventually receive adequate toxicological screening, as do other food additives. However, because many natural colours are so variable in composition and defy firm chemical characterization, the significance of some toxicological evaluations would be very doubtful. This may restrict the usage of natural pigments to those that can be easily characterized. At present HPLC can establish the pigment profile of a large number of plants, fruits, juices or wines, but no methods exist for the complete quantitative analysis of pigment components. The situation is complicated by the fact that reproducibility of HPLC chromatographic

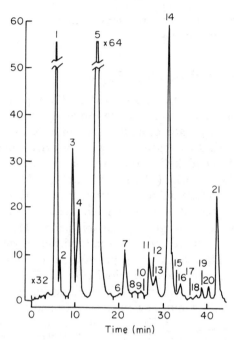

**Fig. 10.5** Separation of the anthocyanins of Cabernet Sauvignon grapes. Abbreviations: Dp, delphinidin; Cy, cyanidin; Pt, petunidin; Pn, peonidin; Mv, malvidin; Gl, monoglucoside; Ac, acetate; Coum, *p*-coumarate; Caf, caffeoate. Peaks: 1, Dp-3-Gl; 2, Cy-3-Gl; 3, Pt-3-Gl; 4, Pn-3-Gl; 5, Mv-3-Gl; 6, unknown; 7, Dp-3-Gl-Ac; 8, Cy-3-Gl-Ac; 9, unknown; 10, unknown; 11, Pt-3-Gl-Ac; 12 Mv; 13, Pn-3-Gl-Ac; 14, Mv-3-Gl-Ac; 15, unknown; 16, Dp-3-Gl-Coum; 17, Cy-3-Gl-Coum; 18, Mv-3-Gl-Caf; 19, Pt-3-Gl-Coum; 20, Pn-3-Gl-Coum; 21, Mv-3-Gl-Coum. [Redrawn with permission from Wulf and Nagel (1978).]

data is found to be dependent upon the pH of the mobile phase, the operating pressure and the temperature at which the chromatography is performed (Williams, Hrazdina, Wilkinson, Sweeny and Iacobucci, 1978). Thus a most important progression in methodology will be the application of HPLC to the quantitative determination of individual pigment components, and is an area that looks set for significant development in the next few years.

## REFERENCES

Adamovics, J. and Stermitz, F. R. (1976) *J. Chromat.;* **129**, 464.
Aitzetmüller, K. and Arzberger, E. (1979). *Z. Lebensmittelunters. u-Forsch.* **169**, 335.

Andrey, D. (1979). *Mitt. Geb. Lebensmittelunters. Hyg.* **70**, 237.

Attina, M. and Ciranni, G. (1977). *Il Farmaco-Ed. Pr.* **32**, 186.

Bailey, J. E., (1980). *J. Ass. off. analyt. Chem.* **63**, 565.

Bailey, J. E. and Cox, E. A. (1975). *J. Ass. off. analyt. Chem.* **58**, 609.

Bailey, J. E. and Cox, E. A. (1976). *J. Ass. off. analyt. Chem.* **59**, 5.

BFMIRA (1963). *Analyst, Lond.* **88**, 864.

Boley, N. P., Crosby, N. T. and Roper, P. (1979). *Analyst, Lond.* **104**, 472.

Boley, N. P., Bunton, N. G., Crosby, N. T., Johnson, A. E., Roper, P. and Somers, L. (1980). *Analyst. Lond.* **105**, 589.

Braumann, T. and Grimme, L. H. (1979). *J. Chromat.* **170**, 264.

Chudy, J., Crosby, N. T. and Patel, I. (1978). *J. Chromat.* **154**, 306.

Cox, E. A. (1980). *J. Ass. off. analyt. Chem.* **63**, 61.

Cox, E. A. and Reed, G. F. (1981). *J. Ass. off. analyt. Chem.* **64**, 324.

Dolinsky, M. and Stein, C. (1962). *J. Ass. off. agric. Chem.* **45**, 767.

Eskins, K., Scholfield, C. R. and Dutton, H. J. (1977). *J. Chromat.* **135**, 217.

Gilhooley, R. A., Hoodless, R. A., Pitman, K. G. and Thomson, J. (1972). *J. Chromat.* **72**, 325.

Graichen, C. (1975). *J. Ass. off. analyt. Chem.* **58**, 278.

Hajibrahim, S. K., Tibbetts, P. J. C., Watts, C. D., Maxwell, J. R. and Eglinton, G. (1978). *Analyt. Chem.* **50**, 549.

Hurst, W. J., McKim, J. M. and Martin, R. A., Jr (1981). *J. Fd Sci.* **46**, 419.

Iriyama, K., Yoshiura, M. and Shiraki, M. (1978). *J. Chromat.* **154**, 302.

Jandera, P. and Churacek, J. (1980). *J. Chromat.* **197**, 181.

Jones, A. D., Hoar, D. and Sellings, S. G. (1978). *J. Chromat.* **166**, 619.

Knox, J. H. and Laird, G. R. (1976). *J. Chromat* **122**, 17.

Lawrence, J. F., Lancaster, F. E. and Conacher, H. B. S. (1981). *J. Chromat.* **210**, 168.

Lehmann, G., Collet, P., Hahn, H. G. and Ashworth, M. R. F. (1970). *J. Ass. off. analyt. Chem.* **53**, 1182.

Marmion, D. M. (1975). *J. Ass. off. analyt. Chem.* **58**, 719.

Marmion, D. M. (1977). *J. Ass. off. analyt. Chem.* **60**, 168.

Martin, G. E., Tenenbaum, M., Alfonso, F. and Dyer, R. G. (1978). *J. Ass. off. analyt. Chem.* **61**, 908.

Masiala-Tsobo, C. (1979). *Analyt. Lett. A* **12**, 477.

Passarelli, R. J. and Jacobs, E. S. (1975). *J. chromatogr. Sci.* **13**, 155.

Puttemans, M. L., Dryon, L. and Massart, D. L. (1980). *Analyt. chim. Acta* **113**, 307.

Puttemans, M. L., Dryon, L. and Massart, D. L. (1981). *J. Ass. off. analyt. Chem.* **64**, 1.

Schwartz, S. J. and von Elbe, J. H. (1980). *J. agric. Fd Chem.* **28**, 540.

Schwartz, S. J. Woo, S. L. and von Elbe, J. H. (1981). *J. agric. Fd Chem.* **29**, 533.

Shoaf, W. T. (1978). *J. Chromat.* **152**, 247.

Singh, M. (1974a). *J. Ass. off. analyt. Chem.* **57**, 219.

Singh, M. (1974b). *J. Ass. off. analyt. Chem.* **57**, 358.

Singh, M. (1975). *J. Ass. off. analyt. Chem.* **58**, 48.

Singh, M. (1977a). *J. Ass. off. analyt, Chem.* **60**, 173.

Singh, M. (1977b). *J. Ass. off. analyt, Chem.* **60**, 1067.

Singh, M. and Adams, G. (1979). *J. Ass. off. analyt. Chem.* **62**, 1342.

Stanley, R. L. and Kirk, P. L. (1963). *J. agric. Fd Chem.* **11**, 492.

Steuerle, H. (1979). *Z. Lebensmittelunters. u. Forsch.* **16**a, 429.

Stewart, I. (1977). *J. ass. off. analyt. Chem.* **60**.
Takeshita, R., Yamashita, T. and Itoh, N. (1972). *J. Chromat.* **73**, 173.
Uematsu, T., Kurita, T. and Hamada, A. (1979). *J. Chromat.* **172**, 327.
Van Peteghem, C. and Bijl, J. (1981). *J. Chromat.* **210**, 113.
Vincent, K. R. and Scholz, R. G. (1978). *J. agric. Fd Chem.* **26**, 812.
Wilkinson, M., Sweeny, J. G. and Iacobucci, G. A. (1977). *J. Chromat.* **132**, 349.
Williams, M., Hrazdina, G., Wilkinson, M. M., Sweeny, J. G. and Iacobucci, G. A. (1978). *J. Chromat.* **155**, 389.
Wittmer, D. P., Nuessle, N. O. and Haney, W. G., Jr (1975). *Analyt. Chem.* **47**, 1422.
Wolfgang, F. (1978). *Dt. LebensmittRdsch.* **74**, 263.
Wulf, L. W. and Nagel, C. W. (1976). *J. Chromat.* **116**, 271.
Wulf, L. W. and Nagel, C. W. (1978). *Am. J. Enol. Vitic.* **29**, 42.
Yanuka, Y., Shalon, Y., Weissenberg, E. and Nir-Grosfeld, I. (1963). *Analyst, Lond.* **88**, 872.

# 11 Determination of Mycotoxins†

## D. C. HUNT
Laboratory of the Government Chemist, London, U.K.

## I INTRODUCTION

Mycotoxins are *secondary* chemical metabolites which can be produced by certain moulds during their growth. An indication of their toxicity can be obtained from their $LD_{50}$ values (i.e. the amount, usually expressed in milligrams per kilogram of bodyweight, required to kill 50% of a group of animals). Bullerman (1979), lists the $LD_{50}$ value of various mycotoxins ranging from aflatoxin $B_1$ at 0.36 mg kg$^{-1}$ for ducklings to penicillic acid at 600 mg kg$^{-1}$ for mice. Some of these chemicals, in addition to being toxic, have also been shown to be carcinogenic in animals, hence the interest in their detection in foodstuffs at the level of parts per thousand million.

In general, mycotoxins are relatively high molecular weight compounds containing one or more oxygenated alicyclic rings. The identification of individual compounds and their metabolites presents the analyst with a difficult problem, since over one hundred such compounds are known, and any individual compound is likely to be present in minute concentrations in a highly complex organic matrix. Until recently, the preferred methods for the assay of mycotoxins have involved the use of thin layer chromatography (TLC) for the final quantification. Detection is facilitated by the fact that some mycotoxins fluoresce under ultraviolet light and others can be visualized by various spray reagents.

High performance liquid chromatography (HPLC) has certain advantages over TLC, many of which are relevant to the analysis of mycotoxins. In addition to the advantages of speed of analysis, greater efficiency, increased sensitivity and better reproducibility, there is a greater safety factor for handling toxic chemicals, particularly in the case of preparative work as the toxins always remain in solution.

## II AFLATOXINS

Aflatoxins are the most important of the mycotoxins, and interest in their analysis dates back to the early 1960s following the outbreak of an unknown disease (called "Turkey X" disease) which resulted in the death of over 100 000 turkey poults (Blout, 1961). The cause was eventually traced to one of the feed components, mould-contaminated peanut meal, which contained aflatoxins. They can be formed during the growth on various foodstuffs of the common storage mould *Aspergillus flavus*, from which they derive their name.

The chemical structure of the aflatoxins is based on difurancoumarin and, although a number of different aflatoxin metabolites are known, interest is usually focused on the four main aflatoxins, which are $B_1$, $B_2$, $G_1$, $G_2$, and the so-called "milk toxin", aflatoxin $M_1$. The structures of these main aflatoxins are illustrated in Fig. 11.1.

Before a food sample can be assayed for its aflatoxin content, either by TLC or by HPLC, the toxin must first be extracted with a solvent and the resulting extract cleaned up using suitable techniques such as liquid/liquid partitioning or open column chromatography, on absorbants such as silica gel, alumina or Florisil, in order to remove any interfering co-extracts. The method most often used in the United Kingdom for the analysis of animal feedingstuffs for aflatoxins is contained in the *Official Journal of the European Communities* (1976) and is similar to the "CB" method of the Association of Official Analytical Chemists (1980) for peanuts and peanut products. Essentially, the methods consist of extracting 50 g of a ground homogeneous sample with 250 ml of chloroform and 25 ml of water in the presence of kieselguhr, by shaking in a stoppered flask for 30 min. After filtering, an aliquot is mixed with twice its volume of *n*-hexane and cleaned

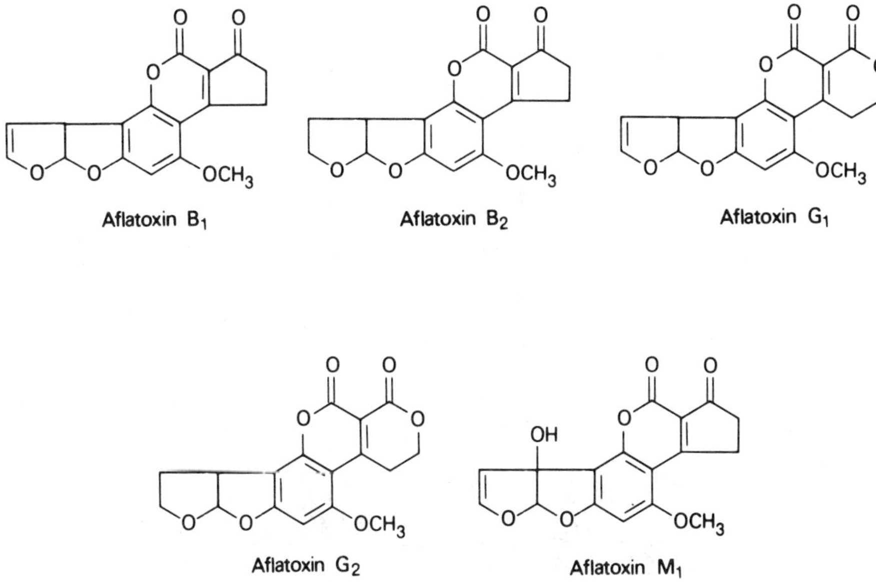

Aflatoxin B₁          Aflatoxin B₂          Aflatoxin G₁

Aflatoxin G₂          Aflatoxin M₁

**Fig. 11.1**  Structures of five main aflatoxin metabolites.

up on a chromatographic column containing 10 g of silica gel topped with 5 g anhydrous sodium sulphate. After washing the loaded column with 100 ml of diethyl ether to remove lipids and other interfering compounds, the aflatoxins are eluted using 150 ml of a chloroform–methanol (97 : 3 v/v) mixture. This final eluate is concentrated to dryness and the residue redissolved in a suitable solvent ready for assay by TLC. Although these methods use TLC for the final quantitation, this step may be replaced by normal phase HPLC. Figure 11.2 shows the chromatogram obtained with two extracts of naturally contaminated peanuts extracted by the "CB" method and subjected to normal phase HPLC. The 20 cm column contains LiChrosorb SI 60 (5 $\mu$m) and the mobile phase consists of 10% (v/v) acetic acid in water-saturated chloroform. Detection is by fluorescence using an excitation wavelength of 360 nm and an emission wavelength of 425 nm. The two samples contain 6 $\mu$g kg$^{-1}$ (p.p.b.) and 316 $\mu$g kg$^{-1}$ of aflatoxin B₁ respectively; the system is capable of detecting 1 ng of aflatoxin B₁. It has been observed that the strength of the fluorescence emission signals given by aflatoxins B₁ and G₁ are greater when they are bound to silica gel than when they are in solution. Advantage can be taken of this behaviour to enhance the response of the HPLC system, by packing the flow cell with silica gel. Most flow cells can easily be dry-packed with silica gel of 30–50 $\mu$m size, and full procedural details are contained in the original

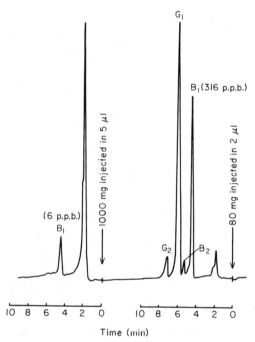

**Fig. 11.2**  Normal phase HPLC of peanut extracts.

paper of Panalaks and Scott (1977). Use of the packed flow cell restricts the mobile phase, in the main, to chlorinated solvents, since these have a suitable refractive index to make the packed cell transparent. Similarly, the mode of chromatography is restricted to normal phase HPLC. Each packed cell must be individually calibrated and occasionally repacked, since over a period of time the background fluorescence builds up, leading to a noisy base-line. Nevertheless, the fluorescence signal of aflatoxin $B_1$ can be considerably increased using this technique, leading to a 10- to 20-fold increase in sensitivity, so that the limit of detection is lowered to 50 pg. This refinement has been used by Pons (1979), for the determination of aflatoxins in corn.

It is very desirable in any analysis for aflatoxin to confirm the peak identity, and this can be achieved, for aflatoxins $B_1$ and $G_1$, by converting them to their corresponding hemiacetals ($B_{2a}$, $G_{2a}$) using acid hydrolysis with trifluoroacetic acid (Fig. 11.3), followed by further chromatography. The increased polar nature of the hemiacetals over their parent compounds leads to increased elution times on normal phase systems, but the opposite is true with the reversed phase mode. Aflatoxins can be separated on an

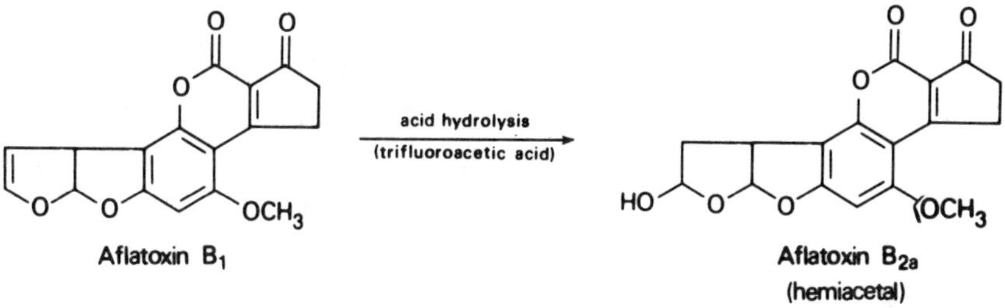

**Fig. 11.3** Formation of aflatoxin $B_{2a}$.

ODS column using an aqueous methanol mobile phase, but in such polar solvents the fluorescence of aflatoxins $B_1$ and $G_1$ is reduced, though that of $B_2$ and $G_2$ is unaffected. In the work of Takahashi (1977) for the determination of aflatoxins in wine using a reversed phase system, it was shown that by their conversion to hemiacetals, not only can their presence be chemically confirmed, but their sensitivity is increased such that about 100 pg can be detected. Furthermore, by using on-column concentration, 500 $\mu$l of sample extract can be loaded on to the analytical column to aid sensitivity. Beebe (1978) used both of these techniques in the determination of aflatoxins in green coffee, corn, raw peanuts, pistachio nuts and cottonseed meal, whilst Gregory and Manley (1981) have described a similar procedure to determine aflatoxins in animal tissues, dairy products and eggs. Figure 11.4 shows the determination of aflatoxins in peanuts by reversed phase HPLC using 500 $\mu$l loadings of extract both before and after the hemiacetal conversions.

Diebold, Karni, Zare and Seitz (1979a) have used a helium–cadmium laser to induce fluorescence in aflatoxin $B_1$. The beam, which operates at 325 nm, is focused on to a 4 $\mu$l droplet of eluate from the end of the analytical column, before impinging on a small spike and running to waste. The droplet serves as a windowless cell, and the resulting fluorescence is observed by a photomultiplier connected to a lock-in amplifier in phase with the laser modulation. It is claimed that 750 fg of aflatoxin $B_1$ can be detected as aflatoxin $B_{2a}$.

Confirmation of aflatoxin $B_1$ can also be carried out by the formation of an iodine derivative. Full details are given by Davis and Diener (1980), who chromatographed aflatoxin $B_1$ on an ODS column both before and after treatment with iodine. The elution time of the iodine derivative is virtually half that of the parent mycotoxin, and a 25-fold increase in fluorescence sensitivity is obtained.

Aflatoxin $M_1$ is a hydroxylated metabolite of aflatoxin $B_1$ (Fig. 11.1) and can appear in the urine and milk of animals which have been fed aflatoxin-

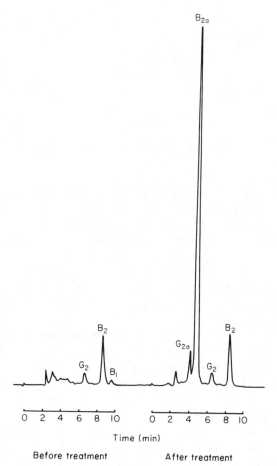

**Fig. 11.4** Reversed phase HPLC of peanuts, and the hemiacetal conversion of aflatoxins $B_1$ and $G_1$. Column, ODS Spherisorb; mobile phase, water–acetonitrile–methanol (15 : 3 : 2, v/v/v); injecting solvent, water–acetonitrile (9 : 1, v/v); detector, fluorescence (ex. 362 nm, em. 425 nm).

contaminated feed, thus it is important to be able to analyse milk products for aflatoxin $M_1$. Similar HPLC techniques are used for aflatoxin $M_1$ assay as are used for aflatoxin $B_1$, the main difference being in the mobile phase, which needs to be suitably modified to elute the more polar $M_1$ with a reasonable retention time. Normal phase HPLC with a silica gel-packed fluorescence flow cell has been used by Blanc (1980) for the analysis of a range of milk products including liquid milk, powdered milk, yogurt and cheese, whilst Winterlin, Hall and Hsieh (1979) have used the technique of on-column concentration with their reversed phase system to achieve a sensitivity of 0.1 $\mu$g l$^{-1}$ for unconverted aflatoxin M$_1$. In a similar way to aflatoxins B$_1$ and G$_1$, aflatoxin M$_1$ may be converted to its hemiacetal (M$_{2a}$), and this is the method favoured by Gregory and Manley (1981).

## III  OCHRATOXIN A

This mycotoxin may be produced by a number of fungi including *Aspergillus ochraceus*. It can occur in cereals, and animals which are fed rations contaminated with ochratoxin A can accumulate the unchanged mycotoxin in their kidneys.

The chemical structure of ochratoxin A, shown in Fig. 11.5, includes a carboxylic acid group which is used in the chemical confirmation of the mycotoxin, and a phenol group which can lead to poor chromatography. Suppression of ionization of the phenol group is achieved by acidification of the mobile phase with a suitable organic acid.

The mycotoxin can be determined in cereals by reversed phase HPLC using the method of Josefsson and Möller (1979), who used two 25 cm columns packed with Spherisorb ODS and a mobile phase of methanol–water (70 : 30, v/v) with the pH adjusted to 4.5 with acetic acid. Detection was by fluorescence with excitation at 340 nm and emission at 470 nm. Under acid conditions, ochratoxin A will fluoresce green, but under alkaline conditions a more intense blue fluorescence is obtained. Unfortunately, ochratoxin A is completely retained on an HPLC column with alkaline mobile phases and, furthermore, it is unwise to use solvents having a high pH with silica-based HPLC packings, since the packing material is liable to dissolve. It is possible to overcome these problems by chromatographing ochratoxin A under acid conditions and then use the simplest form of post-column derivatization by introducing a stream of

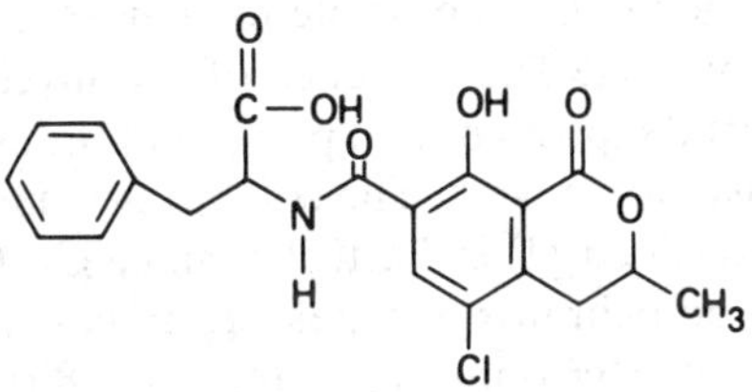

Ochratoxin A

**Fig. 11.5**   Structure of ochratoxin A.

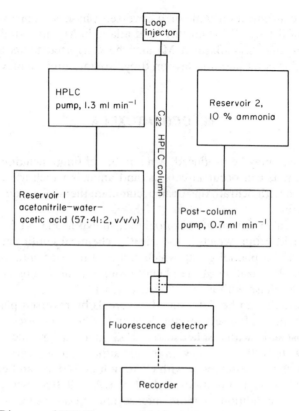

**Fig. 11.6**  Diagram of HPLC system with post-column derivatization to determine
ochratoxin A.

dilute ammonium hydroxide into the eluate after it emerges from the
column and before it enters the detector (see also Chapter 4). The work of
Hunt, Philp and Crosby (1979) on the determination of ochratoxin A in
pigs' kidneys uses such a system, which is shown diagrammatically in Fig.
11.6. Using this arrangement it is possible to achieve a 10-fold increase in
sensitivity for ochratoxin A. The presence of the mycotoxin can be confirmed by switching off the post-column pump and observing the reduction
in the height of the mycotoxin peak. The column packing used for the work
on pigs' kidneys is $C_{22}$ (docosyl) in Partisil 5, since $C_{18}$ (ODS) is only able
partially to separate the ochratoxin A peak from co-extracted material.

In a later paper, Hunt, McConnie and Crosby (1980) describe a method
for the chemical confirmation of ochratoxin A in which the mycotoxin
present in an extract is converted to its methyl ester by using a boron

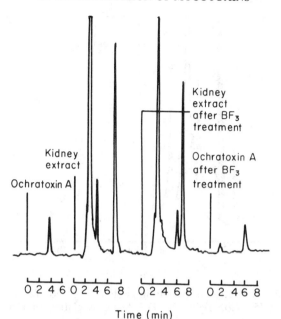

Kidney
extract
after BF₃
treatment

Ochratoxin A
after BF₃
treatment

Kidney
extract

Ochratoxin A

0 2 4 6   0 2 4 6 8   0 2 4 6 8   0 2 4 6 8

Time (min)

**Fig. 11.7** Chromatogram of extract of pig kidney, containing 5 $\mu$g kg$^{-1}$ of ochratoxin A; before and after derivatization. [Redrawn with permission from Hunt *et al.* (1980). Crown copyright.]

trifluoride-catalysed esterification with methanol. The chromatogram obtained using the post-column system and the chemical confirmation is illustrated in Fig. 11.7.

## IV  PATULIN

Patulin, which can occur in fruit, is a metabolite of several fungi, primarily species of *Penicillium* and *Aspergillus*. The mycotoxin absorbs light at 276 nm and this is the usual method for its detection in HPLC following either normal or reversed phase chromatography. Most of the survey work on patulin has been carried out on apple juice or apple products, and one rapid method for the determination of patulin in apple juice with reversed phase HPLC is that of Möller and Josefsson (1980) which uses Spherisorb ODS with distilled water as the mobile phase. The determination of patulin in apple juice is relatively straightforward with few interfering substances but with concentrated apple products co-extracted material causes far

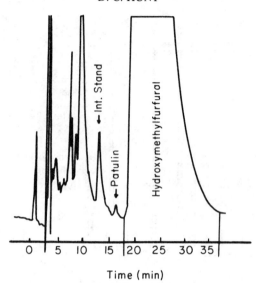

**Fig. 11.8** HPLC detection of patulin in apple butter. [Redrawn with permission from Ware (1975). Copyright © 1975 The Association of Official Analytical Chemists.]

more problems. Such a product is apple butter, which contains considerable amounts of 5-hydroxymethylfurfural, which absorbs in the ultraviolet region of the spectrum. A normal phase HPLC system has been described by Ware (1975) in which, by using a Zorbax Sil column and a mobile phase of 2,2,4-trimethylpentane–diethyl ether–acetic acid (750 : 250 : 0.5, v/v/v), patulin elutes before 5-hydroxymethylfurfural instead of occurring on the trailing edge as with previous HPLC systems (Fig. 11.8).

## V ZEARALENONE

Zearalenone, which is also known as F-2 toxin, may occur in cereals and can be produced by several species of *Fusarium*. The HPLC assay of zearalenone may be carried out using either normal or reversed phase chromatography and detection can be by either ultraviolet absorption or fluorescence. The majority of HPLC methods favour the use of reversed phase chromatography and Scott, Panalaks, Kanhere and Miles (1978) found that, although successful chromatography could be achieved for zearalenone using a Spherisorb silica packing, certain corn products gave interferences which could only be removed by connecting a second column of Spherisorb ODS to the first silica column.

Although ultraviolet absorption at 236 nm will give good sensitivity for zearalenone, fluorescence detection is to be preferred since a greater degree of selectivity is obtained, with co-extractants no longer interfering to the same extent. Ware and Thorpe (1978) used peak height ratios as confirmation of the mycotoxin during the determination of zearalenone in corn using fluorescence detection. The system consisted of a Spherisorb ODS column, a mobile phase of methanol–water (58 : 42, v/v) and a fluorescence detector with an emission cut-off filter of 418 nm and variable wavelength excitation. The height of the zearalenone peak in the sample extract was measured at excitation wavelengths of 236, 254, 274 and 314 nm, and their ratios compared with a standard solution of the mycotoxin.

As with aflatoxins, Diebold, Karni and Zare (1979b) have used laser fluorimetry for the determination of zearalenone by HPLC, but the increase in sensitivity over conventional detectors is not so spectacular as in the case of aflatoxins since the wavelength of the helium–cadmium laser at 325 nm is a considerable distance from the optimum wavelength for excitation of zearalenone at 236 nm. The detection limit for zearalenone is 300 pg compared with less than 1 pg for aflatoxin $B_1$.

## VI  OTHER MYCOTOXINS

Other mycotoxins which have been analysed by HPLC include roquefortine, citrinin, sterigmatocystin, penicillic acid and rubratoxin. All these compounds can be chromatographed on a reversed phase system using a suitable mobile phase.

Ware, Thorpe and Pohland, (1980), describe a method for the determination of roquefortine in blue cheese, the HPLC separation being obtained on a LiChrosorb RP-2 column using a mobile phase of water–methanol (50 : 50, v/v) plus 0.05 M monobasic ammonium phosphate. Detection of levels down to 16 ng $g^{-1}$ was obtained and confirmed by using two detectors, UV at 312 nm and an electrochemical detector, connected in series.

Marti, Wilson and Evans (1978) have determined citrinin in corn and barley using an ODS column with a mobile phase of 0.25 N orthophosphoric acid–acetonitrile (50 : 50, v/v) and detected the mycotoxin by fluorescence, exciting with UV light of 325–385 nm and recording the emission above 451 nm. With a similar system, but a mobile phase of 0.1 M potassium dihydrogen phosphate–acetonitrile (7 : 5, v/v) and UV detection at 254 nm, Stack, Nesheim, Brown and Pohland (1976) determined sterigmatocystin in corn and oats, whilst Kingston, Chen and Vercellotti (1976) found that a Partisil PAC column gave superior separations of sterigmatocystin and other metabolites of *Aspergillus versicolor* to that obtained with $\mu$-Porasil silica gel. The mobile phase used with the amino/ cyano packing was hexane–chloroform–acetic acid (65 : 35 : 1, v/v/v) and detection was by UV absorption at 254 nm.

Phillips *et al.* (1981) chromatographed penicillic acid on a $\mu$-Bondapak $C_{18}$ column using a mobile phase of acetonitrile–water (60 : 40, v/v) and UV detection at 254 nm, and also produced a pre-column derivative for penicillic acid by reacting it with diazomethane. Using the same HPLC conditions, the sensitivity of the diazomethane derivative is 4 ng, an increase of approximately two and a half times over the parent mycotoxin.

An HPLC system has been developed by Engstrom, Richard and Cysewski (1977) which will resolve a mixture of seven mycotoxins

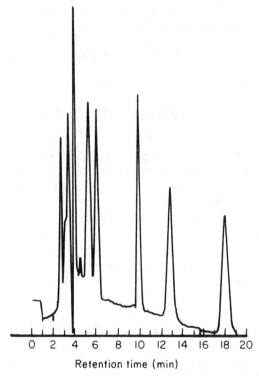

Retention time (min)

**Fig. 11.9** HPLC detection of seven mycotoxins. [Redrawn with permission from Engstrom *et al.* (1977). Copyright © 1977 The American Chemical Society.]

(aflatoxin $B_1$, aflatoxin $G_1$, ochratoxin A, zearalenone, rubratoxin B, patulin, and penicillic acid). The system, which will also resolve trichothecin from the other six mycotoxins in the absence of ochratoxin A, roseotoxin B from the other six mycotoxins in the absence of rubratoxin B, and sterigmatocystin from the other six mycotoxins in the absence of zearalenone, uses a $\mu$-Bondapak $C_{18}$ column and a mobile phase of acetonitrile–water–glacial acetic acid (45 : 55 : 2, v/v/v) with UV detection at 254 and 365 nm (Fig. 11.9). Such a system, which is capable of separating a number of different mycotoxin standards, could well form the final step in a multi-mycotoxin system, but the problem with food samples is the presence of comparatively large quantities of UV-absorbing co-extracted material in the final solution which can interfere with the detection of compounds of interest. One way of producing a very clean extract has been demonstrated by Hunt, Bourdon and Crosby (1978), in which a wide range of food samples including breakfast cereals, nuts, meat, baked beans and milk products are subjected to a clean-up procedure involving semipreparative HPLC on an alumina column before the final assay of the mycotoxins on an analytical HPLC column of silica gel. Improvements may be effected by use of solvent programming, column switching and UV/ fluorescence detectors in tandem.

# REFERENCES

Association of Official Analytical Chemists (1980). *Official Methods of Analysis*, 13th edn, Association of Official Analytical Chemists, Washington, D.C., Chap. 26, p. 26.026.

Beebe, R. M. (1978). *J. Ass. off. analyt. Chem.* **61**, 1347.

Blanc, M. (1980). *Industries Alimentaires et Agricoles*, **1980**, 893.

Blout, W. P. (1961). *J. Brit. Turkey Fedn*, **9**, 52.

Bullerman, L. B. (1979). *J. Fd Protection*, **42**, 65.

Davis, N. D. and Diener, U. L. (1980). *J. Ass. off. analyt. Chem.* **63**, 107.

Diebold, G. J., Karni, N., Zare, R. N. and Seitz, L. M. (1979a). *J. Ass. off. analyt. Chem.* **62**, 564.

Diebold, G. J., Karni, N. and Zare, R. N. (1979b). *Analyt. Chem.* **51**, 67.

Engstrom, G. W., Richard, J. L. and Cysewski, S. J. (1977). *J. agric. Fd Chem.* **25**, 833.

Gregory, J. F. and Manley, D. (1981). *J. Ass. off. analyt. Chem.* **64**, 144.

Hunt, D. C., Bourdon, A. T. and Crosby, N. T. (1978). *J. Sci. Fd Agric.* **29**, 239.

Hunt, D. C., Philp, L. A. and Crosby, N. T. (1979). *Analyst, Lond.* **104**, 1171.

Hunt, D. C., McConnie, B. R. and Crosby, N. T. (1980). *Analyst, Lond.* **105**, 89.

Josefsson, E. and Möller, T. (1979). *J. Ass. off. analyt. Chem.* **62**, 1165.

Kingston, D. G. I., Chen, P. N. and Vercellotti, J. R. (1976). *J. Chromat.* **118**, 414.

Marti, L. R., Wilson, D. M. and Evans, B. D. (1978). *J. Ass. off. analyt. Chem.* **61**, 1353.

Möller, T. E. and Josefsson, R. (1980). *J. Ass. off. analyt. Chem.* **63**, 1055.

*Official Journal of the European Communities* 15 April 1976, No. L. 102/9–18.

Panalaks, T. and Scott, P. M. (1977). *J. Ass. off. analyt. Chem.* **60**, 583.

Phillips, D. T., Ivie, G. W., Heidelbaugh, N. D., Kubena, L. F., Cysewski, S. J., Hays, A. W. and Witzel, D. A. (1981). *J. Ass. off. analyt. Chem.* **64**, 162.

Pons, W. A., Jr (1979). *J. Ass. off. analyt. Chem.* **62**, 586.

Scott, P. M., Panalaks, T., Kanhere, S. and Miles, W. F. (1978). *J. Ass. off. analyt. Chem.* **61**, 593.

Stack, M. B., Nesheim, S., Brown, N. L. and Pohland, A. E. (1976). *J. Ass. off. analyt. Chem.* **59**, 966.

Takahashi, D. M. (1977). *J. Ass. off. analyt. Chem.* **60**, 799.

Ware, G. M. (1975). *J. Ass. off. analyt. Chem.* **58**, 754.

Ware, G. M. and Thorpe, C. W. (1978). *J. Ass. off. analyt. Chem.* **61**, 1058.

Ware, G. M., Thorpe, C. W. and Pohland, A. E. (1980). *J. Ass. off. analyt. Chem.* **63**, 637.

Winterlin, W., Hall, G. and Hsieh, D. P. H. (1979). *Analyt. Chem.* **5**, 1873.

# 12  Determination of Amino Acids and Peptides

## A. P. WILLIAMS

National Institute for Research in Dairying, Shinfield, Reading, U.K.

## I  INTRODUCTION

Analysis of foods to determine their total amino acid composition has long been of importance nutritionally and there is a wealth of such data for traditional foods (Paul and Southgate, 1978). These authors reported that between 1967 and 1978 over 500 new foods were proposed by food scientists for inclusion in McCance and Widdowson's Tables of Food Composition, suggesting a continuing demand for such analyses. In addition there are an increasing number of other applications of amino acid analysis. New areas include the detection of adulteration of foods and beverages (Pechanek, Blaicher, Pfannhauser and Woidich, 1980), the detection of potentially toxic amino acids or peptides produced by new food processing technologies (Woodward and Alvarez, 1967). Peptide analysis has also been used increasingly in the detection of adulterants (Bailey, 1976) and in investigations into the production of flavours in foods (Schormüller, 1968).

Although there are many techniques for the analysis of amino acids, the use of automatic amino acid analysers using ion-exchange liquid chromatography (IEC) is the most popular by far. Many high performance liquid chromatography (HPLC) procedures for the estimation of amino acids have been published in recent years. Most of these achieve high speed (1 h or less), high resolution and high sensitivity by the use of microparticulate stationary phases and the subsequent need for high pressures. By this definition many of the older IEC methods would not qualify as high performance liquid chromatography so that only the more recent methods, post-1970, will be discussed here. The historical development of amino acid analysis and the problems associated with the analysis of pure proteins have been comprehensively reviewed by Hamilton (1966), Hare (1977) and Robinson (1978a). Many of these problems are exacerbated when these techniques are applied to food proteins and these will be discussed in this chapter.

## II  DETERMINATION OF AMINO ACID COMPOSITION

### A  Preparation of Protein Hydrolysates

The determination of the total amino acid composition of food proteins first necessitates hydrolysis of the protein into its constituent amino acids. Acids, alkalis or enzymes may be used for this but Hill (1965), in a review of the hydrolysis of pure proteins, concluded that the best method is treatment with 6 M HCl at 110 °C for 24 h. With the exception of methionine and cystine, which are oxidized to methionine sulphone and cysteic acid prior to hydrolysis, and tryptophan for which alkaline hydrolysis is necessary, the use of 6 M HCL remains the method of choice in most laboratories. Generally a finely ground sample (passing through a 40 mesh sieve) of the food containing about 10 mg N is hydrolysed with 400 ml 6 M HCl, under reflux, at 110 °C for 24 h. After cooling the solution is filtered and diluted to 500 ml and a 5 ml aliquot evaporated to dryness using a rotary evaporator. The residue may then be dissolved in a buffer suitable for loading onto the column of an amino acid analyser. Unfortunately it is generally agreed that the hydrolysis stage is still a major source of error in the amino acid analysis of food (Paul and Southgate, 1978). There are many variations of this procedure (Porter, Westgarth and Williams, 1968) and there have been many investigations (Otterburn and Sinclair, 1973; Davies and Thomas, 1973; Spitz, 1973; Savoy, Heinis and Seals, 1975), designed to improve and standardize this procedure for foods and feeds. The most comprehensive investigation has recently been published by Mason, Bech-Andersen and Rudemo in a series of papers (Mason *et al.*, 1979; 1980a,b; Bech-Andersen *et al.*, 1979a,b 1980; Rudemo *et al.*, 1979, 1980). Factors studied included the influence of the ratio of 6 M HCl to nitrogen, the use of sealed tubes versus open-reflux, the influence of hydrolysis time and stability of amino acids to oxidation with performic acid (with and without the addition of phenol as a halogen scavenger) prior to hydrolysis with 6 M HCl. It was concluded (Mason *et al.*, 1980b) that in routine feed analyses all amino acids, with the exception of tyrosine and tryptophan, could be satisfactorily determined in samples (0.1–1.0 g) containing 10 mg N oxidized with a performic acid/hydrogen peroxide reagent (containing 5 mg phenol ml$^{-1}$) for 16 h at 0 °C. Excess oxidizing agent was degraded with sodium pyrosulphite; formic acid was removed by rotary evaporation; and the residue was hydrolysed with 50 ml 6 M HCl (containing 50 mg phenol) under reflux at 110 °C for 23 h with argon slowly bubbling through the solution. After cooling, the pH of the unfiltered solution was adjusted to 2.2 with 7.5 M NaOH and the mixture diluted to 200 ml with pH 2.20, 0.2 M sodium citrate buffer. Insoluble material was removed by passing through a membrane filter before chromatography. This new procedure has optimized hydrolysis conditions and has advantages in speed and cost since most amino acids can be determined in one hydrolysate. A large scale collaborative trial organized by the EEC to evaluate this method has recently been completed. Its results are awaited with interest, although the problems encountered in tryptophan and tyrosine analyses remain.

## B  Preparation of Samples for Determination of Free Amino Acids

It is sometimes necessary to determine the concentration of free amino acids in beverages. Such samples often contain protein and although they can be analysed without "clean-up", particularly if a guard column is used, it is usual to deproteinize first. Methods of deproteinizing have included the use of picric acid (Stein and Moore, 1945), sulphosalicylic acid (SSA) (Hamilton, 1962), high speed centrifugation (Gerritsen, Rehberg and Waisman, 1965), ultrafiltration (Van Stekelenburg and Desplanque, 1966) and ion-exchange resin (Reid, 1966). The most commonly used method is precipitation with 3% (m/v) SSA (5 volumes : 1 volume of sample solution) followed by centrifugation to remove the precipitated protein. With suitable adjustment of pH the sample is ready for analysis. This procedure is popular because of its simplicity but it is not without its critics. Gerritsen *et al.* (1965), Reid (1966) and Knipfel, Christensen and Owen (1969) reported better precision for amino acid analysis carried out on picric acid-deproteinized blood plasma than for SSA-deproteinized blood plasma. Knipfel *et al.* (1969) and London (1966) suggested that this was due to impaired resolution if more than 200 mg of SSA were added to the column. However, Hamilton (1962), Block, Markovs and Steele (1966) and Ohara and Ariyoshi (1979) found that SSA was as good as other protein precipitants and until a simpler method, which avoids acid precipitation altogether, is found this will probably be the preferred method.

Ethanol (Ohara and Ariyoshi, 1979) and methanol (Hill, Walters, Wilson and Stuart, 1979) appear to be satisfactory protein precipitants and the latter results in a solution suitable for pre-column derivatization with *o*-phthalaldehyde/ethanethiol before HPLC. Some clinical studies suggest that traditional methods of deproteinization result in interferences with HPLC which can be avoided by ultrafiltration (Hartwick, Van Haverbeke, McKeag and Brown, 1979; Green and Perlman, 1980). There are occasions when the free amino acids in solid food samples have been estimated, e.g. in cocoa beans where free amino acids have been shown to constitute one of the major components of the precursors of chocolate aroma (Rohan and Stewart, 1966). These authors reported a procedure for the extraction of free amino acids from cocoa beans. Finely ground beans were blended with water and the suspension filtered. The filtrate was passed through a column of strongly acidic cation-exchange resin which acted as a guard column. The resin was washed with propan-2-ol to remove adsorbed flavonoids and then the amino acids were eluted with 4 N $NH_4OH$. When the eluent reached pH 10 it was evaporated to dryness and the residue redissolved for amino acid analysis without further treatment. This procedure could be used for other solid foods.

## C  Amino Acid Analysis

Spackman, Stein and Moore (1958) reported the development of an automatic analyser for separating and quantitatively determining amino acids in protein hydrolysates in 24 h. This ion-exchange chromatography (IEC) method is still the major method used in amino acid analysis, although there have been many modifications and improvements since 1958, which have resulted in the reduction of the analysis time to 2 h or less, without loss of resolution and with much increased sensitivity (nanomole to picomole level). The initial method used crushed sulphonated, divinylbenzene cross-linked, polystyrene ion-exchange resins with irregular, fairly large particles. The major improvements in performance have been mainly due to the studies of Hamilton, Bogue and Anderson (1960), who recognized the importance of using spherical resins with small, better defined particle sizes in achieving faster flow rates without loss of resolution. However, a further 5 years elapsed before advantage was taken of this commercially when Benson and Patterson (1965) published a procedure for the determination of amino acids in protein hydrolysates in 4 h by the use of spherical resins of mean particle size $22 \pm 6 \mu m$ (neutral and acid amino acids) or $15 \pm 6 \mu m$ (basic amino acids). Further improvements in spherical resins and attention to the other physical parameters, such as column dimensions, considered by Hamilton et al. (1960) to be important in the chromatographic process, led to reduction of this time to 2 h (Hubbard, 1965) and then to 63 min (Ertinghausen, Adler and Reichler, 1969). Further notable developments were due to Piez and Morris (1960) who introduced a single-column system with gradient instead of stepwise buffer elution which, although not reducing the speed of analysis at that time, was important for further advances. Although sensitivity is not usually a problem in food analysis the replacement of ninhydrin in the detection system by fluorescent reagents such as fluorescamine (Udenfriend et al., 1972) and o-phthalaldehyde (Benson and Hare, 1975) considerably improved the sensitivity and has been another major advance. However, since proline and hydroxyproline do not react with these fluorescent reagents, it is usually necessary to oxidize these amino acids by the post-column introduction of hypochlorite or $N$-chlorosuccinimide into the effluent stream (Felix and Terkelsen, 1973). Strangely there has been little change in the sodium citrate buffer systems proposed by Spackman, Stein and Moore (1958) apart from the use of lithium citrate buffers (Benson, Gordon and Patterson, 1967) for the analysis of free neutral and acidic amino acids in physiological and other fluids, which is still the preferred method today. A versatile gradient elution buffer system has been developed by Murren, Stelling and Felstead (1975) and is commercially available in the Chromaspek amino acid analyser (Rank Hilger, Margate, England). A pH gradient is derived by the accurate mixing of only two buffers, one acid (pH 2.2, 0.2 N sodium citrate) and the other alkaline (pH 11.5, 0.2 N sodium borate). Analysis of protein hydrolysates can be completed in less than 1 h using this system, which considerably simplifies buffer preparation.

These developments are perhaps best exemplified by the changes experienced in the author's own laboratory since 1964 and shown in Table 12.1. Also included for comparison are details of the conditions used, in what is generally considered to be the most advanced IEC system, the

**Table 12.1** Typical procedures for the analysis of amino acids in protein hydrolysates by ion-exchange chromatography (1964–79)

| System | Time (h) | Column (cm) | Resin particle size ($\mu$m) | Buffer composition | | Temperature (°C) |
|---|---|---|---|---|---|---|
| | | | | pH | Na$^+$ (N) | |
| Two-column† (1964) | | | | | | |
| Acidic and neutral | 20 | 0.6 × 150 | 40 | 3.25/4.25 | 0.20 | 50 |
| Basic | 4 | 0.6 × 15 | 25–30 | 5.28 | 0.35 | 50 |
| Two-column‡ (1969) | | | | | | |
| Acidic and neutral | 4 | 0.8 × 70 | 13 | 3.25/4.25 | 0.20 | 40/60 |
| Basic | 2 | 0.8 × 15 | 13 | 5.28 | 0.35 | 60 |
| One-column§ (1979) | | | | | | |
| Acidic | | | | 3.25 | 0.18 | |
| Neutral | 1.8 | 0.6 × 35 | 11 | 3.90 | 0.30 | 49/61 |
| Basic | | | | 4.75 | 1.6 | |
| One-column¶ (1976) | | | | | | |
| Acidic, neutral and basic | 0.8 | 0.18 × 48 | 8 | 3.21/3.25/4.25/9.45 | 0.2 | 56/65/75 |

†Evans Electroselenium Ltd, Halstead, England.
‡Jeol 5AH (Jeol Co. Ltd, Tokyo, Japan).
§LC2000 (Biotronik, Frankfurt, West Germany).
¶D500 (Durrum Instruments Co., Palo Alto, USA).

Durrum D500 amino acid analyser (Hewett and Forge, 1976), which has reduced the total analysis time to 0.5 h for hydrolysates and to less than 5 h for physiological fluids. Most commercial amino acid analysers currently available are capable of analysing hydrolysates in less than 1 h using two or three sodium buffer systems and physiological fluids in less than 5 h using four or five lithium buffer systems. Woodham (1974), in a survey of 36 commercially available amino acid analysers, highlighted the problems encountered and was severely critical of the amount of time lost through break-downs. This situation might well have been exacerbated by the much higher pressures used in instruments developed since 1974. Robinson (1978b), in a review of automatic amino acid analysers, concluded that incomplete resolution and loss of precision was a problem in some accelerated systems due to the compression of the chromatogram. Clearly, since many more unusual amino acids and amino acid derivatives formed during new food technologies, such as lysinoalanine, are being continually discovered, the problems of resolution assume even greater importance. Details of current commercial analysers have been published, e.g. the Rank Chromaspek (Bailey and Marks, 1977), the Hitachi 835 (Fujita, Takeuchi and Ganno, 1979), the LKB 4400 (Dilley, 1980) and the Carlo Erba 3A29 (Rossi and Trisciani, 1980). Ersser (1979a,b) has recently reviewed the current techniques and commercial instruments available and warns that unequivocal separation of naturally occurring amino acids in complex mixtures has not been achieved by IEC.

In compiling their tables of the amino acid composition of common foods, Paul and Southgate (1978) considered that although some analyses had been carried out by microbiological assay, for the vast majority IEC was still the method used. However, in recent years a number of HPLC procedures have been reported for the analysis of amino acids in standard mixtures and in hydrolysates of pure proteins. Unlike the ninhydrin detection system used in IEC, derivatization has usually been carried out before chromatography using phenylthiohydantoin (PTH), dansyl, 2-4-dinitrophenyl or o-phthalaldehyde derivatives. Improvements in column packing materials have resulted in increased resolution and the technique has emerged as the most suitable method for the identification of PTH-amino acids produced by the Edman sequence analysis of proteins and peptides. For example, Zimmerman, Appela and Pisano (1977) described the resolution of 20 PTH-amino acids in less than 20 min using reversed phase HPLC on a 25 cm × 0.46 cm Zorbax ODS column with a mobile phase of acetonitrile in 0.01 M sodium acetate (pH 4.5). The conditions, temperature and flow rate, required for this were rather critical. It is apparent from the plethora of papers on the subject that many laboratories have difficulty in obtaining adequate resolution of all 20 amino acids in a single run. Somack (1980) concluded that many laboratories had difficulty in reproducing published procedures due to confusion in establishing the precise chromatographic positions of the acidic and basic amino acid PTH derivatives, in particular those of aspartic and glutamic acids and histidine and arginine. Godtfredsen and Oliver (1980) summarized the experimental conditions and separations reported for synthetic mixtures of PTH-amino acids by HPLC in 35 papers published since 1973. The majority of analyses were performed using columns of reversed phase support material (nine different commercially available materials were used). It was concluded that reproducible, rapid and efficient separation of synthetic mixtures may be achieved on any of the commercially available hydrocarbon, $C_{18}$ and $C_8$ bonded, support materials using gradient elution with aqueous sodium acetate buffers and either acetonitrile or methanol. However, these studies were for synthetic mixtures, many containing far less than 20 amino acids, and although the application of these HPLC techniques to sequence studies is widespread there appears to have been little evidence of their use in analysing food or feed protein hydrolysates (Sloman, Foltz and Yersanian, 1981).

It is therefore all the more surprising to find that the first application of HPLC to food and beverage analysis has been in the determination of free amino acids. Hurst and Martin (1980) used o-phthalaldehyde derivatives with reversed phase HPLC to determine the free amino acids in cocoa beans in 35 min. However, resolution of certain amino acids was poor, and only 13 out of a possible 20 amino acids were detected. Schuster (1980) reported a method for the determination of free amino acids in intravenous solutions, orange juice, raspberry juice, wine and beer in 30 min without clean-up or derivatization. Separation was achieved on a column of 5 $\mu$m LiChrosorb $NH_2$ with an acetonitrile/phosphate buffer and amino acids were detected by UV absorbance at 200 nm. Twenty amino acids could be resolved in a standard mixture but the maximum number detected in any of the samples was 13. Whether the remaining amino acids were absent or incompletely resolved is unclear but high precision was reported for replicate analyses. Free amino acids in intravenous solutions have also been determined by HPLC (Alexander, Haddad, Low and Maitra, 1981) using a new copper tubular electrode as a potentiometric detector coupled to reversed phase HPLC. No sample pretreatment was required other than filtration. Although only 11 of the 15 amino acids present in the intravenous solution were resolved, no attempt was made to improve on this since the detection technique was still considered to be under development. Conrad (1979) reported the analysis by HPLC of free amino acids as their dansyl derivatives in unfermented beer. Resolution for leucine, phenylalanine and isoleucine was incomplete and the analysis time was 1 h.

## III  INDIVIDUAL AMINO ACIDS

### A  Methionine and Cystine

The sulphur amino acids are nutritionally very important because they are the limiting amino acids in many foods. Although they can be estimated by IEC they are difficult to measure because they undergo oxidation during the normal acid hydrolysis conditions used for most other amino acids. Methionine is oxidized to methionine sulphoxide and then to methionine sulphone, and cystine and cysteine are oxidized to cysteic acid. Under normal conditions variable amounts of these compounds are formed and although they can be detected after IEC the peaks are too small for accurate quantification. To overcome this problem controlled oxidation to methionine sulphone and cysteic acid using performic acid (Moore, 1963) is carried out prior to acid hydrolysis followed by IEC using a shortened standard neutral and acidic programme, since these oxidation products are eluted early in the chromatogram. Numerous other methods have been developed and the problems associated with them reviewed by Friedman and Noma (1975) and Walker, Kohler, Kuzmicky and Witt (1975).

Although the performic acid oxidation method of Moore (1963) is widely used for foods and feeds, the precision obtained in collaborative trials has been poor (Williams, Hewitt and Cockburn, 1979; Williams, 1981). The reason for this is not clear, although a contributory factor must be the low concentrations of methionine and cystine in foods. Incomplete oxidation is also likely to be a major source of error since Lipton, Bodwell and Coleman (1977) have reported that oxidation with hydrogen peroxide in the presence of HCl results in the conversion of only 75% of L-cystine to cysteic acid with lanthionine sulphoxide, lanthionine sulphone and a number of unidentified products also being formed. Attempts have been made to overcome this problem for cystine by converting it to a derivative that would be more stable to acid hydrolysis. Cavins and Friedman (1970) reported that 4-vinylpyridine met these requirements. Treatment with this reagent, after reduction with $\beta$-mercaptoethanol, converts cystine and cysteine to S-(4-pyridylethyl)-L-cysteine which, after hydrolysis, can be analysed by IEC using a basic amino acid programme. Results for the cystine content of a variety of foods analysed by this method were in good agreement with the performic acid oxidation method (Friedman, Krull and Cavins, 1970; Davies and Thomas, 1973) but it does not appear to have been widely adopted. Mercaptoethanol has also been used in an alternative method for methionine estimation (Keutmann and Potts, 1969) since its addition to the 6 M HCl before hydrolysis prevents oxidation of methionine to methionine sulphone. At the present time the method described by Mason *et al.* (1980b), in which it is claimed that all the amino acids, with the exception of tyrosine and tryptophan, can be determined in one hydrolysate by prior oxidation with performic acid in the presence of phenol, appears to be promising for the determination of the sulphur amino acids in food and feed samples.

There have been few applications of HPLC to the determination of sulphur amino acids in foods. Saetre and Rabenstein (1978) have reported the determination of free cystine and glutathionine in fruit. Fruit juice was acidified with $H_3PO_4$, centrifuged and filtered before analysis on a column of Zipax SCX with 0.5% $H_3PO_4$ as the mobile phase. Detection was with a mercury-based electrochemical detector and analysis time was 6 min. Under these conditions only the thiol compounds are detected. Free methionine in model food systems fortified with methionine has been estimated by O'Keefe and Warthesen (1978) and Tufte and Warthesen (1979) using HPLC. Following deproteinization with trichloroacetic acid or methanol and filtration, pre-column derivatization was by dansylation. Analysis was completed in about 5 min by reversed phase HPLC using a mobile phase of acetonitrile–0.10 M phosphate buffer (pH 7.0) on a $\mu$-Bondapak $C_{18}$ column and UV detection at 254 nm. It is doubtful if either of these HPLC methods for the estimation of free sulphur amino acids could be successfully applied to food hydrolysates until the hydrolysis problems have been solved.

## B  Tryptophan

Tryptophan cannot be estimated with any degree of precision (Williams, 1981) and this is abundantly clear from the multitude of methods that have been published; see Friedman and Finley (1975) for a review of these methods. Most of the problems are probably related to the hydrolysis stage since tryptophan is extensively degraded during the acid hydrolysis used for most other amino acids. Unlike the cases of cystine and methionine, ammonia is the only recognized product and it is normal to use alkaline or enzymic hydrolysis for this amino acid. Subsequent analysis can be by colorimetric or fluorimetric assay with or without a chromatographic stage. Several methods have used IEC and tryptophan is eluted from an ion-exchange column as a well resolved peak before lysine using a standard basic amino acid programme. This type of procedure was used by Knox, Kohler, Palter and Walker (1970) to estimate tryptophan in a variety of foods and feeds after hydrolysis with barium hydroxide. Comparison with two other methods showed good agreement for the samples tested.

Since 1977 there have been several HPLC methods published for the estimation of free tryptophan in biological materials (De Vries, Koski,

Egberg and Larson 1980). These authors have developed a procedure for the determination of protein-bound tryptophan in food products such as yeasts, soya and citrus products based on the method used by Krstulovic, Brown and Rosie (1977) for serum samples. After enzymic hydrolysis with pronase the samples were passed through a column of $\mu$-Bondapak $C_{18}$ using acetonitrile–sodium acetate buffer at pH 4.0 as the mobile phase. Tryptophan was eluted after 4 min and detected fluorimetrically without derivatization (i.e. using the native fluorescence of tryptophan). The chromatograms usually contained only one other peak, well resolved from tryptophan, and the results obtained compared favourably with a colorimetric method (Spies, 1967). Jones, Hitchcock and Jones (1981) have also recently published a method which uses the native fluorescence of tryptophan as a means of detection. Hydrolysis was carried out with 6 M NaOH in the presence of maltodextrin to prevent tryptophan degradation. After neutralization with 6 M HCl and filtration samples were chromatographed on a column of LiChrosorb SI 100 and eluted with a mobile phase of sodium acetate–acetic acid and the tryptophan detected fluorimetrically. The chromatogram was free of interfering peaks and took only a few minutes to complete. Good agreement was reported for samples of cereals and feeds when the method was compared with that of Slump and Schreuder (1969).

Although these new HPLC procedures are very rapid and sensitive it is generally considered (Friedman and Finley, 1975; Steinhart, 1979; Lucas and Sotelo, 1980) that the major problem is the hydrolysis procedure and that losses of tryptophan can occur even when alkalis are used. Lucas and Sotelo (1980) carried out an extensive study into the effect of different alkalis, temperatures and hydrolysis times on the determination of tryptophan in pure proteins and foods by IEC. They concluded that lithium hydroxide (4 M) at 145 °C gave the best results, but that the hydrolysis time was critical. The best time for pure proteins was 4 h and 8 h for food proteins. Alkaline treatment of foods during processing also leads to the production of new amino acids such as lysinolalanine and ornithinoalanine (Nashef et al., 1977), which are eluted close to tryptophan in IEC procedures and care should be taken in the interpretation of chromatograms.

## C  Lysine and Available Lysine

The protein quality of mixed diets for man or animals is often limited by the lysine content. Cereals supply a large proportion of the protein but are seriously deficient in lysine and considerable effort has been devoted to breeding high lysine strains which require rapid mass-screening techniques for the analysis of lysine. Unfortunately the classical IEC techniques are too time-consuming for the repetitive analysis of a single amino acid and several methods have been published to reduce the time taken for lysine analysis. Most involve the use of multiple columns (Bell and Mason, 1970; Dennison, 1976) to reduce the analysis time from 3 h to less than 1 h but suffer from being complex manual rather than automated methods. Any advantages in time saving seem to be offset by the labour intensity of these methods.

HPLC methods would seem to have more to offer for the rapid determination of lysine in food products. Warthesen and Kramer (1978) have published a method for the estimation of free lysine in lysine-fortified foods such as flour, dough and bread. The free lysine was extracted with trichloracetic acid and didansyl lysine prepared by derivatization with dansyl chloride. Separation was by reversed phase HPLC on a column of $\mu$-Bondapak $C_{18}$ with an acetonitrile–0.01 M $Na_2HPO_4$ (pH 7.0) mobile phase with UV detection at 254 nm. Analysis time was 15 min and good recoveries and reproducibility were claimed in studies of lysine losses during baking. A similar method has been described for the estimation of lysine in pharmaceutical products, but with pre-column derivatization using 1-fluoro-2,4-dinitrobenzene (FDNB) (Muhammad and Bodnar, 1980). Peterson and Warthesen (1979) have used the method of Warthesen and Kramer (1978) to determine the lysine content in hydrolysates of soya, casein, gluten and zein. Good agreement with published values obtained by IEC procedures was obtained.

It has long been recognized that the nutritive value of foods subjected to heating during processing or to adverse storage conditions might be less than would appear from measurements of total lysine in hydrolysates of the foods (Carpenter, 1960). Under such conditions lysine can become nutritionally unavailable if its free $\varepsilon$-amino group reacts with, for example, carbohydrates, forming bonds which are resistant to the enzymes of the digestive tract. The most widely used method (Carpenter, 1960) for the estimation of available lysine uses FDNB to react with the free $\varepsilon$-amino groups of lysine and then the DNP-lysine, is measured, spectrophotometrically after acid hydrolysis. This method is laborious and problems arise with foods containing high levels of carbohydrate due to the formation of interfering compounds, so that there have been many attempts to find a more suitable method. One such approach has been to measure the available lysine indirectly by the so called "difference" method (Roach, Sanderson and Williams, 1967). The difference between the total lysine content after acid hydrolysis and the residual lysine content measured after acid hydrolysis of the sample reacted with FDNB is assumed to be the available lysine content. Total and residual lysine were measured by IEC using a standard programme for basic amino acids. This method is still fairly time-consuming and can suffer from interference from other peaks when the residual lysine peak is small, but it is considered to compare favourably with other methods for estimating available lysine (Milner and Westgarth, 1973; Holsinger and Posati, 1975; Rayner and Fox, 1978).

Hurrell and Carpenter (1974) have shown that the method of Roach *et al*. (1967) does not measure available lysine accurately in foods containing early Maillard products, i.e. those formed when proteins are heated in the presence of reducing sugars under mild conditions. However, this may not be a problem in commercial foods (Rayner and Fox, 1978). Hurrell and Carpenter (1976) have also suggested the use of a dye-binding method for the rapid estimation of available and total lysine in food and feed proteins which might eventually replace the spectrophotometric and chromatographic methods.

Attempts have also been made to separate DNP-lysine from interfering compounds by IEC and to use this as a measure of available lysine (Posati, Holsinger, DeVilbiss and Pallansch, 1974). Again the major disadvantage is the time taken for the analysis (40 min) and recently Peterson and Warthesen (1979) have proposed an HPLC method which can accomplish the analysis of available lysine in 15 min. The filtered hydrolysate of the dinitrophenylated protein containing DNP lysine is passed through a column of $\mu$-Bondapak $C_{18}$ with a mobile phase of acetonitrile 0.01 M acetate buffer at pH 4.0. The DNP lysine was well resolved from dinitrophenol and other DNP amino acids but the method gave lower results for the available lysine content of casein, soy and gluten/glucose mixtures compared with results obtained by FDNB method of Carpenter (1960), possibly due to the interference of other DNP compounds in this latter method.

## D  3-Methylhistidine

3-Methylhistidine is an analogue of histidine and is found mainly in skeletal muscle. Methylation of the histidine occurs after its incorporation into the peptide chains of actin and myosin. Since, after the catabolism of these proteins, the 3-methylhistidine is not recycled but quantitatively excreted in urine, it has been proposed as an index of muscle protein turnover. Considerable interest in its estimation has developed for this reason, but it has also been used to determine the meat content of food products (Hibbert and Lawrie, 1972). The use of non-meat proteins of vegetable or microbial origin to replace animal protein for economic reasons is extensive. Although such practices may have a minimal effect on the nutritive quality, texture and flavour of the product, their detection is of considerable interest for consumer protection. Hibbert and Lawrie (1972) proposed the use of 3-methylhistidine to measure meat protein content since it was considered to be a characteristic constituent of muscle protein. In studies to validate this theory, Rangeley and Lawrie (1976) used an IEC method based on a standard programme for the analysis of basic amino acids. Analysis took 8 h to complete for samples of beef, lamb, pork, whale meat, egg, milk and soya. Although 3-methylhistidine was only found in the meat samples its concentration was much more variable in pork than in beef or lamb. This could have been due to the presence of free 3-methylhistidine and the authors proposed to carry out further studies before unilaterally recommending the method as an index of meat content. The problems of analysing 3-methylhistidine by conventional IEC are the long analysis time and its low concentration relative to other amino acids. Ward (1978) managed to reduce the analysis time to 3 h and to overcome the latter problem by using ninhydrin-$o$-phthalaldehyde instead of ninhydrin. Skurray and Lysaght (1978), however, reduced the analysis time from 8 h to 10 min by using HPLC. After pre-column derivatization to form dinitro-phenyl amino acids, the neutral and acidic DNP-amino acids were extracted with ethyl ether. DNP-methyl histidine was estimated on a column of Spherisorb silica (5 $\mu$m) with water as the mobile phase. Electronic intergration of the peak areas from a 250 nm UV detector reduced the experimental error from 11.7 to 1.9% compared with the IEC method of Rangeley and Lawrie (1978). Application of the method to meat pies and hamburgers revealed 35–96% and 20–30% meat contents respectively.

HPLC has also been used to estimate free 3-methylhistidine and related compounds in fish muscle in 80 min (Abe, 1981). A column of Zipax SCX was eluted with 12–30 mM $KH_2PO_4$ followed by UV detection at 210 nm without derivatization. Excellent resolution of 3-methylhistidine, histidine, 1-methylhistidine, carnosine, anserine and balenine was reported. However, this method has not yet been applied to protein hydrolysates or to samples other than fish muscle. Other HPLC methods for the estimation of 3-methylhistidine have recently been published (Wassner, Schlitzer and Li, 1980; Friedman and Smith, 1980; Ward, Miller and Hawgood, 1981) but have been developed specifically for urine and plasma analysis. None of these methods involve prior removal of the neutral and acidic amino acids as used by Skurray and Lysaght (1978).

## E  Lysinoalanine

In recent years much interest has centred upon the formation of lysinoalanine (LAL), $N^\varepsilon$-D,L-[2-amino-2-carboxyethyl]-L-lysine, during alkaline or heat treatment of food proteins, due to reports concerning its possible toxic effects (Woodward and Alvarez, 1967; Woodward and Short, 1973). Since LAL is formed by an interaction of the $\varepsilon$-amino group of the lysine residues and the double bond of dehydroalanine, there may also be nutritional problems due to a loss of available lysine (De Groot and Slump, 1969).

Although lysinoalanine toxicity may be species specific to rats since other species have failed to exhibit renal lesions when given synthetic LAL (De Groot, Slump, Feron and Van Beek, 1976), the possible effect on the nutritional quality of the alkali-treated food is considered to be important. Alkaline treatment of food proteins is increasingly used in the food industry, for example to isolate protein from soya flour, to prepare textured proteins, to peel fruit and vegetables and to destroy aflatoxin in peanuts. Besides LAL, other new amino acids such as ornithinoalanine, lanthionine and $\beta$-aminoalanine may also be formed by alkaline treatment of proteins (Nashef et al., 1977).

The determination of LAL in foods is a classic example of the problems encountered in amino acid analysis. It is usually present in very low concentrations compared with other amino acids, for example, Sternberg, Kim and Schwende (1975) and Haagsma and Slump (1978) estimated the LAL content of a wide range of foods, both with and without alkaline treatment, and reported that levels as low as $10–15$ mg LAL (kg protein)$^{-1}$ could be measured in gelatin and fresh milk. With one or two exceptions most of the foods analysed contained much less than 1 g LAL (kg protein)$^{-1}$, which is considerably lower than the range of amino acid concentrations, from 8 g cystine (kg protein)$^{-1}$ to 219 g glutamic acid (kg protein)$^{-1}$, found in cows' milk (Williams, Bishop, Cockburn and Scott, 1976). This presents problems, since in achieving measurable quantities of LAL it is necessary to overload the analytical system with other amino acids. Slump (1977) reported that 10 mg of hydrolysed protein is the maximum that can be loaded on to the column of an amino acid analyser without causing blockages in the reaction coil.

Another problem is that LAL is a basic amino acid and is eluted in a region where good resolution is difficult to achieve because of the close proximity of glucosamine, galactosamine, tryptophan, lysine, histidine, hydroxylysine, pyridosine, ornithine and ornithinoalanine (Fig. 12.1). Most of the published methods for the estimates of LAL (Bohak, 1964; Ziegler, Melchert and Lurken, 1967; Robson, Williams and Woodhouse, 1967; De Groot and Slump, 1969; Sternberg et al., 1975; Slump, 1977; Fujimaki, Haraguchi, Abe, Homma and Arai, 1980; Sketty and Kinsella, 1980) have been based on standard IEC programmes for the separation of natural basic amino acids. Although these programmes have been slightly modified to achieve adequate resolution, there is evidence (Haagsma and Slump, 1978; Raymond, 1980) that these procedures can lead to incorrect identification of LAL or overestimation due to interference from other

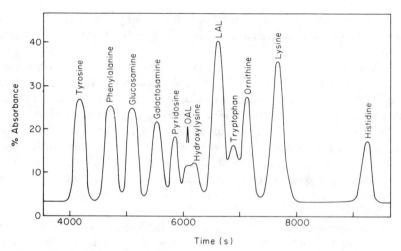

**Fig. 12.1** Chromatography of lysinoalanine (LAL) and ornithinoalanine (OAL) in the presence of some other amino acids. Column, 50 cm × 0.9 cm, Aminex A-4 resin; buffer, 0.61 M sodium – 0.2 M citrate, pH 4.50. [Redrawn with permission from Slump (1977).]

amino acids and amino sugars, particularly hydroxylysine, pyridosine, ornithinoalanine and galactosamine, Haagsma and Slump (1978) used the method of Slump (1977) but chromatographed a sample of each hydrolysate twice, once at a column temperature of 49 °C and once at 57 °C. This was necessary because of interference, probably by pyridosine, at 57 °C with heated milk products. Analysis of such samples at 49 °C resulted in satisfactory resolution of LAL. For some products the reverse was true and for foaming agents adequate resolution could not be obtained at either temperature. Haagsma and Slump (1978) suggested that great care should be taken with unknown samples and that besides carrying out chromatography at two column temperatures the results should be checked against thin layer chromatography. Similar observations were made by Raymond (1980) who used the basic column of a Durrum D500 amino acid analyser eluted with 0.35 M sodium citrate buffer at pH 5.30 to determine LAL in a variety of foods. If LAL alone was estimated the analysis could be completed in 30 min at a sensitivity of 50 $\mu$g (g sample)$^{-1}$. Although a column temperature of 52 °C was satisfactory for most samples, it was necessary to increase this to 65 °C for certain samples, for example meat products, possibly because of the presence of hydroxylysine, which was not resolved from LAL at 52 °C. Certain milk products such as casein also created problems due to interference from several unidentified peaks.

Another solution to the problem would be to carry out a complete amino acid analysis using a single column procedure. This has been suggested by Slump (1977) and Erbersdobler, Holstein and Lainer (1979) but although this has been tried for standard mixtures it does not appear to have been used for food samples, possibly because such an analysis takes about 3 h to complete instead of less than 1 h. Analysis of single amino acids usually creates the problem of whether to sacrifice accuracy for speed. Again HPLC could provide a possible answer and Wood-Rethwill and Warthesen (1980) have described a method using dansylation followed by separation on a $\mu$-Bondapak $C_{18}$ column with a mobile phase of acetonitrile and 0.01 M phosphate buffer, pH 7.0 with detection at 254 nm. Analysis of a hydrolysate of alkaline-treated casein took 16 min and resolution, from interfering compounds, such as the dansyl derivatives of lanthionine and lysine, was good. However, the position of other amino acids formed during alkaline treatment of protein, such as ornithinoalanine was not checked. The quantity of dansyl chloride used was also critical.

## IV  DETERMINATION OF PEPTIDES

Automated peptide chromatography by IEC was first suggested by Spackman et al. (1958) for protein structure studies and the many problems and developments since then have been reviewed by Jones (1970) and Herman and Vanaman (1977). Early studies used essentially the same equipment as for amino acid analysis even to the extent of using the same non-volatile sodium citrate buffers. Unfortunately such buffers lead to high salt concentrations in the effluent which created problems, and these were replaced by volatile buffers based mainly on pyridine. The fact that most samples contained amino acids as well as peptides created problems with resolution and identification, particularly since peptides have much lower colour yields when reacted with ninhydrin. Catravas (1964) attempted to overcome this problem by splitting the effluent stream into two parts, one of which reacted with ninhydrin in the normal way and the other was hydrolysed with NaOH at 105 °C before reaction with ninhydrin. By the use of time delay coils the absorbance of the hydrolysed and non-hydrolysed effluents were measured simultaneously and therefore the presence and size of the peptides in the unhydrolysed chromatogram could be detected. Jones (1970) and Benson (1976) have described simplified accelerated versions of this procedure. Johnson (1979) has further modified the procedure by using a new sulphonic acid ion-exchange resin with pyridine/acetate buffers which provides excellent resolution of neutral, acidic and basic peptides in about 14 h.

Gel filtration using cross-linked dextrans such as Sephadex has also been used to advantage when large peptides or proteins are present since these tend to block ion-exchange resin columns or are eluted together at the beginning of the chromatogram. Indeed this technique has been much used in the study of bitter peptides in cheese since they are considered to be important in the development of cheese flavour (Schormüller, 1968). In common with many foods the estimation of peptides in cheese is complicated by the presence of large amounts of free amino acids and peptides of widely different molecular weights and Polzhofer and Ney (1972) and Rothebühler, Waibel and Solms (1979) have concluded that the best method involves the use of copper–Sephadex columns. Similar methods have been used for the analysis of peptides in beer wort (Clapperton, 1971) and in an enzymic hydrolysate of casein used for intravenous feeding (Sampson and Barlow, 1980). These methods are based on the charge differential, at alkaline pH, between Cu(II) complexes of amino acids (no net charge) and peptides (negative net charge). The amino acid complexes are then separated from the peptide complexes by elution with different mobile phases. Subsequent analysis of the peptides is performed after removal of copper and salt.

In the UK meat products must contain a minimum quantity of meat and the increasing use of vegetable protein to replace the more expensive meat protein has led to the need for methods for detecting their presence in such products. Bailey (1976) has suggested the use of peptide analysis to assess the soya protein content of meat products. After partial hydrolysis of the samples with trypsin the peptides were analysed in just over 3 h, on a modified IEC analyser using a cation-exchange resin and pH 5.47, 0.2 N sodium citrate buffer as the mobile phase followed by detection with ninhydrin. A peptide (molecular weight < 5000 daltons) unique to soya, was discovered, the concentration of which was used to determine the soya content of various products. Attempts are being made to apply this procedure to the determination of other non-meat proteins. Bailey's (1976) method has been modified (Llewellyn, Dean, Sawyer, Bailey and Hitchcock, 1978) by removal of material with molecular weights above 1000 daltons from the trypsin digest, by ultrafiltration to improve resolution and to prevent column blocking due to the presence of large peptides or undigested protein. Reimerdes (1980) has proposed a similar method for detecting milk and other proteins added to foods. Tryptic digests of $\beta$-casein, $\beta$-lactoglobulin, and soya isolate gave characteristic peptide patterns which after ultracentrifugation could be separated by HPLC using reversed phase Lichrosorb RP-8 and RP-2 columns. Elution with water, 0.1% NaCl and methanol was followed by UV detection at 280 or 205 nm.

Enzymic hydrolysis is used to increase the solubility of food proteins to render them suitable for incorporation into liquid foods and beverages. Such liquid protein diets are extensively used in the treatment of obesity but there have been problems of flavour due to the formation of bitter peptides and a number of deaths have also been attributed to the use of these diets (Roland, Mattis, Kiang and Alm, 1978). Grundel, O'Dell, Pirisino and Prosky (1981) have recently analysed 21 commercially available liquid protein products for amino acids, carbohydrates and peptides but failed to establish any links between these compounds and the deaths. The amino acids were analysed by conventional IEC, the peptides by gel filtration chromatography using Sephadex G25 with 0.1 M $CaCl_2$ as the mobile phase. Analysis took about 6 h and complex chromatograms were obtained. No free amino acids were present and there was considerable variation between samples in the molecular weight distribution of the peptides. Amino acid analysis of the hydrolysed samples were also very variable and the concentrations were often quite different from the amounts shown on the product labels.

The effect of heat on the nutritive value of food proteins is well known. Although the mechanism of this heat damage is not completely understood, it is believed to be due to the formation of enzyme-resistant cross-linkages which reduce the digestibility of the protein. Of these cross-linkages one of the most likely appears to be that formed by the reaction between the ε-amino group of lysine and either the carboxyl groups of aspartic and glutamic acids or the amide groups of asparagine and glutamine. There has been much interest in the significance of the lysine-containing isopeptides formed, ε-N-(β-L-aspartyl)-L-lysine and ε-N-(γ-L-glutamyl)-L-lysine (Hurrell, Carpenter, Sinclair, Otterburn and Asquith, (1976) and consequently in methods for the rapid estimation of these compounds (Otterburn and Sinclair, 1976; Weder and Scharf, 1981). Otterburn and Sinclair (1976) used an IEC method (Spackman *et al.*, 1958) but with gradient elution using lithium buffers instead of normal sodium buffers to achieve improved resolution. Resolution of the isopeptides from the common amino acids was achieved in 7 h for heated proteins, but the pH of the mobile phase and the quality of the resins were very critical. Recently Weder and Scharf (1981) improved this procedure and were able to reduce the analysis time to 4 h (2 h if only the isopeptides were required) by the use of a standard single column programme for amino acid analysis using sodium citrate buffers. Apart from the differences in resins and mobile phases the only other major differences which could account for these improvements seemed to be the instruments. It is doubtful whether the term high performance should be applied to the analysis of peptides by IEC and gel filtration since resolution is still inadequate and analyses time-consuming. However, HPLC does not yet appear to have been used in the direct estimation of peptides in food protein samples although several methods have been published (Molnar and Horvath, 1977; Kroeff and Pietrzyk, 1978; Crommen, 1979; Meek 1980) for use in sequence studies.

## V  THE PRECISION OF AMINO ACID ANALYSIS

The precision of an analytical method is its capacity to provide the same result on repeated application to the same sample and is usually described by a statement of the mean value obtained and the coefficient of variation (standard deviation/mean). The precision between laboratories can be tested by collaborative trials and since 1951 there have been 12 trials for the amino acid analysis of protein hydrolysates (foods and feeds) and one on the determination of free amino acids (Williams et al., 1980). Results from these trials have been grouped for ease of comparison, as follows: standard mixtures (Bureau of Biological Research, 1951; Bender, Palgrave and Doell, 1959; Matthias, 1964; Porter et al., 1968; Williams et al., 1980), protein hydrolysates, 1964–8 (Matthias, 1964; Weidner and Eggum, 1966; Wünsche, 1967; Porter et al., 1968), protein hydrolysates, 1971–9 (Knipfel, Aitken, Hill, McDonald and Owens, 1971; Cavins, Kwolek, Inglett and Cowan, 1972; Kreienbring and Wünsche, 1974, 1976; Mikoska, 1974; Buraczewska, 1979), protein hydrolysates 1981–to date (Kreienbring, 1981). Results, expressed as the means of the coefficients of variation (CV) of laboratory mean values are given in Table 12.2.

The precision obtained for standard mixtures of amino acids (mean CV 6.0%) is close to that claimed by manufacturers of automatic amino acid analysers. The trials carried out in the 1960s used high voltage electrophoretic, paper chromatographic, microbiological and manual and automatic IEC techniques. The precision was much poorer than for standards (mean CV 12.9%), particularly for proline, tyrosine, histidine, cystine, methionine and tryptophan. It was clear that the best method was automatic IEC and this has been the method used in all subsequent trials. Consequently precision improved during the 1970s (mean CV 7.8%), although problems still existed for proline, tyrosine, histidine, cystine, methionine and tryptophan. Five trials have recently been instigated only one of which has been completed (Kreienbring, 1981). Overall precision (mean CV 7.3%) showed little improvement and precision for proline, tyrosine and histidine was still poor, although there was an encouraging improvement for cystine and methionine. The consistently poor precision obtained for proline, tyrosine and histidine is probably due to the relatively

**Table 12.2** Comparison of published results of collaborative trials on the analysis of amino acids in standard mixtures, protein hydrolysates and blood plasma [results are expressed as the means of coefficients of variation (%) of laboratory means]

| Year: | 1951–80 | 1964–8 | 1971–9 | 1981– | 1980 |
|---|---|---|---|---|---|
| | Standard mixture | Protein hydrolysates | Protein hydrolysates | Protein hydrolysates | Blood plasma |
| Aspartic acid | 5.8 | 10.3 | 6.5 | 6.3 | — |
| Threonine | 5.4 | 9.7 | 8.1 | 7.5 | 21.5 |
| Serine | 5.3 | 10.8 | 8.4 | 7.4 | 19.0 |
| Glutamic acid | 5.4 | 10.8 | 6.6 | 6.5 | 59.5 |
| Proline | 7.3 | 13.1 | 8.9 | 8.0 | — |
| Glycine | 4.5 | 11.0 | 6.9 | 6.7 | 14.3 |
| Alanine | 5.9 | 10.7 | 7.7 | 5.7 | 22.0 |
| Valine | 6.7 | 10.9 | 6.9 | 6.3 | 17.5 |
| Isoleucine | 7.6 | 13.0 | 7.5 | 6.4 | 12.0 |
| Leucine | 5.2 | 8.6 | 6.3 | 6.2 | 15.0 |
| Tyrosine | 5.8 | 20.2 | 9.6 | 18.8 | 15.0 |
| Phenylalanine | 4.5 | 18.7 | 6.4 | 7.8 | 8.3 |
| Lysine | 4.9 | 12.5 | 8.8 | 6.6 | 24.0 |
| Histidine | 9.2 | 18.2 | 10.1 | 11.6 | 28.5 |
| Arginine | 6.7 | 15.5 | 7.6 | 7.0 | 18.8 |
| Mean | 6.0 | 12.9 | 7.8 | 7.3 | 21.2 |
| Cystine | — | 18.6 | 12.9 | 5.1 | — |
| Methionine | 7.8 | 22.9 | 14.4 | 4.5 | 57.0 |
| Tryptophan | — | 41.5 | 16.7 | — | — |

small sizes of the peaks obtained for these amino acids after chromatography. Cystine, methionine and tryptophan cannot be measured with any degree of precision even when alternative methods are used for their estimation (Williams et al., 1979; Westgarth and Williams, 1974).

Possible reasons for poor precision in amino acid analysis are inexperience of collaborators, incomplete resolution and incorrect identification of peaks, reagent impurities, variations in resins, analysers and standards, sample preparation and storage, sample loading and data processing. Probably a combination of all of these factors is to blame and the situation has been aggravated by the advent of faster analysis times, automatic sample loading and data handling. The preparation of hydrolysates must also be an important factor since the precision obtained for standard mixtures has generally been good. Indeed, in the one trial carried out on free amino acids (Williams et al., 1980), in which extremely poor precision was reported for most amino acids (mean CV 21.2%) in samples of blood plasma, satisfactory results were obtained for a standard mixture analysed at the same time. It was concluded that the deproteinizing procedure was partially to blame, although the wide range of amino acid concentrations in plasma and the problems of peak resolution caused by the presence of many more amino acids and related compounds than in protein hydroly-sates must have contributed to the poor precision.


## VI  CONCLUSION


Although automated, high performance ion-exchange chromatography of amino acids has been in use for more than 20 years there are still doubts about the precision and reliability of the methods. Paul and Southgate (1978) were most critical of published data on foods because most authors omitted details of their hydrolysis conditions and failed to give estimates of hydrolytic losses. They considered hydrolysis to be the most critical part of amino acid analysis and this is probably true, although the criticism that few laboratories carry out serial hydrolyses to determine the optimum length of hydrolysis is unwarranted. Tristram and Smith (1963) defined the ideal analysis as one in which the hydrolysis of the protein should be carried out for 20, 40, 70 and 140 h and the results averaged or extrapolated to provide the best results. This statement is undoubtedly true for pure proteins but for food and feed proteins, particularly those with a high carbohydrate content, there is evidence to suggest that a single hydrolysis time of 24 h is justified for the majority of amino acids (Rudemo et al., 1980). Exceptions were isoleucine, valine and serine for which the 24 h values should be increased by 6% to correct for losses during hydrolysis. Other factors influencing the precision of amino acid analysis by IEC have been discussed in Section V. Of these, the current tendency towards faster and faster analyses must be a major factor since resolution is often compromised in favour of speed (Young and Yamamoto, 1973).

Ion-exchange chromatography remains the preferred method in most laboratories and there is little evidence to suggest that other HPLC systems have been adopted for the determination of total amino acids in food hydrolysates. However, there have recently been many publications reporting the use of HPLC for the determination of individual amino acids. It is in this area that the high speed and resolving power of HPLC would seem to have its greatest potential in the future. Many of the HPLC methods for individual amino acids have been developed by Warthesen and his colleagues at the University of Minnesota and one would anticipate that any further advances in the application of HPLC to the analysis of total amino acids in foods might come from this group. The precision of HPLC methods of analysis does not appear to have been evaluated by collaborative trials to the same extent as IEC methods of amino acid analysis. Indeed only one trial (McSharry and Mahn, 1979) appears to have been carried out for any HPLC method. In this trial the CV values between laboratory means for 14 laboratories, estimating folic acid in six samples, ranged from 1.4 to 2.1%. This is exceptional and illustrates the great potential of HPLC, but only two components had to be resolved and hydrolysis and deproteinization were not involved. In view of the problems associated with the HPLC analysis of pure proteins and standard mixtures reported by Somack (1980) it will be essential that, when eventually applied to food analysis, careful evaluation of the method by collaborative trials is carried out.

## REFERENCES

Abe, H. (1981). *Bull. Jap. Soc. Sci. Fish.* **47**, 139.
Alexander, P. W., Haddad, P. R., Low, G. K. C. and Maitra, C. (1981). *J. Chromat.* **209**, 29.
Bailey, F. J. (1976). *J. Sci. Fd Agric.* **27**, 827.
Bailey, F. J. and Marks, A. J. (1977). *Lab. Equip. Digest* **15**, 51.
Bech-Andersen, S., Rudemo, M. and Mason, V. C. (1979a). *Z. Tierphysiol. Tierernähr. Futtermittelk.* **41**, 248.
Bech-Andersen, S., Mason, V. C. and Rudemo, M. (1979b). *Z. Tierphysiol. Tierernähr. Futtermittelk.* **41**, 265.
Bech-Andersen, S., Rudemo, M. and Mason, V. C. (1980). *Z. Tierphysiol. Tierernähr. Futtermittelk.* **43**, 57.
Bell, J. A. and Mason, V. C. (1970). *J. Chromat.* **46**, 317.
Bender, A. E., Palgrave, J. A. and Doell, B. H. (1959). *Analyst, Lond.* **84**, 526.
Benson, J. R. (1976). *Analyt. Biochem.* **71**, 459.
Benson, J. R. and Hare, P. E. (1975). *Proc. natn. Acad. Sci. U.S.A.* **72**, 619.
Benson, J. V., Jr and Patterson, J. A. (1965). *Analyt. Chem.* **37**, 1108.
Benson, J. V., Jr, Gordon, M. J. and Patterson, J. A. (1967). *Analyt. Biochem.* **18**, 228.
Block, W. D., Markovs, M. E. and Steele, B. F. (1966). *Proc. Soc. exp. Biol. Med.* **122**, 1089.
Bohak, Z. (1964). *J. biol. Chem.* **239**, 2878.
Buraczewska, L. (1979). *Roczn. PZH* **30**, 613.
Bureau of Biological Research (1951). *Cooperative Determination of the Amino Acid Content and the Nutritive Value of Six Selected Protein Food Sources*, Rutgers University, New Brunswick, N.J.
Carpenter, K. J. (1960). *Biochem. J.* **77**, 604.
Catravas, G. N. (1964). *Analyt. Chem.* **36**, 1146.
Cavins, J. F. and Friedman, M. (1970). *Analyt. Biochem.* **35**, 489.

Cavins, J. F., Kwolek, W. F., Inglett, G. E. and Cowan, J. C. (1972). *J. Ass. Off. analyt. Chem.* **55**, 686.

Clapperton, J. F. (1971). *J. Inst. Brew.* **77**, 177.

Conrad, E. C. (1979). In *Liquid Chromatographic Analysis of Food and Beverages*, Vol. 2 (G. Charalambous, ed), Academic Press, New York and London, p. 237.

Crommen, J. (1979). *Acta pharm. Suec.* **16**, 111.

Davies, M. G. and Thomas, A. J. (1973). *J. Sci. Fd Agric.* **24**, 1525.

De Groot, A. P. and Slump, P. (1969). *J. Nutr.* **98**, 45.

De Groot, A. P., Slump, P., Feron, V. J. and Van Beek. L. (1976). *J. Nutr.* **106**, 1527.

De Vries, J. W., Koski, C. M., Egberg, D. C. and Larson, P. A. (1980). *J. agric. Fd Chem.* **28**, 896.

Dennison, C. (1976). *Lab. Pract.* **25**, 81–82.

Dilley, K. J. (1980). *Int. Lab.* **10**, 79.

Erbersdobler, H. E., Holstein, B. and Lainer, E. (1979). *Z. Lebensmittelunters. u-Forsch.* **168**, 6.

Ersser, R. (1979a). *Lab. Equip. Digest* **17** (6), 61.

Ersser, R. (1979b). *Lab. Equip. Digest* **17** (8), 61.

Ertinghausen, G., Adler, H. J. and Reichler, A. S. (1969). *J. Chromat.* **42**, 355.

Felix, A. M. and Terkelsen, G. (1973). *Archs Biochem. Biophys.* **157**, 177.

Friedman, M. and Finley, J. W. (1975). In *Protein Nutritional Quality of Foods and Feeds*, Vol. I (M. Friedman, ed.), Marcel Dekker, New York, p. 423.

Friedman, M. and Noma, A. T. (1975). In *Protein Nutritional Quality of Foods and Feeds*, Vol. I (M. Friedman, ed.), Marcel Dekker, New York, p. 521.

Friedman, M., Krull, L. H. and Cavins, J. F. (1970). *J. biol. Chem.* **245**, 3868.

Friedman, Z. and Smith, H. W. (1980). *J. Chromat.* **182**, 414.

Fujimaki, M., Haraguchi, T., Abe, K., Homma, S. and Arai, S. (1980). *Agric. biol. Chem.* **44**, 1911.

Fujita, K., Takeuchi, S. and Ganno, S. (1979). In *Liquid Chromatographic Analysis of Foods and Beverages*, Vol. 1 (G. Charalambous, ed.), Academic Press, New York and London, p. 81.

Gerritsen, T., Rehberg, M. L. and Waisman, H. A. (1965). *Analyt. Biochem.* **11**, 460.

Godtfredsen, S. E. and Oliver, W. A. (1980). *Carlsberg Res. Commun.* **45**, 35.

Green, D. J. and Perlman, R. L. (1980). *Clin. Chem.* **26**, 796.

Grundel, E., O'Dell, R. G., Pirisino, J. and Prosky, L. (1981). *J. agric. Fd Chem.* **29**, 188.

Haagsma, N. and Slump, P. (1978). *Z. Lebensmittelunters. u-Forsch.* **167**, 238.

Hamilton, P. B. (1962). *Ann. N.Y. Acad. Sci.* **102**, 55.

Hamilton, P. B. (1966). *Adv. Chromat.* **2**, 3.

Hamilton, P. B., Bogue, D. and Anderson, R. A. (1960). *Analyt. Chem.* **32**, 1782.

Hare, P. E. (1977). *Methods Enzymol.* **47**, 3.

Hartwick, R. A., Van Haverbeke, D., McKeag, M. and Brown, P. R. (1979). *J. Liquid Chromat.* **2**, 725.

Herman, A. C. and Vanaman, T. C. (1977). *Methods Enzymol.* **47**, 220.

Hewett, G. E. and Forge, G. O. (1976). *Fedn. Proc. Fedn Am. Socs exp. Biol.* **35**, 1382.

Hibbert, I. and Lawrie, R. A. (1972). *J. Fd Technol.* **7**, 333.

Hill, D. W., Walters, F. H., Wilson, T. D. and Stuart, J. D. (1979). *Analyt. Chem.* **51**, 1338.

Hill, R. L. (1965). *Adv. Protein Chem.* **20**, 37.

Holsinger, V. H. and Posati, L. P. (1975). In *Protein Nutritional Quality of Foods and Feeds*, Vol. I (M. Friedman, ed.), Marcel Dekker, New York, p. 479
Hubbard, R. W. (1965). *Biochem. Biophys. Res. Commun.* **19**, 679.
Hurrell, R. F. and Carpenter, K. J. (1974). *Br. J. Nutr.* **32**, 589.
Hurrell, R. F. and Carpenter, K. J. (1976). *Proc. Nutr. Soc.* **35**, 23A.
Hurrell, R. F., Carpenter, K. J., Sinclair, W. J., Otterburn, M. S. and Asquith, R. S. (1976). *Br. J. Nutr.* **35**, 383.
Hurst, W. J. and Martin, R. A., Jr (1980). *J. agric. Fd Chem.* **28**, 1039.
Johnson, P. (1979). *J. chromatog. Sci.* **17**, 406.
Jones, A. D., Hitchcock, C. H. S. and Jones, G. H. (1981). *Analyst, Lond.* **106**, 968.
Jones, R. T. (1970). *Methods biochem. Anal.* **18**, 205.
Keutmann, H. T. and Potts, J. T., Jr (1969). *Analyt. Biochem.* **29**, 175.
Knipfel, J. E., Christensen, D. A. and Owen, B. D. (1969). *J. Ass. off. analyt. Chem.* **52**, 981.
Knipfel, J. E., Aitken, J. R., Hill, D. C., McDonald, B. E. and Owens, B. D. (1971). *J. Ass. off. analyt. Chem.* **54**, 777.
Knox, R., Kohler, G. O., Palter, R. and Walker, H. G. (1970). *Analyt. Biochem.* **36**, 136.
Kreienbring, F. (1981). *Die Nährung* **25**, 1.
Kreienbring, F. and Wünsche, J. (1974). *TagBer. dt. Akad. LandwWiss. Berl.* **124**, 19.
Kreienbring, F. and Wünsche, J. (1976). *TagBer dt. Akad. LandwWiss. Berl.* **142**, 15.
Kroeff, E. P. and Pietrzyk, D. J. (1978). *Analyt. Chem.* **50**, 502.
Krstulovic, A. M., Brown, P. R. and Rosie, D. M. (1977). *Analyt. Chem.* **49**, 2237.
Lipton, S. H., Bodwell, C. E. and Coleman, A. H., Jr (1977). *J. agric. Fd Chem.* **25**, 624.
Llewellyn, J. W., Dean, A. C. Sawyer, R., Bailey, F. J. and Hitchcock, C. H. S. (1978). *J. Fd Technol.* **13**, 249.
London, D. R. (1966). In *Techniques in Amino Acid Analysis*, Technicon Instruments, Chertsey, Surrey, p. 38.
Lucas, B. and Sotelo, A. (1980). *Analyt. Biochem.* **109**, 192.
McSharry, W. O. and Mahn, F. P. (1979). *J. pharm. Sci.* **68**, 241.
Mason, V. C., Bech-Andersen, S. and Rudemo, M. (1979). *Z. Tierphysiol. Tierernahr. Futtermittelk.* **41**, 226.
Mason, V. C., Rudemo, M. and Bech-Andersen, S. (1980a). *Z. Tierphysiol. Tierernahr. Futtermittelk.* **43**, 35.
Mason, V. C., Bech-Andersen, S. and Rudemo, M. (1980b). *Z. Tierphysiol. Tierernahr. Futtermittelk.* **43**, 146.
Matthias, W. (1964). *TagBer. dt. Akad. LandwWiss. Berl.* **64**, 7.
Meek, J. L. (1980). *Proc. natn. Acad. Sci. U.S.A.* **77**, 1632.
Mikoska, F. (1974). *TagBer. dt. Akad. LandwWiss. Berl.* **124**, 47.
Milner, C. K. and Westgarth, D. R. (1973). *J. Sci. Fd Agric.* **24**, 873.
Molnar, I. and Horvath, C. (1977). *J. Chromat.* **142**, 623.
Moore, S. (1963). *J. biol. Chem.* **238**, 235.
Muhammad, N. and Bodnar, J. A. (1980). *J. Liquid. Chromat.* **3**, 529.
Murren, C., Stelling, D. and Felstead, G. (1975). *J. Chromat.* **115**, 236.
Nashef, A. S., Osuga, D. T., Lee, H. S., Ahmed, A. I., Whitaker, J. R. and Feeney, R. E. (1977). *J. agric. Fd. Chem.* **25**, 245.
Ohara, I. and Ariyoshi, S. (1979). *Agric. biol. Chem.* **43**, 1473.

310    A. P. WILLIAMS

O'Keefe, L. S. and Warthesen, J. J. (1978). *J. Fd Sci.* **43**, 1297.
Otterburn, M. S. and Sinclair, W. J. (1973). *J. Sci. Fd Agric.* **24**, 929.
Otterburn, M. S. and Sinclair, W. J. (1976). *J. Sci. Fd Agric.* **27**, 1071.
Paul, A. A. and Southgate, D. A. T. (1978). *McCance and Widdowson's The Composition of Foods*, 4th revised edn, HMSO, London.
Pechanek, V., Blaicher, G., Pfannhauser, W. and Woidich, H. (1980). *Chromatographia* **13**, 421.
Peterson, W. R. and Warthesen, J. J. (1979). *J. Fd Sci.* **44**, 994.
Piez, K. A. and Morris, L. (1960). *Analyt. Biochem.* **1**, 187.
Polzhofer, K. P. and Ney, K. H. (1972). *Tetrahedron* **28**, 1721.
Porter, J. W. G., Westgarth, D. R. and Williams, A. P. (1968). *Br. J. Nutr.* **22**, 437.
Posati, L. P., Holsinger, V. H., DeVilbiss, E. D. and Pallansch, M. J. (1974). *J. Dairy Sci.* **57**, 258.
Rangeley, W. R. D. and Lawrie, R. A. (1976). *J. Fd Technol.* **11**, 143.
Raymond, M. L. (1980). *J. Fd Sci.* **45**, 56.
Rayner, C. J. and Fox, M. (1978). *J. agric. Fd Chem.* **26**, 494.
Reid, R. H. P. (1966). In *Techniques in Amino Acid Analysis*, Technicon Instruments, Chertsey, Surrey, p. 43.
Reimerdes, E. H. (1980). *Lebensmittel. Gerich. Chem.* **34**, 75.
Roach, A. G., Sanderson, P. and Williams, D. R. (1967). *J. Sci. Fd Agric.* **18**, 274.
Robinson G. W. (1978a). In *Amino Acid Determination. Methods and Techniques* (S. Blackburn, ed.), Marcel Dekker, New York and Basel, p. 39.
Robinson G. W. (1978b). In *Amino Acid Determination. Methods and Techniques* (S. Blackburn, ed.), Marcel Dekker, New York and Basel, p. 101.
Robson, A., Williams, J. J. and Woodhouse, J. M. (1967). *J. Chromat.* **31**, 284.
Rohan, T. A. and Stewart, T. (1966). *J. Fd Sci.* **31**, 202.
Roland, J. F., Mattis, D. L., Kiang, S. and Alm, W. L. (1978). *J. Fd Sci.* **43**, 1491.
Rossi, D. and Trisciani, A. (1980). *Int. Lab.* **10**, 31.
Rothebühler, E., Waibel, R. and Solms, J. (1979). *Analyt. Biochem.* **97**, 367.
Rudemo, M. Mason, V. C. and Bech-Andersen, S. (1979). *Z. Tierphysiol. Tierernahr. Futtermittelk.* **41**, 254.
Rudemo, M., Bech-Andersen, S. and Mason, V. C. (1980). *Z. Tierphysiol. Tierernahr. Futtermittelk.* **43**, 27.
Saetre, R. and Rabenstein, D. L. (1978). *J. agric. Fd Chem.* **26**, 982.
Sampson, B. and Barlow, G. B. (1980). *J. Chromat.* **183**, 9.
Savoy, C. F., Heinis, J. L. and Seals, R. G. (1975). *Analyt. Biochem.* **68**, 562.
Schormüller, J. (1968). *Adv. Fd Res.* **16**, 231.
Schuster, R. (1980). *Analyt. Chem.* **52**, 617.
Sketty, J. K. and Kinsella, J. E. (1980). *J. agric. Fd. Chem.* **28**, 798.
Skurray, G. R. and Lysaght, V. A. (1978). *Fd Chem.* **3**, 111.
Sloman, K. G., Foltz, A. K. and Yersanian, J. A. (1981). *Analyt. Chem.* **53**, 242R.
Slump, P. (1977). *J. Chromat.* **135**, 502.
Slump, P. and Schreuder, H. A. W. (1969). *Analyt. Biochem.* **27**, 182.
Somack, R. (1980). *Analyt. Biochem.* **104**, 464.
Spackman, D. H., Stein, W. H. and Moore, S. (1958). *Analyt. Chem.* **30**, 1190.
Spies, J. R. (1967). *Analyt. Chem.* **39**, 1412.
Spitz, H. D. (1973). *Analyt. Biochem.* **56**, 66.
Stein, W. H. and Moore, S. (1945). *J. biol. Chem.* **211**, 915.
Steinhart, H. (1979). *Arch. Tierernähr.* **29**, 211.
Sternberg, M., Kim, C. Y. and Schwende, F. J. (1975). *Science, N.Y.* **190**, 992.

Tristram, G. R. and Smith, R. H. (1963). *Adv. Protein Chem.* **18**, 227.
Tufte, M. C. and Warthesen, J. J. (1979). *J. Fd Sci.* **44**, 1767.
Udenfriend, S., Stein, S., Bohlen, P., Dairman, W., Leimgruber, W. and Weigele, M. (1972). *Science, N.Y.* **178**, 871
Van Stekelenberg, G. V. and Desplanque, J. (1966). In *Techniques in Amino Acid Analysis*, Technicon Instruments, Chertsey, Surrey, p. 51.
Walker, H. G., Jr, Kohler, G. O., Kuzmicky, D. D. and Witt, S. C. (1975). In *Protein Nutritional Quality of Foods and Feeds*, Vol. 1 (M. Friedmman, ed.), Marcel Dekker, New York, p. 549.
Ward, L. C. (1978). *Analyt. Biochem.* **88**, 598.
Ward, L. C., Miller, M. and Hawgood, S. (1981). *J. Chromat.* **223**, 417.
Warthesen, J. J. and Kramer, P. L. (1978). *Cereal Chem.* **55**, 481.
Wassner, S. J., Schlitzer, J. L. and Li, J. B. (1980). *Analyt. Biochem.* **104**, 284.
Weder, J. K. P. and Scharf, U. (1981). *Z. Lebensmittelunters. u-Forsch.* **172**, 9.
Weidner, K. and Eggum, B. O. (1966). *Acta agric. Scand.* **16**, 115.
Westgarth, D. R. and Williams, A. P. (1974). *J. Sci. Fd Agric.* **25**, 571.
Williams, A. P. (1981). In *Amino Acid Analysis* (J. M. Rattenbury, ed.), Ellis Horwood, Chichester, p. 138.
Williams, A. P., Bishop, D. R., Cockburn, J. E. and Scott, K. J. (1976). *J. Dairy Res.* **43**, 325.
Williams, A. P., Hewitt, D. and Cockburn, J. E. (1979). *J. Sci. Fd Agric.* **30**, 469.
Williams, A. P., Hewitt, D., Cockburn, J. E., Harris, D. A., Moore, R. A. and Davies, M. G. (1980). *J. Sci. Fd Agric.* **31**, 474.
Woodham, A. A. (1974). *Proc. Nutr. Soc.* **33**, 95A.
Wood-Rethwill, J. C. and Warthesen, J. J. (1980). *J. Fd Sci.* **45**, 1637.
Woodward, J. C. and Alvarez, M. R. (1967). *Archs Path.* **84**, 153.
Woodward, J. C. and Short, D. D. (1973). *J. Nutr.* **103**, 569.
Wünsche, J. (1967). *Sitzungsberichte* **16**, 17.
Young, J. L. and Yamamoto, M. (1973). *J. Chromat.* **78**, 221.
Ziegler, K., Melchert, I. and Lurken, C. (1967). *Nature, Lond.* **214**, 404.
Zimmerman, C. L., Appela, E. and Pisano, J. J. (1977). *Analyt. Biochem.* **77**, 569.

# 13 Future Developments in HPLC Relevant to Food Analysis

## R. MACRAE and H. E. NURSTEN

Department of Food Science, University of Reading, U.K.

## I INTRODUCTION

The exponential phase in the development of HPLC is now nearly complete and progress over the next few years will almost certainly be in the form of consolidation and improvement of the basic techniques available. That is not to imply that there will not be significant developments in certain areas, but overall the majority of HPLC systems working in routine analysis in 5 years' time will be remarkably similar to those in use today.

The last few years have seen vast improvements in all aspects of HPLC instrumentation, from the introduction of reliable pulse-free solvent delivery systems to improvements in many detectors, in particular, to variable wavelength ultraviolet detectors, with adequate signal to noise ratios. Several selective detectors have also been introduced, which have proved very valuable in specific areas of analysis, but these have not overcome the problem of the lack of a reliable and sensitive universal detector. It is to be hoped that this is one area where instrumentation will be improved in the near future. Much development has also been carried out in the field of automation and data handling, and the use of HPLC systems under complete microprocessor control will become more widespread, even for non-routine analyses.

The most significant development, in terms of chromatographic resolution, has been the introduction of microparticulate stationary phases and reliable methods of packing these materials to yield efficient chromatographic columns. It is unlikely that the present most commonly used particle size (5 $\mu$m) will be reduced significantly, although some phases with particle sizes down to 3 $\mu$m are becoming commercially available. There is theoretical justification for reducing the particle size further, but the increased resistance to solvent flow would place severe demands on pumping systems.

The present state of development of HPLC methodology is such that many food components can be successfully chromatographed. Thus, the main development required in the area of food analysis by HPLC is not in the chromatographic separation, but rather in the stages of extraction and in sample preparation. The limiting or time-consuming stage in many analyses is not the HPLC determination, which can be automated, but the preceding stages, of which there may be several and which are difficult to automate. Applications of complete analytical systems, such as the Technicon FAST-LC, to food analysis are potentially very great, but much development work needs to be carried out. There are a number of techniques in the early stages of their development, which could have important consequences for food analysis. It is very difficult to decide which of these will become the bases of reliable methods in food analysis, but three areas of development, namely microbore HPLC, size exclusion chromatography and mass spectrometry as a detection system, are considered potentially to offer significant advantages over existing techniques. They are discussed in more detail below (Sections II–IV).

## II  NARROW BORE HPLC

### A  Introduction

The idea of using narrow bore columns for HPLC is by no means novel; indeed, some of the earlier published separations employed columns considerably narrower than those now in common use. Earlier work with bonded-layer (pellicular) stationary phases often made use of columns up to 1 m in length and only 1 mm in diameter, as for example in the separation of purine alkaloids (Wu and Seggia, 1972). Why then should interest now be returning to the use of narrow bore columns, when they have obviously been tried and tested and subsequently abandoned in preference to the wider bore columns that are prevalent today? The earlier work with narrow bore columns used the most efficient chromatographic material then available, that is to say, pellicular stationary phases. However, it was soon realized that this material had certain disadvantages, many of which were overcome by porous microparticulate phases. Difficulties were subsequently encountered in packing efficient narrow columns with microparticulate material and in many cases intrumentation was not available to overcome the increased column resistance to solvent flow or to handle the small injection and detection volumes that are required. These factors led to the development of sophisticated slurrying techniques for packing efficient, short, and comparatively wide microparticulate columns (Majors, 1972). These columns 10–25 cm in length and 4.5 mm in diameter, packed with 5–10 $\mu$m stationary phases, are the mainstay of current HPLC. The same packing techniques can be extended to pack efficient narrow bore columns (Scott and Kucera, 1979), which theoretically possess several advantages over their wider counterparts.

One of the major purported advantages of wide bore columns is that they can operate in the infinite diameter mode (Bristow, 1976); that is to say, the applied sample can pass down the centre of the column without contact with the walls, thus avoiding potential problems of wall effects. However, in practice the majority of commercial chromatographic systems do not allow columns to operate in this manner, as there is considerable radial dispersion at the top of the column, allowing the sample to come in contact with the column walls. The fact that high plate counts can be achieved under these conditions implies that in many cases wall effects may not be that great. With microbore columns, sample/column wall contact is inevitable and hence wall effects would be expected to be more apparent.

Narrow bore columns may be divided into three classes, namely microbore packed columns, packed capillary columns and open tubular columns.

## B  Microbore Packed columns

These are the direct counterpart of the currently used 4–5 mm diameter columns but with a reduced diameter of *c*. 1 mm (Knox, 1980). The columns are packed in a similar manner to normal columns (Scott and Kucera, 1979), but general performance, in terms of reduced plate heights, is poorer than with wider columns for a given stationary phase. This is almost certainly a function of the packing, as there is no inherent reason why narrower columns should give poorer performance, unless wall effects are becoming more significant. This is supported by the observation that the reduction in column performance, as compared with wider bore columns, is greater when small particle size stationary phase material is used.

The attraction of microbore columns is not that they are more efficient than conventional columns, in terms of reduced plate heights, but rather that they can be linked together to form a separating system with a high total plate count. Thus, for microbore systems, separating power is approximately proportional to the total column length, which is not the case with wider bore linked columns. The reasons for this difference are not clear, although Scott and Kucera (1979) have suggested that it may be due to increased thermal effects caused by the higher solvent pressures, the heat generated being less able to dissipate in wider bore columns. They also suggested that, as the permeability across the column is not homogeneous, an "aggravated multipath term" occurs as the columns are linked together, which again should be less serious with narrower columns. An example of

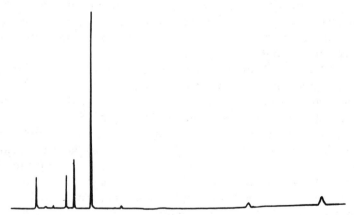

**Fig. 13.1**  Chromatogram  of  bergamot  oil  ($N$ = 160 000).  Column, 10 m × 1 mm i.d.; packing, Partisil 20; mobile phase, ethyl acetate–*n*-heptane (3 : 97, v/v); sample volume, 0.5 $\mu$l; flow rate, 38 $\mu$l min$^{-1}$. [Redrawn with permission from Scott and Kucera (1979).]

the high plate counts that can be realized by linked columns is shown in Fig. 13.1. Analyses requiring columns with very high plate counts also require long elution times. This is an unfortunate consequence of the use of long columns which will only be overcome with improvement in column packing procedures to give shorter columns with similar plate counts.

## C Packed Capillary Columns

These columns have received considerably less attention than microbore backed columns, indeed, there are very few groups actively working in this area. Novotny (1980) has described a technique by which columns with a bore of 50–200 $\mu$m may be prepared by redrawing packed 0.5–2 mm bore heavy-walled glass tubing. The columns prepared in this way have a low particle to column diameter ratio of about 2–3, which is considerably less than other types of packed column. Columns with high plate counts, up to 100 000, can be prepared, but column efficiency (reduced plate height) is lower than with wider bore packed columns.

## D Open Tubular Columns

Open tubular columns, with the liquid stationary phase coated on the wall of capillary tubing, are now well established in gas–liquid chromatography after the pioneering work of Golay (1959). However, there is a large difference in the rate of diffusion in the liquid and the gaseous phases and this initially led to poor success. Theoretical considerations of capillary column liquid chromatography by Knox and Gilbert (1979) have suggested that optimum column diameters would be in the range 10–20 $\mu$m. It was also concluded that under these conditions capillary column HPLC would be considerably more rapid than packed column HPLC where plate numbers in excess of 30 000 were required. At present, mainly for technical reasons, column diameters are usually $c$. 50 $\mu$m and thus the inherent advantages of the capillary system have yet to be realized. When capillary columns are tightly coiled, centrifugal forces may become significant giving rise to a secondary flow and increased mass transfer (Tsuda and Novotny, 1978; Tijssen, 1978). This results in decreased reduced plate heights at high solvent velocities (Hofman and Halasz, 1979). This beneficial effect with capillary columns is in contrast to that reported for packed microbore columns (Scott and Kucera, 1979).

## E Instrumental Requirements

The full potential of narrow bore systems, irrespective of whether they are packed microbore columns or open tubular capillary columns will only be

**Table 13.1**  Instrumental requirements

|  | Conventional | Microbore | Capillary |
|---|---|---|---|
| Pumps | 1–2 ml min$^{-1}$ | 10–50 $\mu$l min$^{-1}$ | <1 $\mu$l min$^{-1}$ |
| Injectors | 10–20 $\mu$l | 1–2 $\mu$l | c. 20 nl |
| Detectors (cells) | 5–10 $\mu$l | 1–2 $\mu$l | 100 nl |

realized when suitable accompanying instrumentation is available. The small column dispersion in these systems demands both small injection volumes and small detection cells. The inherent benefits of narrow bore systems are immediately lost when the sample cannot be applied to the column in a discrete band and the components in the eluate cannot be detected without significant band-broadening. The magnitude of the problem of instrumental requirements is shown in Table 13.1. In the case of capillary columns, the injection and detection volumes are extremely small and very specialized techniques are required (Novotny, 1980). At present, no suitable equipment is commercially available. In the case of packed microbore columns, some commercial instrumentation is available (Karasek, 1977) and, in addition, certain systems used with conventional columns can be modified to produce the low flow rates and small detection cell volumes required (Scott and Kucera, 1979).

## F  Implications for Food Analysis

Microbore chromatography is potentially a powerful technique for the food analyst. For certain areas of analysis it should offer distinct advantages over conventional columns. Individual columns can be linked to provide a separating system capable of high resolution, although with consequent increase in analysis time. This increase in resolving power may be important for the study of complex mixtures, although in many published methods it is quite clear that adequate resolution can be achieved with conventional columns. However, there are a number of research areas where characterization of a complex system has been hampered by inadequate chromatographic resolution. The chromatogram reproduced in Fig.13.2 shows the presence of a vast number of components, many of which are not adequately separated for the purposes of characterization. The use of an extended microbore, or even capillary, system of sufficient resolving power could well facilitate research of this kind.

A more important attribute of narrow bore systems as far as the food analyst is concerned is the increase in mass sensitivity which can be achieved. In narrow bore systems, the sample is applied to the column in

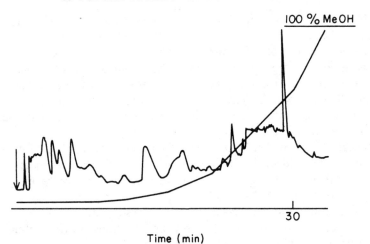

**Fig. 13.2**  Chromatogram of extract from model browning reaction. (Reproduced by courtesy of R. O'Reily.)

smaller volumes and the components also eluted in smaller volumes than in the case with wider columns. Thus, as most detectors are concentration dependent, there will be a significant increase in sensitivity. The magnitude of this sensitivity increase is related to the change in cross-sectional area, thus a change from a 4.6 mm diameter column to one of 1 mm diameter should result in approximately a 20-fold increase in mass sensitivity. In many areas of food analysis lack of sensitivity is a problem, for example with pesticide residues, and it is envisaged that narrow bore systems will prove useful in such areas. It must be remembered, however, that in many cases it is not an absolute lack of sensitivity that is the problem but a lack of sensitivity over interfering peaks, and it is in these latter instances that improvements in sample clean-up are required, rather than in instrumentation.

The reduction in column diameter associated with narrow bore columns also means that the mobile phase flow rate will be reduced as compared with conventional columns for a given linear velocity. This leads directly to a massive saving in solvents, again approximately in the ratio of the column cross-sectional areas. Thus, a change in column diameter from 4.6 to 1 mm will result in a 20-fold saving in solvents. This may not be of vital importance for comparatively cheap solvents, such as methanol or hexane, but even in these cases the saving in a laboratory where several units are in continuous use will be considerable. For more expensive solvents, such as acetonitrile, and indeed additives, such as ion-pair reagents, the saving will be more apparent. In addition to savings in solvents, there will also be a considerable saving in stationary phases, allowing the more widespread use of materials, such as silver-loaded silica. It is unlikely that these savings will be reflected in the cost of commercially available microbore columns, but for those chromatographers who pack their own columns this may be an important consideration. A further consequence of the reduction in column diameter is that high linear velocities can be more readily obtained with modest flow rates, and, with efficient chromatographic material, this should lead to very rapid analyses.

The extremely small volumes in which components are eluted from capillary columns would suggest that direct interfacing with mass spectrometry should be feasible. The small volumes of solvent entering the mass spectrometer source could easily be removed by pumping arrangements used for gas chromatography/mass spectrometry systems.

The theoretical advantages of narrow bore systems are well defined, but how long it will take before they make a significant contribution to food analyses remains to be seen.


## III  SIZE EXCLUSION CHROMATOGRAPHY

### A  Introduction

HPLC techniques have been applied successfully to the analysis of the majority of important food components. However, only recently have these been extended to include macromolecular species such as polysaccharides and proteins. Traditional exclusion chromatography of water-soluble biopolymers employed soft hydrophilic gels, such as synthetic polyacrylamides (Bio-Gel) or dextrans (Sephadex). These materials are insufficiently rigid to be used under high pressures. Thus, large particle sizes are employed to allow elution under low pressures, with a consequent limitation on chromatographic efficiency. Another disadvantage of the polysaccharide-based gels is that they are attacked by certain microorganisms and enzymes. Despite their limitations, these materials have contributed enormously to research in many areas involving water-soluble biopolymers. The need for rigid and stable microparticulate stationary phases for aqueous size exclusion chromatography has been evident for many years and many materials have been evaluated with varying degrees of success. It is interesting to note that gel permeation chromatography with organic solvents initiated high pressure techniques, but its aqueous phase counterpart has had to wait some two decades to reach the same level of development.

The early work with HPLC of proteins made use of porous silica gel.

However, this material cannot be used without surface modification, because of the presence of a high density of highly polar and weakly acidic silanol groups. These may lead to complicating interactions, such as adsorption, cationic sorption, anionic exclusion, or hydrogen bonding (Barth, 1980). These secondary effects may in fact increase the selectivity of certain separations, but in general a stationary phase that separates solely on the basis of molecular size is to be preferred. Initial attempts to deactivate the surface silanol groups employed materials, such as Carbowax 20M, which met with limited success, for example, in the work of Schechter (1974) on carboxylic acid synthetase. A more stable solution to the problem was subsequently found, when the surface of porous silica gel particles, or controlled porosity glass particles, were coated with a layer of Glycophase. This is glycerol covalently bonded via propyl silane to the particle surface. This layer not only reduces the undesirable secondary effects of adsorption, but prevents sensitive biological macromolecules from coming into contact with the inorganic surface, which can lead to protein denaturation. This surface modification has been applied to both controlled porosity glass (e.g. Glycophase CPG) and silica gel (e.g. Synchropak GPC). These materials have been used for the study of proteins, polynucleotides, and polysaccharides (Regnier, 1980). The success of the glycophase materials resulted in the rapid evaluation of a number of covalently bonded hydrophilic phases for use in size exclusion chromatography. The many earlier phases need not concern us here, but the interested reader is referred to two recent reviews (Barth, 1980; Mikes, 1981).

Currently a wide range of high performance stationary phases is available for aqueous size exclusion chromatography. The majority of these are based on porous silica gel, and a representative selection is shown in Table 13.2, together with some of their characteristics. In addition to these inorganic-based materials there is also a number of organic-based packings and a selection of these, and their characteristics, is shown in Table 13.3. It will be noted that the structures of some of these proprietary phases have not been published.

## B  Secondary Effects in Size Exclusion Chromatography

The ideal material for size exclusion chromatography would separate components solely on the basis of molecular size. However, owing to the existence of secondary effects, mentioned in Section II.A, this ideal is not readily attained in practice. In many situations these secondary effects can be important and may result in considerable modification to the pattern of fractionation obtained where size exclusion alone is taking place. It is thus

**Table 13.2** Silica-based packings

| Packing | Pore size (Å) | Particle size (μm) | Fractionation range | Supplier |
|---|---|---|---|---|
| | | | **(a) Unmodified** | |
| *LiChrospher* | | | | 14 |
| Si 100 | 100 | 10 (s) | $<5 \times 10^4 - 8 \times 10^4$(d) | |
| Si 300 | 300 | | $<1.5 \times 10^5 - 3 \times 10^5$ | |
| Si 500 | 500 | | $<3 \times 10^5 - 6 \times 10^5$ | |
| Si 1000 | 1000 | | $<0.6 \times 10^6 - 1.4 \times 10^6$ | |
| Si 4000 | 4000 | | $<2.5 \times 10^6 - 8 \times 10^6$ | |
| *SE series* | | | | 15 |
| SE 60 | 60 | 10 (s) | $2 \times 10^2 - 1 \times 10^4$(d) | |
| SE 100 | 100 | | $5 \times 10^3 - 7 \times 10^4$ | |
| SE 500 | 500 | | $3 \times 10^3 - 5 \times 10^5$ | |
| SE 1000 | 1000 | | $2.5 \times 10^4 - 3 \times 10^6$ | |
| SE 4000 | 4000 | | $10^5 - (>10^7)$ | |
| *Spherosil* | | | | 16 |
| XOA400 | <100 | 7 (s) | $<4 \times 10^4$(d) | |
| XOA200 | 150 | | $<2 \times 10^5$ | |
| XOA075 | 300 | | $<4 \times 10^5$ | |
| XOB030 | 600 | | $<1 \times 10^6$ | |
| XOB015 | 1000 | | $<1.5 \times 10^6$ | |
| XOB005 | >1500 | | NA | |
| *Zorbax* | | | | 15 |
| PSM60 | 60 | 6 (s) | $2 \times 10^2 - 4 \times 10^4$(d) | |
| PSM1000 | 750 | | $3 \times 10^4 - 6 \times 10^6$ | |
| | | | **(b) Modified** | |
| *Glycerolpropyl* | | | | 17 |
| SynChropak | 100 | 10 (s) | $1 \times 10^3 - 1 \times 10^5$(a)   $3 \times 10^3 - 3 \times 10^5$(b)   $<5 \times 10^4 - 8 \times 10^4$(d) | |
| [also sold as | 300 | | $3 \times 10^3 - 5 \times 10^5$   $<1.5 \times 10^5 - 3 \times 10^5$ | |

| Packing | Nominal pore size (Å) | Particle size (μm) | Molecular weight range | | | Ref. |
|---|---|---|---|---|---|---|
| Aquapore (18,19) and Bio-Sil GFC (2) | 500 | | $3 \times 10^3 - 2 \times 10^6$ | $1 \times 10^4 - 5 \times 10^6$ | $<3 \times 10^5 - 6 \times 10^5$ | |
| | 1000 | | $1 \times 10^4 - (>10^7)$ | $1 \times 10^5 - 2 \times 10^7$ | $<0.6 \times 10^6 - 1.4 \times 10^6$ | |
| | 4000 | | | | $<2.5 \times 10^6 - 8 \times 10^6$ | |
| LiChrosorb DIOL | 100 | 10 (i) | | $1 \times 10^4 - 1 \times 10^5$(b) | | 14 |
| LiChrospher DIOL | 100 | 10 (s) | | (Similar to SynChropak) | | 14 |
| | 500 | | | | | |
| | 1000 | | | | | |
| | 4000 | | | | | |

*Polyether [Si(RO)$_n$CH$_3$] (extract structure unknown)*

| Packing | Nominal pore size (Å) | Particle size (μm) | Molecular weight range | | | Ref. |
|---|---|---|---|---|---|---|
| μ-Bondagel | | | | | | |
| E125 | 125 | 10 (i) | $<3 \times 10^4$(a) | $2 \times 10^3 - 5 \times 10^4$(d) | | 13 |
| E 500 | 500 | | $<5 \times 10^5$ | $5 \times 10^3 - 5 \times 10^5$ | | |
| E 1000 | 1000 | | | $5 \times 10^4 - 2 \times 10^6$ | | |
| E Linear | Blend | | | $1 \times 10^2 - 2 \times 10^6$ | | |

*Unknown coatings*

| Packing | Nominal pore size (Å) | Particle size (μm) | Molecular weight range | | | Ref. |
|---|---|---|---|---|---|---|
| Protein I-125 | 125 | 10 (i) | | $2 \times 10^3 - 8 \times 10^4$(b) | | 13 |
| Protein I-250 | 250 | | | $1 \times 10^4 - 5 \times 10^5$ | | |
| TSK G2000SW | | 10 ± 2 (s) | $<2 \times 10^4$(a) | $<4 \times 10^4$(b) | | 2, 9, 10, 20 |
| TSK G3000SW | | 10 ± 2 | $<1.5 \times 10^5$ | $4 \times 10^4 - 4 \times 10^5$ | | |
| TSK G4000SW | | 13 ± 3 | $<6 \times 10^5$ | | | |

*Suppliers*

1. Pharamacia Fine Chemicals (Piscataway, N.J.) 2. Bio-Rad Laboratories (Richmond, Calif.) 3. LKB Instruments, Inc. (Rockville. Md) 4. Aldrich Chemical Company (Milwaukee, Wisc.) 5. Koch-Light Laboratories Ltd (Aylesbury, Bucks) 6. Lachema (Brno, Czechoslovakia) 7. Perkin-Elmer Corp. (Norwalk, Conn.) 8. Showa Denko K.K. (Tokyo, Japan) 9. Variar Associates (Walnut Creek, Calif.) 10. Toya Soda Manufacturing Company Ltd (Tokyo, Japan) 11. Electro-Nucleonics, Inc. (Fairfield, N.J.) 12. Pierce Chemical Company (Rockford, Ill.) 13. Waters Associates (Milford, Mass.) 14. E. Merck (EM Laboratories, Elmsford, N.Y.) 15. DuPont Instrument Products Division (Wilmington, Del.) 16. Rhone Poelence (France) 17. SynChrom. Inc. (Linden, Ind.) 18. Brownlee Labs. Inc. (Santa Clara, Calif.) 19. Chromatix (Sunnyvale, Calif.) 20. Altex Scientific, Inc. (Beckman, Berkeley, Calif.)

*Abbreviations*

(a) Polysaccharides (aqueous mobile phase)  (b) Proteins (aqueous mobile phase)  (c) Polyethylene glycols (aqueous mobile phase)  (d) Polystyrene (THF mobile phase)  (i) Irregularly shaped particles  (s) Spherical shaped particles  NA Not available
Data abstracted with permission from Barth (1980).

**Table 13.3**  Organic-based packings

| Packing | Pore size (Å) | Particle size (μm) | Fractionation range | pH stability | Supplier |
|---|---|---|---|---|---|
| *Poly(hydroxyethyl methacrylate)* | | | | | |
| Spheron | | | | | |
| P40 | 25–40, 40–63, 63–100 (wet) | | $2 \times 10^4$–$6 \times 10^4$(a) | 1–12 | 6 |
| P100 | | | $7 \times 10^4$–$2.5 \times 10^5$ | | |
| P300 | | | $2.6 \times 10^5$–$7 \times 10^5$ | | |
| P1000 | | | $8 \times 10^5$–$5 \times 10^6$ | | |
| P100000 | | | $<10^8$ | | |
| *Methacrylate glycerol copolymer* | | | | | |
| Shodex OHpak | NA | 10 (s) | $<4 \times 10^5$ | 4–12 | 7,8 |
| B-804 | | | | | |
| *Sulphonated poly(styrene divinylbenzene)* | | | | | |
| Shodex Ionpak | | 10 (s) | | 2–11 | 7,8 |
| S-801 | 55 | | $<1 \times 10^3$(a) | | |
| S-802 | 100 | | $<5 \times 10^3$ | | |
| S-803 | 160 | | $<5 \times 10^4$ | | |
| S-804 | 220 | | $<5 \times 10^5$ | | |
| S-805 | 350 | | $<5 \times 10^6$ | | |
| *Hydroxylated gel (extract structure unknown but contains* | | | | | |
| *—$CH_2CHOHCH_2O$— groups)* | | | | | |
| TSK G1000PW | NA | 10 ± 2 | $<1 \times 10^3$(c) | 2–12 | 9, 10 |
| TSK G2000PW | | 10 ± 2 | $<5 \times 10^3$ | | |
| TSK G3000PW | | 13 ± 2 | $<2 \times 10^4$ / $<6 \times 10^4$(a) | | |
| TSK G4000PW | | 13 ± 2 | $<3 \times 10^5$ / $<7 \times 10^5$ | | |
| TSK G5000PW | | 17 ± 2 | $<7 \times 10^6$ | | |
| TSK G6000PW | | 17 ± 2 | $<3 \times 10^7$ | | |

Abbreviations and suppliers as in Table 13.2.
Data abstracted with permission from Barth (1980).

important to be able to gain some information, preferably qualitative and quantitative, on these secondary effects in a particular column so that performance can be predicted. The fact that the majority of successful separations is now being carried out on proprietary columns, of ill-defined composition and structure, increases the need for column characterization of this nature. This problem has been faced in a recent comprehensive paper (Pfannkoch, Lu, Pegnier and Barth, 1980), where deviations from ideality are studied with a range of high and low molecular weight probes. In ideal size exclusion chromatography, the elution volume of a component $(V_e)$ can be defined by the equation

$$V_e = V_0 + K_D V_p,$$

where $V_0$ is the column void volume, $V_p$ is the support pore volume and $K_D$ is the distribution coefficient. Thus low molecular weight probes which do not associate with the column material would have a $K_D$ value of 1. Anionic species, which could be affected by ion exclusion, would be expected to have $K_D$ values, below 1, provided that this secondary effect were signific-

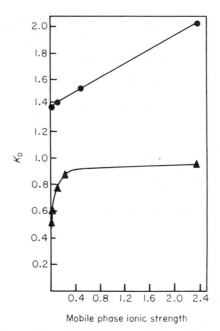

**Fig. 13.3** Dependence of the distribution coefficient $(K_D)$ of small molecules on mobile phase ionic strength. Mobile phase, pH 7.05 phosphate buffer of indicated ionic strength. ▲, citrate; ●, phenylethanol. [Redrawn with permission from Pfannkoch *et al.* (1980).]

ant. Similarly hydrophobic interactions would produce $K_D$ values greater than 1. The effect of ionic interactions would be expected to decrease with increasing ionic strength of the mobile phase, while the reverse would be predicted for hydrophobic interactions. These effects are illustrated for a typical ionic and hydrophobic probe in Fig. 13.3. The optimum ionic strength for the mobile phase will therefore depend on the nature of the components to be separated, but in general is found to be in the region 0.1–0.6 M.

The resolution $(R_s)$ between two components may be defined as on p. 4 in terms of elution parameters and peak widths. However, for size-exclusion chromatography it is more useful to define the specific resolution factor $R_{sp}$, which takes into account the molecular weight ratio of the components $(M_1 * M_2)$:

$$R_{sp} = R_s \frac{1}{\log(M_1/M_2)}.$$

This parameter, which provides a useful indication of column performance, is given in Table 13.4 for some of the most frequently used high performance size exclusion columns. Values for plate numbers are also recorded and it should be noted that the actual value obtained is highly dependent on the test compound employed, which therefore should be specified in all cases. The fractionation range of a column is simply

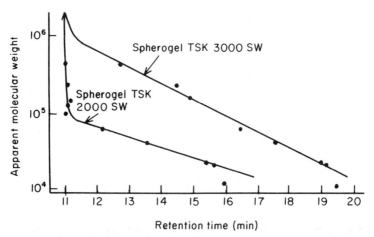

**Fig. 13.4** Protein molecular weight calibration curves for Spherogel TSK-2000 SW and TSK-3000 SW. (Reproduced by courtesy of Altex Scientific, a subsidiary of Beckman Instruments Inc.)

**Table 13.4** Resolution and fractionating limits of various columns with proteins

| Columns | $L_c$ (cm) | $V_i/V_0$ | $K_D$ (ova)† | $N$ (ova)† | $K_D$ (gt)‡ | $N$ (gt)‡ | $R_M$§ | $R_{sp}$ | Fractionating limits (log of molecular weight) |
|---|---|---|---|---|---|---|---|---|---|
| TSK SW 3000 | 30.5 | 1.35 | 0.61 | 2488 | 1.04 | 9216 | 2.00 | 3.32 | 3.45–6.20 |
| TSK SW 2000 | 30.5 | 0.95 | 0.35 | 886 | 1.07 | 6770 | 2.46 | 2.57 | 3.15–5.25 |
| SynChropak GPC 300 | 24.5 | 1.45 | 0.72 | 848 | 1.01 | 4200 | 3.03 | 2.08 | 4.15–6.70 |
| Waters I-125 | 25.0 | 0.92 | 0.35 | 1070 | 1.06 | 4947 | 3.09 | 2.04 | 2.90–5.65 |
| SynChropak GPC 100 | 25.0 | 1.23 | 0.58 | 620 | 1.01 | 2079 | 3.21 | 1.65 | 3.50–5.80 |
| LiChrosorb Diol | 30.0 | 0.63 | 0.33 | 418 | 1.16 | 1764 | 11.06 | 0.84 | 2.90–5.65 |

†"ova" specifies the protein ovalbumin.
‡"gt" specifies the peptide glycyltyrosine.
§$R_M$ is minimum molecular weight ratio to give resolution ($R_s$) of unity.
A mobile phase velocity of 0.33 mm s$^{-1}$ was used in measurements of $N$, $R_M$ and $R_{sp}$.
Data abstracted with permission from Pfannkoch et al. (1980).

determined by plotting retention time, or elution volume, against molecular weight for a number of standards. Such a process is illustrated in Fig. 13.4 for two types of TSK column. A completely smooth curve will not be obtained with protein standards, due to differences in molecular shape and also secondary effects. Smoother curves may be realized with standards of more similar chemical structure, for example a series of dextrans, but protein calibration is more realistic and is the most common method.

## C  Applications of Aqueous Size Exclusion Chromatography

### 1  *Protein Separations*

The majority of published papers on the use of high performance size exclusion chromatography for protein separations appear to have concentrated on characterization of the columns, rather than on their application to specific problems. The power of modern propriety materials is evident from the simple chromatogram of protein standards, such as that illustrated in Fig. 13.5. However, resolution of this quality still requires low flow rates with a long analysis time. The application of these columns to a range of biological problems will prove their worth, and a few such applications

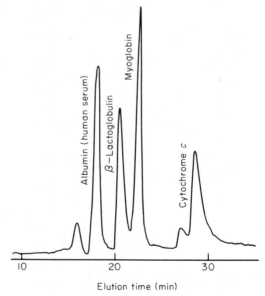

**Fig. 13.5**  Example of separation of commercial protein mixture. (Reproduced with permission of Toyo Sada Mfg Co., Japan.)

have appeared recently, for example the purification of hormone receptors and the fractionation of muscle contractile proteins (Somack, 1980) and the separation of human serum proteins (Schlabach, 1980).

The applications of high performance size exclusion chromatography to food analysis are at present very limited. However, there are many instances where conventional soft gels are being used and these analyses could be potentially improved and the time involved reduced by the introduction of modern columns. One area where this change has been proposed is in the analysis of cheese proteins (Hartman and Persson, 1980). Further areas where high performance size exclusion chromatography would be a viable technique in terms of quantification and speed of analysis include those methods where gel electrophoresis is used to determine proteins. Size exclusion techniques should be of comparable quantitative precision and more rapid, but at present electrophoresis offers greater resolution. In some instances, resolution may not be a limiting factor, as in the determination of whey proteins, and in such cases the greater speed of analysis would be attractive.

## 2   Carbohydrate Separations

The main application of high performance size exclusion chromatography is undoubtedly in the area of protein analysis, but the technique can be used with carbohydrates and other macromolecules. Indeed, some size exclusion columns are more suited to carbohydrates, for example, the TSK Type PW columns. A simple application to the determination of lactose in a milk powder is shown in Fig. 13.6. The speed of the chromatographic stage is in fact considerably less than that achievable with a partition system. Here the attraction of the method is that no deproteination is required, the proteins being eluted rapidly, well away from the peak of interest. Thus the total analysis time is quite reasonable.

A more important application would be the fractionation of higher molecular weight carbohydrates, which cannot be chromatographed by a partition system and require long analysis times with conventional soft gels. Fractions can also be collected for further characterization.

## 3   Sample Clean-up

Size exclusion chromatography is potentially a very powerful technique for cleaning-up samples, by removal of high molecular weight species, prior to HPLC analysis. Automated systems allow fractions to be cut from one chromatographic run and applied directly to a second column (Erni, Krummen and Pellet, 1979) and this technique is ideal for removing unwanted material, which would disturb a high resolution analysis. Any

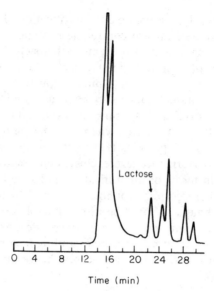

Time (min)

**Fig. 13.6**  Separation of lactose in a non-fat dry milk powder. Proteins are eluted first and do not interfere with lactose determination. (Reproduced by courtesy of Altex Scientific, a subsidiary of Beckman Instruments Inc.)

column material can be used in this "semi-preparative" mode, but size exclusion columns are particularly suitable, as all the components, provided they remain in solution, should be eluted in a reasonable volume. Thus, the columns should not become blocked with strongly retained compounds. An interesting example of column switching of this kind, also known as multidimensional chromatography, is reported by Majors (1980), who determined the vitamins in a protein food supplement. Without primary chromatography, considerable clean-up would be necessary. The fundamental disadvantage is the present enormous cost of such columns, but this it is hoped will change in the future.

## IV  HPLC/MS

### A  General

The mass spectrometer is currently the detector for HPLC which can combine sensitivity, versatility, and universality to the highest degree and, judging by the outstanding advances made with the help of combined gas chromatography and mass spectrometry, (MS) the potential of combined HPLC/MS seems even greater.

The key step is the introduction of the HPLC eluent into the mass spectrometer which normally has to operate under high vacuum. Two methods are currently the front-runners: (1) direct liquid introduction and (2) mechanical transfer.

## 1  Direct Liquid Introduction

Here, a pin-hole allows 1% (i.e. about 0.01 ml min$^{-1}$) eluent to leak directly into the ion source, where the organic component of the eluent forms the reagent gas for the operation of the MS in the CI mode (see Fig. 13.7).

## 2  Mechanical Transfer

Adaptation of the moving wire FID detection system for HPLC to MS has led to moving belts of stainless steel or polyvinide, which can carry about 1 ml min$^{-1}$ towards the mass spectrometer (Fig. 13.8). Most of the solvent is removed by an infrared heater, before the belt takes the sample and residual solvent into the mass spectrometer via vacuum locks. A heater then vaporizes the sample close to the ion source and the belt is cleaned by a further heater, before its return to pick up further eluate. The amount of solvent entering the ion source is insufficient to act as reagent gas, so the mass spectrometer can be operated either in the EI or the CI mode, and the selection of reagent gas is not determined by the HPLC conditions. High proportions of polar eluents, such as water, reduce sensitivity, but use of microbore HPLC/MS can counteract this (Lant, Westwood and Games, 1981).

## B  Applications to Food Science

Although the significance of the technique has been recognized for a number of years (Nursten, 1979) very little progress has yet been made. In a review of mass spectrometry in food science (Harman, 1979) none of the 175 references is quoted for its concern with HPLC/MS and indeed more than 95% of the studies introduced samples by gas chromatography.

HPLC/MS has been applied to classes of compounds which occur in food, such as aflatoxins, alkaloids, amino acids and peptides, essential oils, lipids, nucleosides and nucleotodes, sugars, and tannins (McFaden, 1980). Clearly, much, much more remains to be done in developing such methods and in applying them to the analysis of foodstuffs.

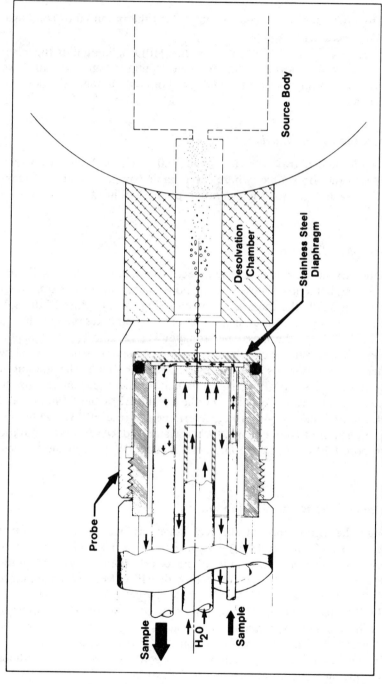

**Fig. 13.7** Tip of HPLC/MS probe for direct liquid introduction. [Reproduced with permission from Melera (1979).]

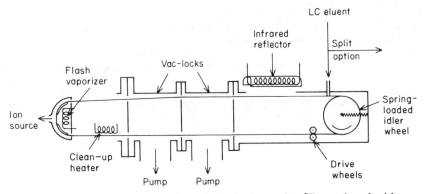

**Fig. 13.8**  HPLC/MS interface using mechanical transfer. [Reproduced with permission from Millington (1980).]

## REFERENCES

Barth, H. G. (1980). *J. chromatogr. Sci.* **18**, 409.

Bristow, P. A. (1976). *Liquid Chromatography in Practice*, HETP Press, Wilmslow, Cheshire, p. 31.

Erni, F., Krummen, K. and Pellet, A. (1979). *Chromatographia*, **12**, 399.

Golay, M. J. E. (1959). In *Gas Chromatography* (D. H. Detsy, ed.) Butterworth, London, p. 36.

Harman, (1979). In *Mass Spectrometry*, Vol. 5 (R. A. W. Johnstone, ed.), The Chemical Society, London, p. 211.

Hartmann, A. and Persson, B. (1980). *Sci. Tools* **27**, 57.

Hofman, K. and Halasz, I. (1979). *J. Chromat.* **173**, 211.

Karasek, F. W. (1977). *Res. Devlmt* **28**, 42.

Knox, J. H. (1980). *J. chromatogr. Sci.* **18**, 453.

Knox, J. H. and Gilbert, M. T. (1979). *J. Chromat.* **186**, 405.

Lant, M. S., Westwood, S. A. and Games, D. E. (1981). Paper presented at the 12th meeting, British Mass Spectrometry Society, Cambridge, 21–24 September.

McFaden, W. H. (1980). *J. chromatogr. Sci.* **18**, 97.

Majors, R. E. (1972). *Analyt. Chem.* **44**, 1722.

Majors, R. E. (1980). *J. chromatogr. Sci.* **18**, 571.

Malera, A. (1979). Technical Paper No. MS-10, Hewlett-Packard Scientific Instruments Division, Palo Alto, Calif.

Mikes, O. (1981). *Ernährung.* **5**, 88.

Millington, D. S. (1980. *New Mass Spectral Techniques for Organic and Biochemical Analysis*, VG-Micromass Ltd, Altrincham, Cheshire.

Novotny, N. (1980). *J. chromatogr. Sci.* **18**, 473.

Nursten, H. E. (1979). In *Progress in Flavour Research* (D. G. Land and H. E. Nursten, eds), Applied Science Publishers, London, p. 337.

Pfannkoch, E., Lu, K. C., Pegnier, F. E. and Barth, H. E. (1980). *J. chromatogr. Sci.* **18**, 430.

Regnier, F. E. (1980). *Analyt. Biochem.* **103**, 1.
Schlabach, T. (1980). Applications Report, Varian Associates, LC-120.
Schechter, I. (1974). *Analyt. Biochem.* **58**, 30.
Scott, R. P. W. and Kucera, P. (1979). *J. Chromat.* **169**, 51.
Somack, R. (1980). Applications Data Sheet, Beckman Scientific Instruments, T1B-BS12, p. 4.
Tijssen, R. (1978). *Sepn Sci. Technol.* **13**, 681
Tsuda, T. and Novotny, M. (1978). *Analyt. Chem.* **50**, 532.
Wu, C.-Yi and Seggia, S. (1972). *Analyt. Chem.* **44**, 149.

# Index

# FOOD SCIENCE AND TECHNOLOGY

## A SERIES OF MONOGRAPHS

Maynard A. Amerine, Rose Marie Pangborn, and Edward B. Roessler, PRINCIPLES OF SENSORY EVALUATION OF FOOD. 1965.

S. M. Herschdoerfer, QUALITY CONTROL IN THE FOOD INDUSTRY. Volume I — 1967. Volume II — 1968. Volume III — 1972.

Hans Reimann, FOOD-BORNE INFECTIONS AND INTOXICATIONS. 1969.

Irvin E. Leiner, TOXIC CONSTITUENTS OF PLANT FOODSTUFFS. 1969.

Martin Glicksman, GUM TECHNOLOGY IN THE FOOD INDUSTRY. 1970.

L. A. Goldblatt, AFLATOXIN. 1970.

Maynard A. Joslyn, METHODS IN FOOD ANALYSIS, second edition. 1970.

A. C. Hulme (ed.), THE BIOCHEMISTRY OF FRUITS AND THEIR PRODUCTS. Volume 1 — 1970. Volume 2 — 1971.

G. Ohloff and A. F. Thomas, GUSTATION AND OLFACTION. 1971.

George F. Stewart and Maynard A. Amerine, INTRODUCTION TO FOOD SCIENCE AND TECHNOLOGY. 1973.

C. R. Stumbo, THERMOBACTERIOLOGY IN FOOD PROCESSING, second edition. 1973.

Irvin E. Liener (ed.), TOXIC CONSTITUENTS OF ANIMAL FOODSTUFFS. 1974.

Aaron M. Altschul (ed.), NEW PROTEIN FOODS: Volume 1, TECHNOLOGY, PART A — 1974. Volume 2, TECHNOLOGY, PART B — 1976. Volume 3, ANIMAL PROTEIN SUPPLIES, PART A — 1978. Volume 4, ANIMAL PROTEIN SUPPLIES, PART B — 1981.

S. A. Goldblith, L. Rey, and W. W. Rothmayr, FREEZE DRYING AND ADVANCED FOOD TECHNOLOGY. 1975.

R. B. Duckworth (ed.), WATER RELATIONS OF FOOD. 1975.

Gerald Reed (ed.), ENZYMES IN FOOD PROCESSING, second edition. 1975.

A. G. Ward and A. Courts (eds.), THE SCIENCE AND TECHNOLOGY OF GELATIN. 1976.

John A. Troller and J. H. B. Christian, WATER ACTIVITY AND FOOD. 1978.

A. E. Bender, FOOD PROCESSING AND NUTRITION. 1978.

D. R. Osborne and P. Voogt, THE ANALYSIS OF NUTRIENTS IN FOODS. 1978.

Marcel Loncin and R. L. Merson, FOOD ENGINEERING: PRINCIPLES AND SELECTED APPLICATIONS. 1979.

Hans Reimann and Frank L. Bryan (eds.), FOOD-BORNE INFECTIONS AND INTOXICATIONS, second edition. 1979.

N. A. Michael Eskin, PLANT PIGMENTS, FLAVORS AND TEXTURES: THE CHEMISTRY AND BIOCHEMISTRY OF SELECTED COMPOUNDS. 1979.

J. G. Vaughan (ed.), FOOD MICROSCOPY. 1979.

J. R. A. Pollock (ed.), BREWING SCIENCE, Volume 1 — 1979. Volume 2 — 1980.

Irvin E. Liener (ed.), TOXIC CONSTITUENTS OF PLANT FOODSTUFFS, second edition. 1980.

J. Christopher Bauernfeind (ed.), CAROTENOIDS AS COLORANTS AND VITAMIN A PRECURSORS:
TECHNOLOGICAL AND NUTRITIONAL APPLICATIONS. 1981.

Pericles Markakis (ed.), ANTHOCYANINS AS FOOD COLORS. 1982.

Vernal S. Packard, HUMAN MILK AND INFANT FORMULA. 1982.

George F. Stewart and Maynard A. Amerine, INTRODUCTION TO FOOD SCIENCE AND TECHNOLOGY, SECOND EDITION. 1982.

Malcolm C. Bourne (ed.), FOOD TEXTURE AND VISCOSITY: CONCEPT AND MEASUREMENT. 1982.

R. Macrae (ed.), HPLC IN FOOD ANALYSIS. 1982.

*In preparation*

Héctor A. Iglesias and Jorge Chirife, HANDBOOK OF FOOD ISOTHERMS: WATER SORPTION PARAMETERS FOR FOOD AND FOOD COMPONENTS. 1982.

John A. Troller, SANITATION IN FOOD PROCESSING AND SERVICE. 1983.

# FOOD SCIENCE AND TECHNOLOGY

## A SERIES OF MONOGRAPHS

Maynard A. Amerine, Rose Marie Pangborn, and Edward B. Roessler, PRINCIPLES OF SENSORY EVALUATION OF FOOD. 1965.

S. M. Herschdoerfer, QUALITY CONTROL IN THE FOOD INDUSTRY. Volume I — 1967. Volume II — 1968. Volume III — 1972.

Hans Reimann, FOOD-BORNE INFECTIONS AND INTOXICATIONS. 1969.

Irvin E. Leiner, TOXIC CONSTITUENTS OF PLANT FOODSTUFFS. 1969.

Martin Glicksman, GUM TECHNOLOGY IN THE FOOD INDUSTRY. 1970.

L. A. Goldblatt, AFLATOXIN. 1970.

Maynard A. Joslyn, METHODS IN FOOD ANALYSIS, second edition. 1970.

A. C. Hulme (ed.), THE BIOCHEMISTRY OF FRUITS AND THEIR PRODUCTS. Volume 1 — 1970. Volume 2 — 1971.

G. Ohloff and A. F. Thomas, GUSTATION AND OLFACTION. 1971.

George F. Stewart and Maynard A. Amerine, INTRODUCTION TO FOOD SCIENCE AND TECHNOLOGY. 1973.

C. R. Stumbo, THERMOBACTERIOLOGY IN FOOD PROCESSING, second edition. 1973.

Irvin E. Liener (ed.), TOXIC CONSTITUENTS OF ANIMAL FOODSTUFFS. 1974.

Aaron M. Altschul (ed.), NEW PROTEIN FOODS: Volume 1, TECHNOLOGY, PART A — 1974. Volume 2, TECHNOLOGY, PART B — 1976. Volume 3, ANIMAL PROTEIN SUPPLIES, PART A — 1978. Volume 4, ANIMAL PROTEIN SUPPLIES, PART B — 1981.

S. A. Goldblith, L. Rey, and W. W. Rothmayr, FREEZE DRYING AND ADVANCED FOOD TECHNOLOGY. 1975.

R. B. Duckworth (ed.), WATER RELATIONS OF FOOD. 1975.

Gerald Reed (ed.), ENZYMES IN FOOD PROCESSING, second edition. 1975.

A. G. Ward and A. Courts (eds.), THE SCIENCE AND TECHNOLOGY OF GELATIN. 1976.

John A. Troller and J. H. B. Christian, WATER ACTIVITY AND FOOD. 1978.

A. E. Bender, FOOD PROCESSING AND NUTRITION. 1978.

D. R. Osborne and P. Voogt, THE ANALYSIS OF NUTRIENTS IN FOODS. 1978.

Marcel Loncin and R. L. Merson, FOOD ENGINEERING: PRINCIPLES AND SELECTED APPLICATIONS. 1979.

Hans Reimann and Frank L. Bryan (eds.), FOOD-BORNE INFECTIONS AND INTOXICATIONS, second edition. 1979.

N. A. Michael Eskin, PLANT PIGMENTS, FLAVORS AND TEXTURES: THE CHEMISTRY AND BIOCHEMISTRY OF SELECTED COMPOUNDS. 1979.

J. G. Vaughan (ed.), FOOD MICROSCOPY. 1979.

J. R. A. Pollock (ed.), BREWING SCIENCE, Volume 1 — 1979. Volume 2 — 1980.

Irvin E. Liener (ed.), TOXIC CONSTITUENTS OF PLANT FOODSTUFFS, second edition. 1980.

J. Christopher Bauernfeind (ed.), CAROTENOIDS AS COLORANTS AND VITAMIN A PRECURSORS: TECHNOLOGICAL AND NUTRITIONAL APPLICATIONS. 1981.

Pericles Markakis (ed.), ANTHOCYANINS AS FOOD COLORS. 1982.

Vernal S. Packard, HUMAN MILK AND INFANT FORMULA. 1982.

George F. Stewart and Maynard A. Amerine, INTRODUCTION TO FOOD SCIENCE AND TECHNOLOGY, SECOND EDITION. 1982.

Malcolm C. Bourne (ed.), FOOD TEXTURE AND VISCOSITY: CONCEPT AND MEASUREMENT. 1982.

R. Macrae (ed.), HPLC IN FOOD ANALYSIS. 1982.

*In preparation*

Héctor A. Iglesias and Jorge Chirife, HANDBOOK OF FOOD ISOTHERMS: WATER SORPTION PARAMETERS FOR FOOD AND FOOD COMPONENTS. 1982.

John A. Troller, SANITATION IN FOOD PROCESSING AND SERVICE. 1983.